THE WORLD ATLAS OF
WHISKY

3RD EDITION

THE WORLD ATLAS OF WHISKY

3RD EDITION

More than 500 distilleries profiled and
480 expressions tasted

DAVE BROOM

André Simon award winner

MITCHELL BEAZLEY

To Jo and Rosie with all my love.

First published in Great Britain in 2024 by Mitchell Beazley,
an imprint of Octopus Publishing Group Ltd
Carmelite House
50 Victoria Embankment
London EC4Y 0DZ
www.octopusbooks.co.uk

An Hachette UK Company
www.hachette.co.uk

Distributed in the US by
Hachette Book Group
1290 Avenue of the Americas
4th and 5th Floors
New York, NY 10104

Distributed in Canada by
Canadian Manda Group
664 Annette St, Toronto
Ontario, Canada M6S 2C8

ISBN: 978-1-78472-673-7

A CIP catalogue record for this book is available from the British Library.

Printed and bound in Malaysia.

10 9 8 7 6 5 4 3 2 1

Publisher: Alison Starling
Art Director: Juliette Norsworthy
Senior Editor: Leanne Bryan
Copyeditor: Jo Richardson
Designer: Geoff Fennell
Picture Researchers: Giulia Hetherington & Jennifer Veall
Proofreader: Sarah Prior
Assistant Production Manager: Allison Gonsalves
Illustration (pages 16–17): Claire Huntley
Cartography: Original cartography and digital mapping by
Encompass Graphics, Hove, UK

Publisher's note
The publishers will be grateful for any information that will assist them in keeping future editions up to date. Although all reasonable care has been taken in the preparation of this book, neither the publishers nor the author can accept any liability for any consequence arising from the use thereof, or the information contained therein.

CONTENTS

INTRODUCTION

Whisky is not fixed. It moves, oscillates, weaves. It adapts and evolves, and at no time have those changes been more dramatic than in the past decade. We are now in a world where there are more distilleries than at any time in whisky's history, and there are more countries than ever making it. In the time I've taken to write this sentence, a few more whisky distilleries will probably have opened up.

Why this remarkable, unprecedented boom? There is the importance that we put on provenance. We want to know where our food and drink comes from, who made it, even how it was made. That plays right into whisky's zone. These people at this distillery in this place making a spirit that tastes unlike any other whisky on the planet. That's a lot of ticks.

There's also the shift to premium – people drinking less but better – and a general desire for bold, complex flavours both in food and drink. Whisky is not bland or neutral, and neither are people's responses to it. This is a spirit that even in its lightest examples has presence.

On that reading, whisky is in a good place. Distillers around the globe have taken the simple principle of grain, water, yeast, copper and wood and spun a multiplicity of variations on that theme. Not only are there more whiskies to choose from, there are a wider variety of whiskies to choose from.

What might have seemed a worthy but rather staid category is alive, innovative, connected, contemporary. It's also more than just a rise in distilleries or bottlings – it's the shift in thinking that lies behind all of this. One example would be the rethinking of whisky as being an agricultural product rather than a quasi-industrial one.

This has partly been driven by the need to work towards sustainable practices, necessitating closer links with farmers, examination of the best grains to grow, the need to reduce chemical inputs and work to restore damaged peatlands. As those links deepen, so a new way of making whisky emerges, one sensitive to the environment and that also has flavour benefits.

There are fascinating parallels that can be drawn with wine – the two drinks are closer to each other than perhaps realized. The exploration of organic principles (seen most significantly in Continental Europe's distillers) mirrors that of wine, as does a willingness and eagerness to explore older techniques. In wine, it's forgotten varieties, old vines and a re-adoption of lost artisanal techniques. In whisky, it manifests itself as working with lower-yielding barleys, heritage grains, brewer's yeasts, direct-fired stills. The old ways adapted for the 21st century. In many ways, whisky is starting anew.

With it comes a new generation of distillers, and a greater number of women being involved in all aspects of whisky: from production to blending, to owning distilleries and running major firms, as well as selling, retailing, writing and educating. The antediluvian line that whisky is a man's drink is finally being discarded. If you hear anyone suggesting that a whisky show is no place for a woman, or that whisky shouldn't be in their glass, call them out. Whisky has no barriers. We are all whisky.

This is the greatest time to be a whisky lover, or to be starting out on your exploration of the spirit.

There is also no such thing as 'world whisky', which is as patronizing a term as 'world music'. The idea that the old countries and styles are in any way entitled to greater reverence should be dismissed. There's no hierarchy here.

So, plenty of positives. This is the greatest time to be a whisky lover, or to be starting out on your exploration of the spirit. And yet there has to be a word of caution. In terms of style, the 'double new oak' technique is growing, a belated replication of the same approach that gave the world overripe, one-dimensional, overloaded Cabs and Chards. Just when wine is moving away from its muscle-bound recent past, some whiskies appear to be embracing it.

The fact that there has never been so many distilleries suggests that there has never been such demand, but today's drinker is more circumspect than in the past. They will pick and choose. Sales remain significantly behind the increase in capacity. Is this a softening, a downturn or a shallow dip in whisky's roller-coaster 600-year existence? Time alone will tell. In the interim, keep exploring (responsibly)... and tell your friends. Come on in, the whisky's lovely.

HOW THIS BOOK WORKS

I wrote the first edition of *The World Atlas of Whisky* in 2010, and the second edition came out in 2014. In the decade since then, the whisky world has changed utterly. There are now in excess of 3,500 distilleries making whisky, sometimes exclusively, sometimes as part of a portfolio of other spirits. This spells joy for a whisky lover but a challenge for a writer, and I am both of these things.

The scale of the 21st-century whisky revolution means that this book isn't an update but a complete rewrite, with more distilleries, more whiskies, more countries. The approach, however, is the same. At its heart this is a book about flavour, people and place.

It will show you where distilleries are but, more importantly, it will tell you what the whiskies made there taste like and why they taste the way they do. To avoid too much repetition, there is a guide to show you the processes of the four main styles – single malt, Bourbon and rye, single pot still, grain – and also a guide to still types.

The differences between distilleries are fascinating, but so are the shared characteristics. Even with a map, it can be hard to navigate this world. So, to avoid the idea that all whiskies from one region or one country are the same, the tasting notes group whiskies together by flavour. They show connections, contrasts and how regional and international boundaries melt away.

Not every distillery could be profiled or mapped. Forgive me. Those that are featured are the distilleries that, after research (that is to say, tasting) and invaluable assistance from friends around the world, have been identified as helping to define their country's approach.

This is an atlas of the world of whisky flavour and of whisky thinking.

At the risk of offending the shade of the beer-loving, wine-drinking Jane Austen, two words came to mind after months of conversations with distillers and trying their wares. Style and Sensibility.

Style is the manner in which something, in this case whisky, is made, as well as referring to its characteristics. Sensibility, on the other hand, is the sensation that the distiller has in creating their whisky, and the response we have when drinking it.

The idea of style can be taken as something that gives a degree of similarity; 'made in this style' is something that, as humans, we look for. We like patterns, categorization and groupings. It makes the world easier to understand.

However, what is as relevant – and more exciting – in today's whisky world is sensibility, the mindset of today's distillers and how they are moving this thing called whisky forward. The book reflects the growth globally in the past decade, but more importantly it's about why this has happened and what 21st-century whisky looks like. It's not about numbers and statistics. This is an atlas of minds.

Above *Convention is being turned on its head in the new world of whisky.*

WHAT IS WHISKY?

The simple answer to that question is a cereal-based, non-neutral spirit that has been aged in wood. Or, if you want to distil it down further, boiled beer. It's also a question that the world's best distillers have been asking. Their 'What is whisky?' is couched slightly differently. They are asking, 'What can whisky be?' The answer is, within the legal boundaries, anything they want it to be.

Distillers can choose from the family of cereals and the almost infinite varieties within each. They can pick a particular yeast variety, distil in numerous ways, explore the world of smoke, oak and other woods. They can play with temperatures and time.

For the consumer, whisky can be the simple highball, the casual drink or the slow, late-night contemplative dram. It can be the base for a fancy cocktail, or nestle inside a hip flask. It can be sipped absent-mindedly, or dissected. It's versatile. It can be whatever you want it to be.

It can also be geeky (writes the whisky geek) and stimulate a desire to know every detail about how it's made, though to reduce whisky to a process is to miss its point. When we listen to music or see a great piece of art, we react viscerally to what has been created. We can be fascinated by brushstrokes or puzzle over tunings, but our response is driven by our emotional investment. It has spoken to us.

It's the same with a great whisky. It can simply make us happy, or demand a deeper investigation.

I was at a concert with a music-loving whisky friend who has an encyclopedic knowledge of production, but here he was speaking of how he wanted his whisky to come from the hearts of the people who made it and celebrate where it was from, and how that desire was the same as that which had driven the music we'd been listening to.

Folk music, the singer Alasdair Roberts told me once, is a continuum, a passing down of a tradition and how, by that transference, the music is refreshed, revivified. This process of interpretation and reinterpretation is vital and it applies to whisky as well. Even new whisky-making countries are involved in this. They are part of that wide river, taking the traditions, interpreting them in their own way, then passing them on. Changed, but the same.

What is whisky? It's an art, a craft, alchemical. It's the whisper of a flower's scent, the reek of a peat fire. It's the grip of tannin and the ethereal aroma of a spring meadow. It's herbal and lushly fruited; it batters you with spice or soothes you. It's the secret hearts of barley, corn, rye and oats; the yeast working its frothy magic, clouds of vapour. It's sunlight and rain on crops and oak trees. It's molecules creating images in our minds, connecting the aromas in the glass with our past. It links us to place and it links us to each other. It's stimulating and yet dangerous if not treated with respect. It's our wildness captured.

Not bad for boiled beer.

Above *Distillers need to be patient. Whisky takes time.*

SINGLE MALT PRODUCTION

While all single malt distilleries follow the same process, each will have its own particular take on how this is done: the use of speciality malts, peated or smoked malt, the type of yeast, length of ferment, the shape and size of the still, cut points and the type(s) of cask(s) used. Within these variations on the theme lies a significant part of each single malt's distillery character, its DNA. All through the process the distiller makes decisions. Here are the main ones.

1

BARLEY

All single malt is made from malted barley, water and yeast. Most modern barley strains have been bred to have disease resistance and give high yields (litres/US gallons of alcohol per tonne/US ton). There is, however, increasing work being done with older varieties and landraces that are lower-yielding but have more characterful flavour.

2

MALTING

A grain of barley is like a small packet of starch. Malting is fooling the barley into thinking it's time to start growing by steeping it in water and then allowing it to germinate in cool, damp conditions. Enzymes in the barley are triggered that will convert the starch into sugar, which is what the distiller needs. To retain the sugar, the germination process is halted by drying the barley.

3

KILNING

Kilning Option 1 Drying the malted barley over hot air will stop germination but will not impart any flavour.

Kilning Option 2: Peating The second option is to dry the barley over a peat fire. This will give a smoky aroma to the final spirit. Peat is semi-carbonized vegetation that, when burned, gives off a fragrant smoke. The oils (phenols) in the smoke stick to the damp surface of the barley, adding their aromas to it and, ultimately, the whisky. Other smokes (wood, nettles, dried animal dung) are also used.

Kilning Option 3: Speciality malt By raising the kilning temperature, the malt becomes increasingly toasted and caramelized. Distillers can add some amber, crystal, caramel or chocolate malts to the mash to produce a range of different flavours.

4

MILLING

The malt is ground into a coarse flour called grist, comprising 20% husk (which acts as a filter bed), 70% grits and 10% flour. Too much of the last could cause the mash to stick; too much of the first and the yield will drop.

5

MASHING

The grist is mixed with hot water (at 63.5˚C/146.3°F) in a large vessel called a mash tun. As soon as the hot water strikes the grist, the conversion from starch to sugar takes place. The sweet liquor, known as wort, is then drained through the perforated bottom of the mash tun. The process is repeated a further two times with increasingly hot water to extract as much of the sugar as possible. The final 'water' is then retained as the first water of the next mash.

Mashing Option 1: Clear wort If distillers pump the wort slowly from the mash tun, they obtain what is known as clear wort. This will tend to produce a spirit with no great cereal character.

Mashing Option 2: Cloudy wort If distillers wish to produce a spirit with a dry, nutty, cereal character, the wort will be extracted quickly, pulling some solids through from the mash tun.

YEAST

Yeast converts the sugars in the wort into alcohol. It also gives off heat and CO2. A distiller's yeast will be efficient in the conversion process, but won't impart any specific flavour. Brewer's, wine, sake or wild yeast will, however, help to generate specific flavour profiles. Distillers may use one or a combination.

6

FERMENTATION

The wort is then cooled and pumped into a fermenting vessel known as a washback. Too high a temperature might kill the yeast. The washbacks can be made from either wood or stainless steel. Yeast is added and fermentation begins, creating wash.

Fermentation Option 1: Short Fermentation is where flavour is created. The conversion of sugar to alcohol is completed in 48 hours. If distillers take this 'short' option, the final spirit will have a more pronounced malty character.

Fermentation Option 2: Long Yeast dies off after roughly 55 hours. Keeping the wash in the washback for longer increases its acidity and allows lactobacillus to start to work on the wash. This combination of acidity, alcohol and bacteria helps to create fruity compounds known as esters. Long ferments can last over 100 hours.

7

DISTILLATION A
The wash has a strength of around 9% ABV. This is then distilled twice in copper pot stills. The first distillation in a 'wash still' will produce a spirit of 23% ABV called 'low wines', which is then redistilled in a 'spirit still'. This time the distillate is divided into three: foreshots, heart and feints. Only the heart (new make) is retained for maturation. The foreshots and feints are recycled with the next batch of low wines.

Distillation Option 1: Long conversation The longer the conversation between the alcohol vapour and copper, the lighter the final spirit will be. This means tall stills are more likely to produce light spirit than small ones. Also, running a still slowly can extend the conversation.

Distillation Option 2: Short conversation Conversely, the shorter the conversation, the heavier the resulting spirit will be. Small stills or quick distillation will tend to give this character.

DISTILLATION B: CONDENSING
The alcohol vapour is turned back into liquid by passing it through a condensing system containing cold water. Once again the distiller has options that will impact on flavour.

Condensing Option 1: Shell and tube This is a tall cylinder containing a mass of small copper pipes filled with cold water. When the alcohol vapour hits the cold pipes, it turns back to liquid. Because there is a large area of copper, shell and tube condensers will help to 'lighten' a spirit.

Condensing Option 2: Worm tub This, the traditional method of condensing, utilizes a long copper pipe coiled in a tank of cold water. Because there is less copper at play here, worm tubs will tend to produce a heavier and often sulphury new make.

DISTILLATION C: CUT POINTS
When the condensed spirit from the second distillation arrives in the spirit safe, it's divided into three parts: foreshots/heads, heart/middle cut and feints/tails. The point where the distiller cuts from foreshots to spirit and from spirit to feints will also have an effect on flavour.

Cut points Option 1: Early As a spirit is distilled, so its aromas change. The first to appear are light and delicate. If a distiller wishes to make a fragrant whisky, they will cut off spirit early.

Cut points Option 2: Late As distillation continues, so the aromas deepen, becoming more oily and rich – smokiness would be one of those. A distiller wishing to make a heavy spirit will therefore cut late.

8

MATURATION
The new make spirit is then reduced (usually to 63.5% ABV) and placed in oak casks to mature.

There are three processes happening here:
1 Subtractive The cask helps to remove the aggressive new spirit character.
2 Additive The flavour compounds in the cask are extracted by the spirit.
3 Interactive The flavours from the wood and the spirit meld together to increase complexity. Time, the freshness of the cask and the type of oak all have a part to play.

Cask type Option 1: American oak (virgin/ex-Bourbon) American oak (*Quercus alba*) is high in compounds that give aromas reminiscent of vanilla, crème brûlée, pine, eucalyptus, spice and coconut.

Cask type Option 2: Ex-sherry These casks are mostly made from European oak (*Quercus robur*), which imparts aromas of dried fruits, clove, incense and walnut, and a richer colour and more mouth-drying tannin.

Cask type Option 3: Other oaks/woods Other oak species such as sessile/French oak (*Quercus petraea*), mizunara/Japanese oak (*Quercus crispula*), Garryana and Chinkapin are all now being used. Each has its own flavour character. Although Scotch whisky can only be aged in oak, other countries can use a range of woods: chestnut and acacia are the most widely used.

Cask type Option 4: Refill Whisky distillers can use casks many times. The more they are used, the less effect the species of oak has on the whisky. These 'refill' casks are important in allowing distillery character to be shown. In practice, most distillers will use a mix of all options to add complexity to the palette of flavours.

Maturation Option 5: Cask finishing Distillers can give the final flavour of their whisky a twist by 'finishing' it. This involves taking a whisky (normally aged in ex-Bourbon or refill casks) and giving it a short period of secondary ageing in a very active cask that has previously held sherry, port, Madeira, wine etc., imbuing the whisky with some of its character.

Maturation: Time It takes time to mature whisky. Logically, the longer a whisky spends in a cask, the more influence the oak will have on the spirit. Eventually it will dominate the whisky, rendering it impossible to tell what distillery it came from. A very active cask will produce this effect more quickly than one that has been filled many times (and which may give a neutral effect). The age statement on a bottle simply tells how long the youngest whisky has spent in wood; it doesn't indicate how active, or otherwise, the cask has been. Old does not automatically equal good! Temperature will also affect how rapidly a whisky matures.

COPPER

Copper is hugely important in creating whisky's flavour. Because copper holds onto heavy elements, distillers can either prolong or restrict the length of the conversation between the alcohol vapour and the copper to create their desired character.

WATER

Distillers need lots of pure cold water in order to make whisky. Accordingly, finding a ready supply is of vital importance. Most distillers use spring water, but lake, river and even town supplies are also used. While there might be a slight impact on fermentation efficiency, it's not believed that water itself has a major contributory factor in a whisky's final flavour.

9

BOTTLING
The whisky is finally ready for bottling, but there are still some final decisions to make.
Caramel adjustment? The addition of spirit caramel helps standardize colour.
Chill-filtration? This will stop a whisky going cloudy, but reduces mouthfeel.
Strength? Legally, whisky must be a minimum of 40% ABV, but 'cask strength' is becoming more popular.

BOURBON & RYE WHISKEY PRODUCTION

Straight Bourbon and rye whiskey can be made anywhere in the USA and not just Kentucky (or, with a slight variation, Tennessee). Its method of production is controlled by regulations dictating composition of mash bill, maximum strength of new make spirit, minimum barrelling strength, cask type, minimum ageing and bottling strength. Within those parameters, however, distillers can play their own tunes.

1

MASH BILL

Corn (for a fat sweetness). If making Bourbon, this must make up a minimum of 51%.

Malted barley for enzymes to convert starches into sugar (though it's also legal to add enzymes).

Rye (for spice and acidity). If making straight rye whiskey, this must be a minimum of 51%.

OR

Wheat (for delicacy). If making straight wheat whiskey, this must be a minimum of 51%.

The ratio of corn to small grains will have a significant effect on the final whiskey, i.e. the more rye, the spicier the end spirit. Wheated Bourbon will tend to be sweeter. Distilleries will often use more than one mash bill.

2

MILLING

The corn is ground.

3

WATER

Hard limestone water high in mineral content.

4

COOKING

A Corn is cooked at 90–100°C (194–212°F) first to hydrolyse (soften) its starch either in open cookers or pressure cookers. When cooked, the mix is cooled and...**B**...rye/wheat is added at the lower temperature (average 70°C/158°F), then cooked again. The mix is cooled once more to 64°C (147°F). Then... **C**...malted barley is added to convert the starch into sugar.

5

BACKSET/SOURMASH/SETBACK

This is the acidic spent liquid at the end of distillation, which is added to the fermenter, making up as much as 33% of the volume. This adjusts the pH of the fermentation – important when you have alkaline water, as in Kentucky. It also prevents bacterial infection and helps with producing a consistent new make character. The amount of sour used affects the percentage of sugars in the mash, so for a fresher Bourbon, less sour is used. Its use is optional – a 'sweet mash' uses no souring.

6

YEAST

Each distillery will have its own strain or strains of proprietary yeast. These are jealously guarded, as the character of the yeast will have a significant effect on the flavour of the final spirit, helping to promote specific congeners (flavouring elements).

7

FERMENTING

Generally this will take up to 3–4 days, at the end of which the distiller has a beer of around 8% ABV after a shorter ferment, rising to 10–11% ABV for longer.

8

DISTILLING

The beer is then distilled in a single column divided by perforated horizontal plates. It arrives at the top, while hot steam is pumped in at the base. As the beer runs through the plates in the column, it meets the rising steam, which strips off the alcohol. This is then condensed into a spirit (aka the 'white dog').

TENNESSEE WHISKEY

In Tennessee, after distillation the white dog is filtered through vats of hard sugar maple charcoal, which removes harsh elements in the new spirit.

9

SECOND DISTILLATION

This takes place either in a 'thumper' (a vessel containing water through which the alcohol vapour passes, extracting any heavy elements) or a 'doubler', a simple pot still. The final strength of the spirit is very important in flavour creation. By law, the white dog must be no stronger than 80% ABV. Most distillers collect significantly below this. The lower the strength of the white dog, the fuller the character will be.

Above *Charcoal filtration is what differentiates Tennessee whiskey.*

11

POT STILL
Some Bourbons are distilled twice, or three times, in pot stills, or in hybrid stills.

12

WAREHOUSING
The warehousing will have a further effect on the spirit's character. The hotter the temperature in the warehouse, the more intense the interaction between the spirit and the oak will be. Conversely, the cooler the environment, the slower this process will be. This means that the location of the warehouse, the number of floors and what it's made out of (brick, metal, wood) is important in flavour creation. Equally, the location of the barrel within those warehouses will have an impact. Some distillers will rotate barrels to get an even maturation, some 'scatter' their barrels across different warehouses, while others set aside warehouses or floors within warehouses for specific brands. By law Straight Bourbon must be aged for a minimum of two years.

10

MATURATION
The white dog must then be reduced in strength to no more than 62.5% ABV before being aged in new, 200-litre (53-US-gallon), charred oak barrels. The barrel entry proof will have an effect on flavour – the lower the strength, the easier it is to dissolve wood-derived sugars into the spirit, giving greater balance between fruits, caramel and tannins.

SINGLE POT STILL IRISH WHISKEY PRODUCTION

The gold standard of whiskey in the 19th century, single pot still Irish whiskey almost became extinct in the 20th century, but was saved by Irish Distillers Limited (IDL) for its brands and in recent years has seen a remarkable revival thanks to Ireland's new distillers and the research of historian Fionnán O'Connor. Single pot still is a complex world of grains, mash bills and distillation techniques. Here are the basic principles.

1

MALTED BARLEY

Although different barley strains are used, barley is malted in the same fashion as in Scotland in order to activate the enzymes that will convert the starches into sugar. The kilning process is over hot air. Normally no smoke is used but some peated experiments have taken place.

2

UNMALTED BARLEY

The key element in single pot still is the use of unmalted barley. This has been shown to have been common practice since the mid-18th century (see page 126). It's this ratio of malted to unmalted barley that gives single pot still Irish whiskey its distinctive spicy/apple character, as well as enhancing its unctuous/oily texture. The IDL ratio is 60% barley to 40% malt.

3

OTHER GRAINS

Oats, rye and wheat can all be used in the mash bill, either individually or in different combinations. Under current regulations they can only comprise 5% of the mash bill, but there are moves to raise this to 30% – as it was in the 19th century. Oats add a distinct creaminess, but also act as a glue or a flavour enhancer. Rye contributes spice and green herbs, while wheat adds a buttery character.

4

BREWING

If a mash filter is used (see step 6 below), the grains are passed through a hammer mill and pumped into a conversion tank with hot water.

5

MASH TUN

This operates in the same way as with single malt, though the mashing temperature is lower to help convert the unmalted grains.

6

MASH FILTER

Mash filtering is a more efficient method of extracting sugars from wort. The wort is pumped from the conversion tank into the mash filter – a long rectangular machine divided by modules. Each of these has a fine filter sheet on one side, a flexible membrane on the other and a mash chamber in the centre. Both the filter plates and the membrane can be expanded by air pressure. The mash is added along the length of the filter, flooding each chamber. Then the membrane in each module is gently expanded, pushing the wort through the bed of solids into outlets at the top and bottom of the filter, leaving the solids on the filter sheet.

7

FERMENTATION

With yeast – normally distiller's but other strains are being used – added, fermentation starts and lasts over 60 hours in order to activate lactobacillus, creating esters. Controlling temperature between 28°C and 32°C (82.4°F and 89.6°F) also helps with oiliness.

8

DISTILLATION 1

Most single pot still Irish whiskey is triple distilled, but there are some that are double distilled, which works in a similar way to single malt production. In triple distillation, the low wines are collected between 25% (for heavy single pot still) and 40% ABV (for light single pot still).

9

DISTILLATION 2

A mix of the low wines and weak feints is redistilled in the intermediate (feints) still. The distillate is cut into three parts: heads, centre cut and tails. The heads and tails are both collected in the 'weak feints' receiver. The middle cut is collected in the 'strong feints' receiver and used for the third distillation. By varying the cut points, different flavours are either retained or removed. The lower the strength of the middle cut, the heavier the resulting spirit; the higher it is, the lighter it will be.

10

DISTILLATION 3

Again, three parts are isolated. The heads are collected as 'strong feints'. The middle cut will be taken forward to be aged. The tails are split into two. The first cut goes into the strong feints receiver, the last runnings go into the weak feints receiver. By varying all of these cut points, once again different characters are created. The deeper the cut, the heavier and oilier the result.

11

MATURATION

A wide variety of casks are used.

GRAIN WHISKY PRODUCTION

Grain whisky was the result of 19th-century distillers using new technology, specifically in still design, to produce a higher-strength spirit more efficiently. By the mid-century, it was being mixed with single malts to create blended Scotch. While most grain still performs this function, it's gaining a following as a stand-alone style of whisky. This diagram shows both its production and the principles of column still distillation.

1

GRAIN

A selection of cereals can be used in grain whisky production. In Scotland these days, the preferred base grain is wheat (for cost reasons), but in the past it was corn (maize). Each type of grain will give a different final spirit, wheat being generally lighter than corn. Malted barley is also used in the mash bill for its enzymes.

2

COOKING

The base grain is cooked in a pressure cooker in order to hydrolyse the starch. When the cooking process has been completed, high diastatic malted barley (i.e. with a high enzyme content) is added to convert the starch into fermentable sugar. In some countries, enzymes are allowed to be added to aid this conversion (but not in Scotland).

3

FERMENTATION

The wort is pumped into a fermenter where yeast is added and a vigorous fermentation takes place, giving 'wash'. This is normally completed in 48 hours.

4

DISTILLATION OPTION 1A: COFFEY STILL A

Aeneas Coffey designed this still in 1831 and the basic design is still used today. It consists of two tall linked columns (the analyser and the rectifier), each of which is divided internally into compartments by perforated trays. Wash is fed in through a coiled copper pipe running from the top to the bottom of the rectifier. From there the pipe rises to the top of the analyser where the wash is sprayed out over the top plate. It then begins to descend through the compartments while live steam is pumped into the bottom of the analyser. As this rises through the perforations, it meets the descending wash and strips off the alcohol.

5

DISTILLATION OPTION 1B: COFFEY STILL B

The alcohol-rich vapour is piped to the foot of the rectifier where it begins to rise. Because different compounds fractionate at different temperatures, as it moves up the column, the spirit starts separating into its constituent parts, with the heavier alcohols condensing into liquid and lying on the plates. As the vapour rises, so it gains in strength but loses weight. Only the lightest and most fragrant alcohols can rise to the top of the column where they hit a solid plate and are drawn off into a condenser and collected, normally at between 90–94.8% ABV. Though strong, Coffey malts have an oilier character than grain spirit made by the other two options.

6

DISTILLATION OPTION 2: THREE-COLUMN STILL

Here the wash enters the top of the first column (stripping column), which removes the most volatile compounds. The alcohol vapour then passes through an intermediate column and finally to a rectifying column, both of which have ports attached at set points in order to draw off specific congeners. This tends to make a lighter style.

7

DISTILLATION OPTION 3: MULTIPLE COLUMN STILL

After running through two columns, the spirit passes into a hydroselector where it's mixed with hot water. The volatile congeners vaporize and are drawn off from the top of the column, while the diluted solution passes to another set of columns to be redistilled. This gives the lightest and 'cleanest' spirit.

8

MATURATION

Grain spirit is normally aged in first-fill ex-Bourbon barrels, which adds a vanilla/coconut character to the finished whisky.

STILLS

Stills come in a multiplicity of shapes and styles, but they all perform the same function in the production of whisky. They are where in which the low-strength fermented wash is heated. As alcohol boils at a lower temperature than water, so it (along with the flavour compounds previously created) vaporizes and then rises up the still's neck and onwards to be condensed back into liquid.

Distillation is not about creating flavour but is concerned with the process of concentration and selection of the flavours that the distiller desires in the new make.

As the alcohol boils, so the most volatile elements, which are also the lightest in terms of aromatics, are liberated first. As the distillation progresses, so the compounds with higher boiling points, which are also heavier in terms of flavour, are released. Ultimately, it's up to the distiller to decide what to retain for maturation and what to reject.

The stills used for whisky are made from copper. As a reactive metal, it holds onto heavy elements such as naturally occurring sulphur. As the vapour rises, so it comes into contact with the slightly cooler copper surface. When this happens, some of the heavier elements are condensed and fall back into the boiling liquid. This process of internal condensing is known as reflux. The more reflux that takes place, the lighter the spirit will be. A small still will tend to produce a heavier style, while a tall/large one will tend to make a lighter spirit, but distillers can find ways in which to run against type, so don't jump to conclusions!

Although every still operates in essentially the same way, its shape manipulates the way in which the vapour cloud moves and interacts with the copper. By adding certain features (see below), distillers can move the cloud to increase or decrease reflux. Equally, the speed of distillation will also affect the nature of reflux. The faster distillation is run, the less reflux; and conversely, the slower, the greater chance for copper conversation.

How the still is heated can also add character. Direct fire (naked flame) will create a Maillard reaction between dead yeast cells and cereal particles, with the hot copper adding depth.

Here are five of the most commonly used still types, each of which has numerous variations.

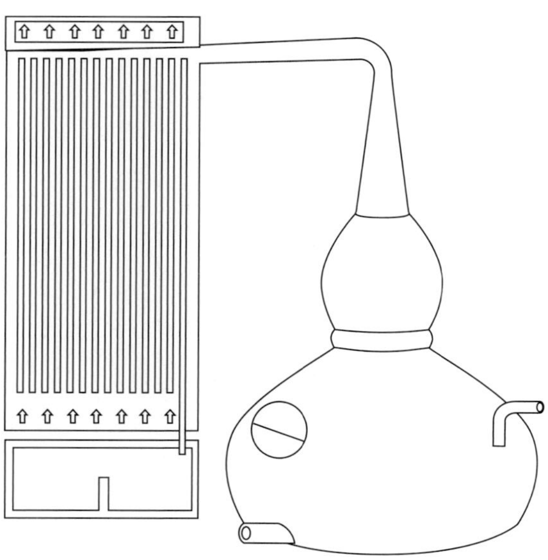

A. POT STILL

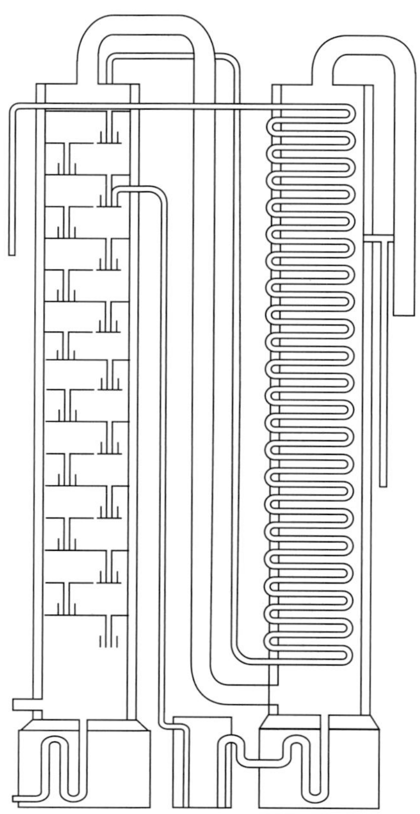

B. COLUMN/COFFEY STILL

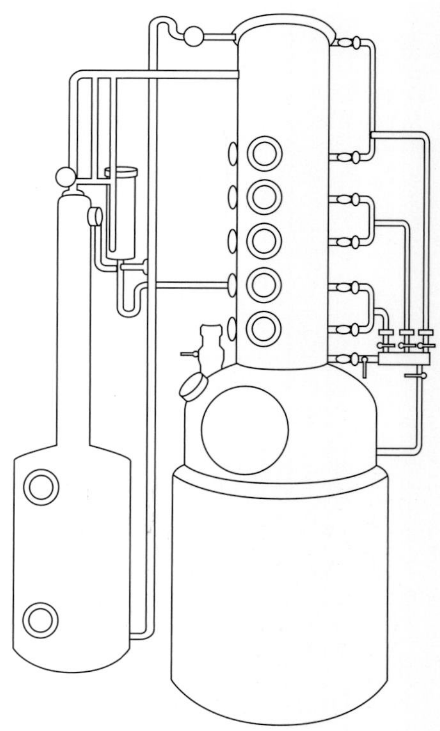

C. HYBRID STILL

POT STILL (A)

The most widely used still type, the pot still basically consists of a broad base, neck and lyne arm that runs into the condensing system, but it can come in many different sizes and shapes. A tall neck will allow for more reflux, while the angle of the lyne arm also has an influence: upward-facing increases the amount of copper and potential reflux; downward-facing and the reflux will stop and a heavier style is more likely to be made.

Pot stills can have a boil ball or bulb between the base and the neck, which helps to increase reflux as well. Purifier pipes can additionally be installed in the lyne arm to carry refluxed elements back into the base. It requires two (sometimes three) distillations or 'passes' to reach the required strength and character.

COLUMN/COFFEY STILL (B)

Pot distillation is a batch process and therefore time-consuming. Column stills were the 19th-century answer to the challenge of making an efficient still that could operate continuously. As long as wash entered at one end, so finished new make would come out the other. The still invented by Aeneas Coffey in 1831 remains the model for this process. Column stills are very tall, so only the lightest (and highest-strength) compounds can be collected as new make. (The operation of column stills is explained on page 15.)

HYBRID STILL (C)

Hybrid stills combine the pot and column approaches. The wash is boiled in the pot and then rises through the neck and into a linked column, which is divided into a series of rectifying plates. In each of these, a mini distillation tales place, with the heavier elements refluxing out and only the lightest being carried up to the next plate where the process is repeated. This allows a distiller to make a desired new make character with a single pass.

In some models, the rectifying plates are contained in the neck of the pot. Because each of the plates in a hybrid still can be engaged or disengaged, reflux can be further manipulated, allowing a higher degree of control than in a pot still.

CHARENTAIS STILL (D)

This type of still – originally designed to produce Cognac – is being widely used in France, but is also appearing around the world. The pot has an exaggerated rounded head (chapeau), which encourages a lot of reflux, as does the elegantly arced swan neck that then runs into a worm tub. Sometimes this pipe runs through another vessel containing the wash for the next distillation, warming it up and saving energy. The majority of Charentais stills are direct-fired.

HOGA STILL (E)

A simplified Charentais shape, this still is closest in style to the alembics used in the early days of distillation. It comprises a fat, rounded base with a smaller head and a swan neck. The Portuguese firm Hoga produces these stills in sizes ranging from 5-litre (1.3-US-gallon) table-top models to 1,000-litre (264-US-gallon) capacity. Most are direct-fired. The flexibility in terms of size and also lower costs make Hogas popular with many start-ups.

D. CHARENTAIS STILL

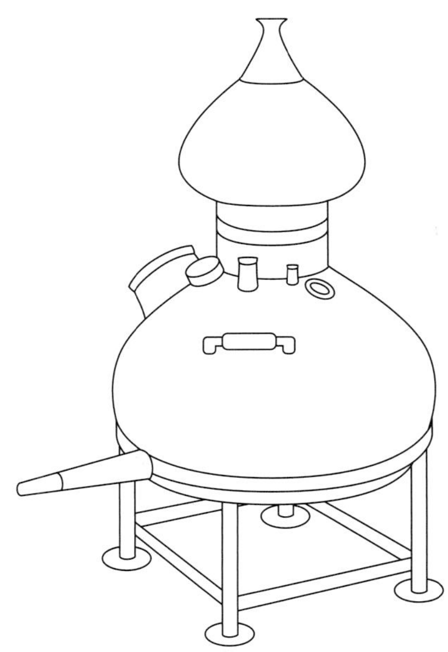

E. HOGA STILL

PLACE

Whisky can be a way of looking at a country: its history, people, stories and thinking. It's a distillation of sensibilities and conditions which creates a spirit that is not just from a place but of it. Some call this terroir, but that concept focuses on the physical environment and its influence on the object in question. There is the father of bioregionalism Patrick Geddes' Valley Section, a diagram of a river valley flowing from mountain to sea, with the occupations of the people in each part showing how land, people and culture are entwined: 'Work, Place, Folk'. The people of the Kaurna country in the Adelaide Plains call it *pangkarra*, how flora, fauna, people and cosmos are connected to a location.

This idea of 'place' begins with those conditions – how geology and weather create a specific environment, which in turn dictates what grows where, while the climate will also impact on the nature and rate of maturation. Then there is the human interpretation of what arises from these conditions – ways of drinking, matching spirit to food culture.

Place is also alive to stories. It is not a single narrative, but many. It is concerned with the where and the how, about the ways in which people live and engage with their location and the manner in which different perspectives create the weave and pattern of a culture. Opinions and approaches vary – but that is the point. It is the engagement that matters.

Whisky's story is not just about what it's made from but where it's made, how it's made and, more interestingly, why it's made that way. It's more than just a liquid in a glass with, god help us, a score attached, but something that has been created by the interplay of geology, weather, farmers, maltsters, distillers, coopers and many more.

Above *Can the mineral content of water help to define distillery character?*

Every country starts from the same point: what grows here, how do we harness it and how does it reflect where we are and who we are? This is what has been picked up by the new distillers, both in 'old' whisky-making countries and new – a heightened awareness of the anchoring role a distillery plays within its community.

Ask a single malt distiller from one of those newer whisky countries what they are making and their immediate response is, 'We're not making Scotch. We're making whisky from here.' The next question is, 'How are you doing this?'

One barrier to achieving this has been the commodity system, which controls cereal production, resulting in distillers only being able to draw from a set of approved varieties whose quality is measured by their yield (how many litres/US gallons of spirit per tonne/US ton), not by their flavour. They are bred to be grown only with the assistance of fertilizers, degrading soil and increasing the release of nitrogen – a death spiral.

The picture is changing, however. Malt distillers and crop scientists (see, for example, page 111, International Barley Hub) are exploring location-appropriate cereal varieties, and rediscovering landraces, while Mexican and US distillers are working with heritage corn and ryes and in doing so preserving them.

Farmers are being paid by the hectare (acre), not the yield, and distillers are sourcing from their immediate surroundings where possible. On the agricultural side, there are more farmers implementing regenerative practices, no-till methods and switching to organic. Place is inextricably linked to sustainable approaches, all of which are ways of retying knots in a relationship between farmer and distiller that has frayed.

It's not about fighting the place and forcing it into an approved way of making whisky, but understanding it in all its peculiarities: what grows here, how to use resources wisely, what flavours can be created. If smoke is desired, then how to minimize the use of peat (and help restore peatlands) or explore new/old smoking techniques taken from local food culture – wood, nettles, seaweed, animal dung.

All of these are answers to that question, 'What can 21st-century whisky be?'

Above *Finding the right cereal variety to suit conditions, and add flavour, also helps to define place.*

HOW TO TASTE WHISKY

We might buy our first glass (or bottle) of a specific whisky because of a recommendation, label or price, but the second time we buy the same whisky it will be because we like its flavour. As we start to explore the wider world of whisky, we continue to do it by flavour – maybe using the experience of that first revelatory dram as our reference point.

You don't have to be an expert to taste or appreciate whisky. Every glass, even of a whisky you're familiar with, will be the first time you've ever tried this whisky. The setting is different; so is the atmosphere, humidity, time of day, company, sound...you. Approach every whisky as something new to be discovered.

The Zen priest Shunryu Suzuki wrote, 'In the beginner's mind there are many possibilities, but in the expert's there are few.' It's important to have a beginner's mind when tasting. None of us are experts. No one knows everything about whisky.

We all pick up the glass and inhale, gently. Scent molecules rise, enter our brain and pictures are triggered. We all know how to smell, but what appears scary about whisky is the idea that you need a different language to appreciate it. You don't. It's a matter of having confidence in your own abilities.

Aroma is memory – it's personal, it's cultural. My frame of reference is Scottish. My aromatic mind map will be different to those of friends in China or the USA, but we are all saying the same thing. So trust your nose, not the 'experts'.

Above & above right *You don't have to be an expert to taste whisky...but it helps.*

Start with asking simple questions.
Light or heavy? Sweet or dry? Smoky or not? Is it fruity? If so, what fruit? Does it remind you of a season or a place? If so, why? That question can often break down the wall.

Then taste.
How does it feel? Then another sip. Let it flow and see how it changes from the tip of the tongue to the middle – where all the elements coalesce – and then the back. It should change, evolve. Listen to it.

Always look for three things:
1 Balance: All the elements are working in equilibrium – sweet balanced by dry, for example.
2 Complexity: Each time you smell and taste, something new appears – the flavour is compelling.
3 Character: It has an individuality. No other spirit has such a complex array of flavours. No other spirit spans the spectrum from whisper light to heavily peated and all points in between.

Don't look at a whisky as being a brand, but as a flavour package. If you understand flavour, you understand whisky.

FLAVOUR CAMPS

As two distilleries next door to each other can make significantly different styles, it makes more sense to group whiskies together within broad flavour camps than by region. The flavour camps indicate the dominant character, so read the words!

Understanding shared characteristics allows you to guide yourself around the fascinating but often daunting world of whisky with confidence. Also, remember that the cask that a whisky has matured in can result in different bottlings from a distillery falling into different flavour camps.

Some whiskies have a star next to them to indicate high quality and classic examples of a style.

All the distillers featured were asked to send two samples of their choosing for tasting. Sadly, Brexit-generated layers of bureaucracy meant that not all managed to reach me. Some distillers, however, sent more. I then picked the examples I felt best highlighted the distillery character.

FLAVOUR CAMPS USED IN TASTINGS

FRAGRANT/FLORAL
Light and delicate, with low oak. Blossom, cut flowers, cut grass, green apple/pear. Fresh and lively.

FRUITY/SOFT
Medium-bodied. Ripe orchard and tropical fruits: peach, apricot, mango, banana, orange, citrus; some vanilla creeping in.

MALTY/NUTTY
Dry. A distinct cereal quality: malt extract, porridge, mixed nuts, straw/hay, rabbit food.

RICH/DEEP
Full-bodied. Savoury, deep, cask-driven aromas (mainly ex-sherry). Dried/fresh black fruits, liquorice, fig, date, incense, resin, meat.

SMOKY/PEATY
Pungent. Woodsmoke, bonfire, tar, seaweed, smoked fish, smoked tea.

OILY/SPICY
Mainly single pot still: leather, oil, green apple, ginger/pepper, citrus, fresh barley.

SPICY/HERBAL
Mainly rye: dill, fennel, rose, clove, allspice, caraway, earthiness.

SWEET/VANILLA
Mainly corn-derived: fat, cornbread, sweetcorn, crème brûlée, vanilla pod, cream, coconut, maple syrup. Some crossover with fruity/soft due to the use of American oak.

OAKY/DRY
Dry and heavy-bodied. Gripping, tannic, stewed black tea, woodshed, varnish, bitter.

Above, right & far right *Humble ingredients, but complex results.*

SCOTLAND

Twenty-first century Scotch is a complex beast. It remains the biggest global player, but one that is now facing up to the reality that many other countries in the world are not only making whisky but doing so to a high standard. Export will be key to the success of the new(er) arrivals – and established markets like the USA and Japan. Scotch no longer has the world to itself. Rather, it is feeling the hot breath of its competitors on its neck. And that is a good thing.

LOOKING BACK

A quick history lesson. At the start of the 20th century, there were three major whisky-producing countries: Scotland, Ireland and America. Prohibition put paid to the latter's expansion, and although production restarted, albeit at smaller scale, in 1933, the industry shut down again for the duration of World War II, effectively meaning virtually no whiskey was made between 1919 (earlier in Kentucky and Tennessee) and 1946.

Most of Ireland gained its independence in 1921 and lost access to the British Empire market. Fatally, its distillers refused to engage with bootleggers in the USA, unlike the Scots. Add in high domestic taxes and an export ban, and the Irish industry imploded.

Scotch struggled in the late 1920s and early 1930s but, though bloodied, was the last man standing. Global hegemony was the result and it held that dominant position for close to a century.

The world wanted whisky and Scotland delivered with its blends. That meant larger distilleries, higher-yielding barley varieties and centralized maltings. It was the right thing to do commercially and was done brilliantly.

THE NEW REALITY

The whisky world, however, is changing. The planet's new distillers proudly (and rightly) all say that they are not making Scotch but whiskies that reflect their location, culture and personality. The consequence of this is an existential crisis for Scotland's distillers. Who are we? What are we? How are we different?

It's not been easy to engage with this new reality and realize that for all your heritage and quality there are now other voices, flavours and approaches all with compelling stories. Scotch at times has seemed trapped within its own mythos, believing naively that since it is the biggest, that means it's the best, that its scale means it's impervious to challenge.

Scale has also had an impact on flavour. Since the 1970s, firms have been continually pushing for ever-higher yields (achieving the maximum litres of alcohol per tonne of barley). Today, the old heads from the production side quietly regret this rush for efficiency that has, in many cases, resulted in a thinning of the resulting spirit.

Right *Scotland's landscape has been a powerful asset in the image of its whiskies.*

There has also been the push to luxury. There is no reason why Scotch shouldn't play in this area – Cognac has been doing it for years – as there clearly is demand. The question as ever is balance. Luxury is one (albeit lucrative) element within a complex category whose strength has always come from delivering quality at all price points. Scotch wasn't built by luxury brands but by blends that offered a reward to all people. Luxury has to be more than packaging and opaque messaging: it has to offer quality, and a justification for the price tag. Ultimately, Scotch has to remain democratic, not elitist.

LOOKING FORWARD

We are at an inflexion point in Scotch's story or, to be precise, at another one, as these changes seem to come at 20-year intervals. In the 1960s, it was the ascendency of blends and satisfying the demands (in volume and flavour) of the US market. By the 1980s, distillers were having to deal with a crash in sales and how to attract new drinkers. By the turn of the 21st century you see the rise of single malt, tapping into the desire for provenance, a widening of the flavour palette and premiumization. What will happen as the 2020s roll on?

That thought reminded me of a chat I had with Emma Walker, Johnnie Walker's master blender, where she spoke of Scotch moving through different eras. The late 19th century saw the start of an Era of Business, while the Era of Engineering started in the 1920s when the industry got to terms with how to make whisky at scale. The Era of Science started in the 1970s as the secrets of whisky were explored.

Above *Glenmorangie's Lighthouse is a beacon for forward-thinking approaches.*

Now? 'It's the Era of Freedom,' she told me. 'We have established businesses, we know how to make it and have an idea of the finer chemistry. Now we can use all of that to play!' To hear that is hugely encouraging. The sense of stasis that was creeping into Scotch has gone and the shape of 21st-century Scotch is beginning to emerge.

It takes two pathways. Volume is clearly still needed because of the importance of blends. That means that not every distillery can be wildly experimental. The old model will therefore remain the same for one side of the modern industry. Single malt might be rising in popularity, but there are still 30 million cases of blended Scotch to be made every year.

The other pathway is dedicated single malt distilleries that play to different rules, which can – and should – push the envelope. It's not right or wrong but it's a choice.

The question now is, what is the new single malt Scotch? The answer to that isn't coming from the mainstream but from the fringes, often geographically: the west and Hebrides, Fife, the Borders. Innovative, smaller distilleries looking at the whiskies coming from around the world and seeing what's there. The world of whisky has learned from Scotch. Now the opposite is happening.

It's about exploring the past and adapting old techniques as well as adopting new ones. It's about having flavour rather than yield as the motivating factor. It's about yeasts and longer ferments, different types of stills, direct fire, new oak types and other cereals such as rye. It's finding a quality- and flavour-driven path that sets them apart and moves Scotch onwards. Crucially, it's about rediscovering and re-establishing links to place, a shift from distillery as 'plant' to the centre of a community.

That identity with place means closer links to farmers, maltings, agronomists and the producers who stand behind the casks. Sustainability sits at the heart of it all, because sustainability means having to be local, and ultimately that is what single malt is about – flavours that can only come from this one spot in the earth and which is made by these people.

We're only at the start of this shift, but in art or music, change is made by smaller movements that have an influence greater than their commercial success. That change, and the necessity for that change, always happens outside the mainstream before seeping in slowly, becoming absorbed and ending up as a new mainstream, by which time two decades have passed and the next shift is taking place on the fringes.

That's what is happening in Scotch – the smaller players finding a new route, influencing the more perceptive larger players. Post-punk whisky making.

It struck me that there are parallels here with the Arts and Crafts Movement of the late 19th century: the elevation of the handmade, the useful, communities of makers, the beauty of craft and its 'honesty', the belief that mass production and homogenization stifles creativity, and beneath that the concept that beauty should be affordable to all.

Scotch is in a good place. There are more distilleries of all sizes, spread across the entire country. The gaps that existed are being filled, not only in a geographical sense but also in terms of production. Scotch had to change; it had to take on board that the world was not the same, and that if it was to retain its position things could not remain the same. That change is underway. New whiskies, a new generation, new thinking.

SCOTLAND'S WHISKY REGIONS

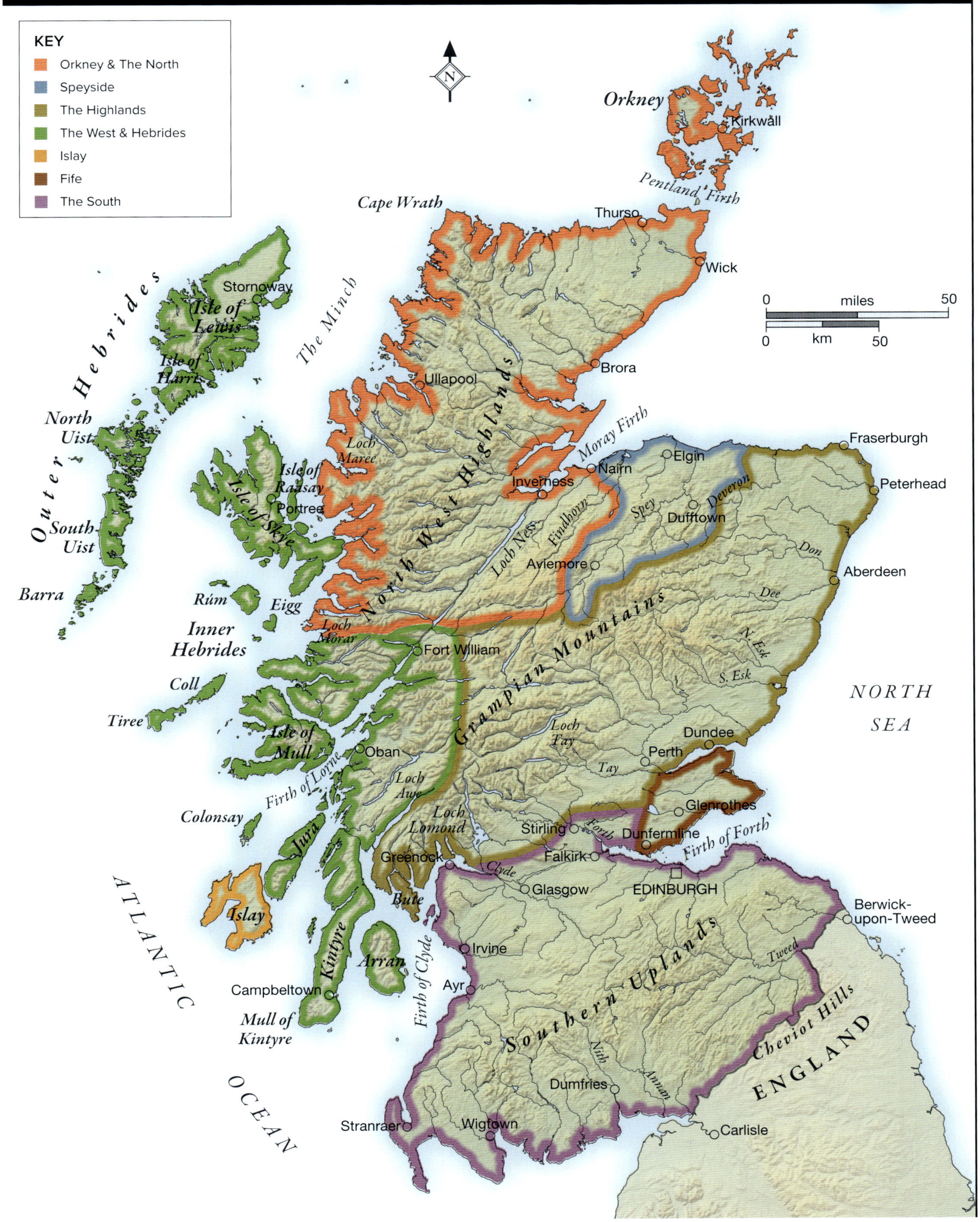

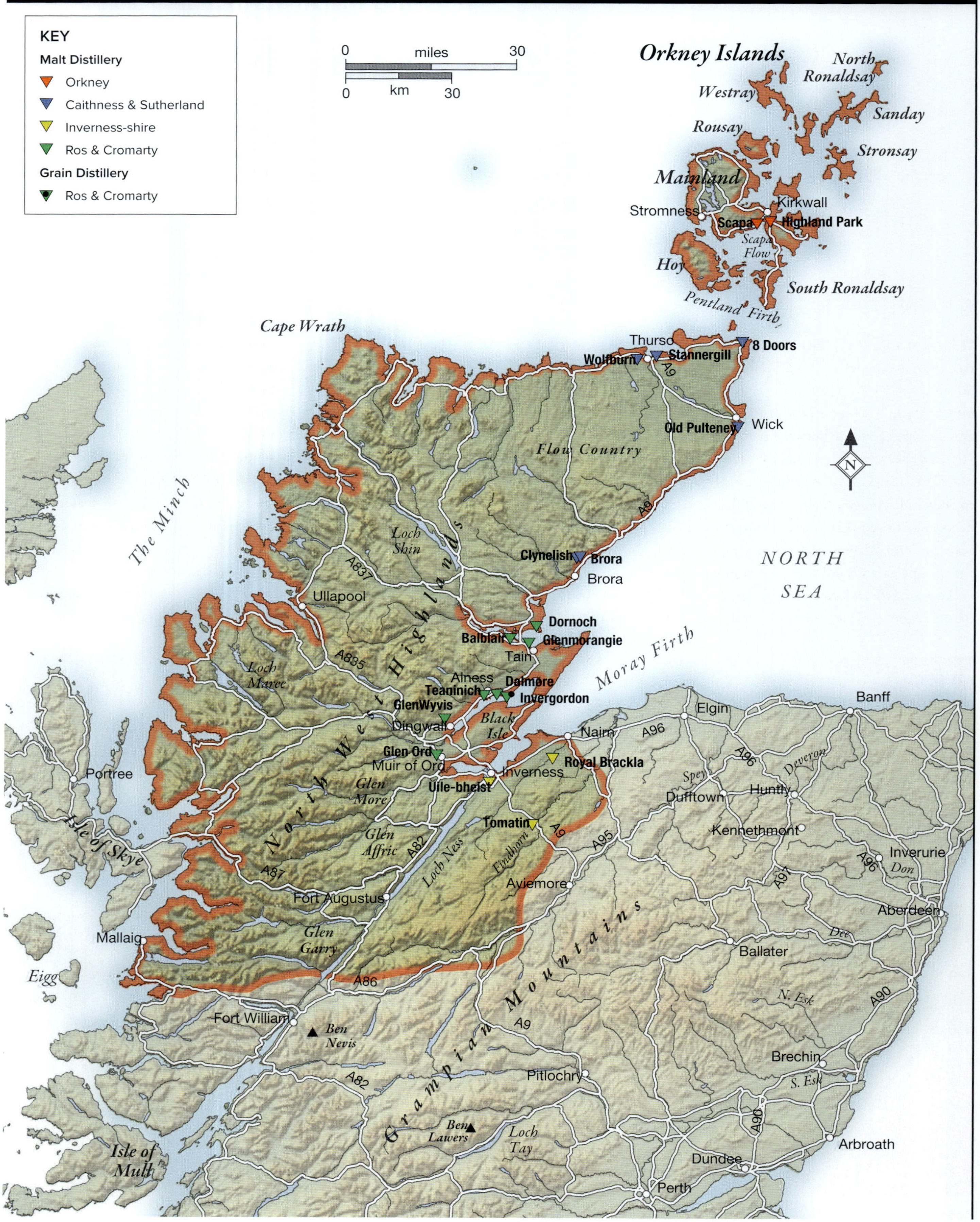
KEY
Malt Distillery
Orkney
Caithness & Sutherland
Inverness-shire
Ros & Cromarty
Grain Distillery
Ros & Cromarty
0 miles 30
0 km 30
Orkney Islands
North Ronaldsay
Westray
Sanday
Rousay
Stronsay
Mainland
Kirkwall
Stromness
Scapa
Highland Park
Scapa Flow
Hoy
South Ronaldsay
Pentland Firth
Cape Wrath
Thurso
8 Doors
Wolfburn
Stannergill
A9
Wick
Old Pulteney
Flow Country
N
The Minch
Loch Shin
A837
Clynelish
Brora
Brora
NORTH SEA
Ullapool
Dornoch
Balblair
Glenmorangie
Tain
Moray Firth
A835
Loch Maree
Alness
Dalmore
Teaninich
Invergordon
GlenWyvis
North West Highlands
Black Isle
Dingwall
Nairn
A96
Elgin
Banff
Glen Ord
Muir of Ord
Royal Brackla
Inverness
Uile-bheist
Portree
Glen More
Spey
Deveron
Dufftown
Huntly
Tomatin
A9
Kennethmont
Isle of Skye
Glen Affric
A82
Loch Ness
Findhorn
A95
A97
Inverurie
Don
A87
Aviemore
Fort Augustus
Aberdeen
Mallaig
Glen Garry
Grampian Mountains
Dee
Ballater
Eigg
A86
N. Esk
A90
Fort William
Ben Nevis
A9
Brechin
A82
Pitlochry
S. Esk
Ben Lawers
Loch Tay
A90
Arbroath
Isle of Mull
Dundee
Perth

ORKNEY & THE NORTH

The far reaches of Scotland, heading on from Inverness to the north coast and beyond to Orkney, both fascinate and confound when it comes to whisky making. There is no single character to be identified – instead, there are oils and waxiness, fresh fruits and density, light malt and grain, and all points in between. It is those particular individualities that matter.

ORKNEY

Orkney isn't Scotland. That much is clear as soon as you either sail past the banded sandstone cliffs of Hoy into the medieval village of Stromness, or fly into Kirkwall Airport. Orkney is low-lying, green and purple; its light is blue and watery. The sea's presence is constant, exerting an uneasy relationship between it and the land. A liminal place.

The dialect is different too, the accent more Nordic, as are many of Orkney's laws. Legally, Orkney and Shetland have only been loaned to Scotland, admittedly since the 15th century, given as security for the dowry of Princess Margaret of Denmark and Norway who was marrying Scotland's King James III. Should the dowry finally be paid, in theory the archipelagoes could become Scandinavian once again.

Barley has been farmed on Orkney since Neolithic times, and in 2007 on the island of Wyre, one of the largest collections of Neolithic charred grains in Scotland was discovered. Archaeologists have concluded that the building in which the grains were found could have acted as a granary for much of Orkney. What were these grains? Naked barley (a variety with no husk) and what we would know as 'normal' barley, landraces naturally adapted to the conditions.

While there is evidence that brewing was underway in the Neolithic, there was no distillation, but without barley and then beer there would be no whisky. Distillers use what grows around them, so Scotland's farmers and early distillers, particularly in the north and west, used those landraces of barley such as bere as well as oats. It is bere that, along with Scotch Common and Scotch Annat barley, made all of Scotland's whisky until the early 19th century.

Plantings of bere had died out by the end of the 19th century with only a few isolated plots remaining. One was on Orkney. Thanks to the Barony Mill at Birsay and the work of Dr Peter Martin at the

Above *Orkney is an ancient land.*

Above *Highland Park's chunky stills help to add subtle weight to its whisky.*

Orkney College University of the Highlands and Islands in Kirkwall, bere was not just saved but in recent times has begun to be grown in commercial quantities and sold to brewers and distillers – most notably Bruichladdich (see page 103).

Church official-cum-smuggler Magnus Eunson would have used bere to make the whisky in the High Park above Kirkwall. In 1798, with a move to a site lower down the hill and a new licensee in David Robertson, **Highland Park** was born.

Where is Orkney in Highland Park? It lies in that thread of smoke that runs through each of the expressions – most apparent in recent years in a 15-year-old release. All of the whisky's peated component is floor malted at the distillery and smoked over hand-cut Orcadian peat from the distillery's Hobbister Moor. You have to dig to find its peculiar qualities. The salt-laden air burns saplings, resulting in a virtually treeless landscape. The flora is heather and mosses that, when compressed for thousands of years and then cut, dried and lit in the distillery's kiln, release a fragrant, slightly green, turfy, floral, heathery smoke. In the 19th century, Highland Park even had a heather shed and sprigs were thrown onto the fire.

In recent times, the kilning regime has changed, with less peat being used to provide the required result. Now only a third of the licensed permitted amount is being cut at Hobbister annually.

The level of smokiness is dependent on the weather. If the day is still and the smoke sits in the kiln for longer, the batch will be smokier. High wind, on the other hand, might pull the smoke through more quickly, meaning lower phenols.

HIGHLAND PARK

HIGHLAND PARK 18 YEAR OLD 43%

Fruits and cereal, and the musky character of Orkney peat. Light chocolate, Darjeeling tea, tweed, old apples. Light struck match, heather blossom and toffee.

FRUITY/SOFT

★ HIGHLAND PARK 25 YEAR OLD 46%

Mature. Cedar/sandalwood, fragrant musky smoke, waxed fruit, medlar, tablet and bookbinder's glue. Savoury and luscious, dried peels, glacé cherry. Enveloping elegance.

SMOKY/PEATY

Above *Sorting peats at Hobbister Moor for Highland Park's kiln.*

Even after each peated batch is blended with unpeated mainland malt, as it is for every expression, there will be small variations. Add in the individual characteristics of each cask, which themselves can accentuate or suppress smoke, and you have complexity. These variations are welcomed, not ironed out by spirits giant and owner Edrington.

The malting barns still work, the kiln still burns. On one visit I was asked if I had noticed anything new since my last visit. I had to confess I hadn't. 'Good,' came the reply, 'that's how we like it.'

That's not to say things haven't been tweaked and changed, including yeast type, cask recipes and barley varieties – there is no longer any bere, for example. Of course things have moved with the times – that's whisky – but the feeling of permanence remains. That's whisky as well.

The air itself helps in the creating of the distillery character. Orkney is mild, albeit windy, (all the distillery's electricity is wind-power generated), and this lack of temperature extremes results in a gentle maturation.

Whisky aged here can easily reach 40 years without becoming too aggressive, and you could argue that the malt only starts to hit peak maturation when it's in its mid- to late teens. This is when the distillate's sweetness becomes more caramelized and fudge-like, the citrus more pronounced, the dried fruits from ex-sherry casks pushing forward and the musky Orcadian smoke that is the ultimate signifier finally exerting itself fully.

Its neighbour **Scapa**, down the hill by the river next to the Flow, is owned by Chivas Brothers. It's home to one of the last Lomond stills (see box, page 81), albeit with its plates removed. For many years, Scapa's fermentation times were, at over 100 hours, the longest in the industry. Now they seem more in line with the recent global trend. That time spent with lactobacillus and the peculiar, copper-rich, oil-drum-like neck of the still helps to promote an oily, super-juicy, peachy style.

Most excitingly, though still in cask at the time of writing, there is a whisky made from Orkney-grown bere. A circle completed.

CAITHNESS & SUTHERLAND

It's only a 90-minute ferry ride from Stromness to the Caithness coast and then, after docking at Scrabster, a 5-minute drive to Thurso and the **Wolfburn** distillery.

The original Wolfburn had a brief existence. William Smith took out a licence to distil in 1820, but it seems the business foundered by the end of the 1830s. There is evidence of a short revival in the 1850s, before it fell silent and into ruin.

In 2011, a new distillery was built close to the pile of stones that was all that remained of the original. On 25 January 2013, the Wolf Burn was once again helping to make whisky.

Above *The Scapa distillery sits on the shores of Scapa Flow.*

There is a 1.1-tonne (1.2-US-ton) mash and clear worts from a very long mashing cycle, and a long ferment of between 72 and 96 hours. There is no automation, and two stills as well as two styles – peated and unpeated.

Wolfburn impressed from the start, light as a new make with a touch of cereal giving a textured grip to the fruity/floral top notes. The use of quarter casks has helped to push maturation, but the extra punch they give (small casks means more extract) is tempered by ex-Bourbon hogsheads and ex-sherry butts. Other fortified wine casks are pressed into service as well.

Now there is a four-strong core range: Northland from quarter casks that ups the cereal and a hint of smoke; Aurora adding dried fruits to the mix; Morven showing the gentle smoke mixing with bright fruits; and higher-strength Langskip balancing ginger with biscuits and dried fruits.

The far north is the most sparsely populated region of the British Isles, effectively cut off from the rest of Scotland by the Flow Country – a vast expanse of blanket peat bog, the most extensive in the world. What seems to be a featureless mix of pools, lochans and peat hags is a vitally important carbon sink, a complex, biodiverse environment where the small details combine to make a huge impact.

WOLFBURN

WOLFBURN 10 YEAR OLD 46%

Nutty, walnut bread, fruit and nut chocolate. More substance on the palate with a lovely, lightly sherried, softly fruity lift in the middle.

FRUITY/SOFT

The fertile straths (valleys) that cut deep into the Flow County from the coastal plain were once densely populated, but were cleared in the 19th century when the people were displaced to make way for sheep and later deer. This resulted in small townships being established on the coast from Wick in the north to Tain in the south. Today, all are connected by rail, making this the only Scottish whisky region where you can visit almost all the distilleries by train (useful if you're having a dram or two).

The cleared people were put to work in a variety of money-making schemes by landowners and Lowland businessmen, of which the most significant was fishing. Nowhere is this more evident than in Wick (from the Old Norse *Vik*, meaning 'bay'), which was once the herring capital of the north.

The fishing industry brought in people. Thirsty people. In the 1840s, when over 1,000 boats were berthed in the harbour, over 800 gallons (3,640 litres/960 US gallons) of whisky were being consumed in Wick every week. Unsurprisingly, a strong temperance movement emerged and from 1922 to 1947 the town was dry, resulting in its sole distillery, Pulteney, closing between 1930 and 1951. Now the fish have gone, but **Old Pulteney** remains.

There was significant moonshining going on in the north in the late 18th and early 19th centuries, being one reliable way people could

Above *Old Pulteney has survived all the vicissitudes of Scotch whisky's convoluted history.*

OTHER CAITHNESS DISTILLERIES

In 2022, the far northern pair were joined by **8 Doors Distillery** at John O'Groats, the most north-easterly point on the British mainland. It's named after the fact that Jan de Groot (after whom the place is named) had a house with eight sides. Ageing is in ex-Montilla casks supplied by Casknolia in southern Spain.

The same year, 5 miles (8km) to the east of Thurso next to Dunnet Bay, Claire and Martin Murray started converting the old Castletown Mill into a distillery where they will make **Stannergill** whisky together with their existing Rock Rose gin.

Above *The new upgraded tun room at Clynelish – where the waxiness starts.*

make ends meet. Legend has it that when James Henderson built his distillery in 1826, he simply enlarged an old illicit pot still. Personally, I think he added a few bells and whistles to the old simple black pot design.

This remains one of the oddest set-ups in Scotland. The wash still is more boil ball than base and has a flat top. The spirit still's lyne arm does a loop before a purifier takes off any heavy elements back into the base. There is massive reflux. Both lyne arms go into worm tubs, which, because of the lack of copper conversation, helps to add weight/oiliness and a leathery/linseed oil note to the new make. Above this funky base, however, are delicate, almost floral elements and a distinct briny touch, something that can be found in some, but by no means all, coastal distilleries.

Part of the International Beverage (IB) portfolio, in 2012 it gave land for the building of a biomass plant using locally sourced wood chips to provide heat for the distillery, 200 homes and the town's hospital.

IB's master blender Stuart Harvey has recently expanded and deepened the firm's range and Old Pulteney now not only has five core expressions but a series of extra-matured releases where the whisky, after spending most of its life in ex-Bourbon casks, spends two to three years in a different type of cask. Harvey believes this gives a more complex end result than short-term 'finishing'.

'Not every cask delivers in six months,' he says. 'For me, it takes that time for the whisky to penetrate the oak and only then can I see what it delivers. It's not just about the previous content but the oak influence as well.'

OLD PULTENEY

OLD PULTENEY 12 YEAR OLD 40%

The freshly washed, briny smell of the morning after a sea storm. Oils, new leather, ham fat, soft pulpy fruit. Unconventional and all the better for that.

FRAGRANT/FLORAL

OLD PULTENEY 15 YEAR OLD 46%

Lively, tangerine, gooseberry and ozone. The vinous side of cider. Oozing fruit juices and slightly oxidized. Brine and pigskin leather.

FRUITY/SOFT

The first of the series was a Pineau des Charentes, which will be followed by ruby port and others. Could 'the manzanilla of the north' end up in a manzanilla sherry cask?

BRORA TOWN

If fishing was the solution imposed on the cleared farmers in the northern straths, then in Strathbrora there were many other options devised by the Marquess of Stafford and his wife the Duchess of Sutherland, who oversaw some of Scotland's most brutal Clearances.

In the town of Brora there was a coalmine, brickworks, tweed weaving and, from 1819, a distillery that the Marquess named Clynelish. By the end of the century, and under significantly better management, it was a considerable success, selling most of its make to private individuals. The arrival of the railway allowed whisky and salmon to be transported overnight to London.

In 1896 it was expanded and in 1930 was absorbed into the Distillers Company Limited's (DCL) malt division, Scottish Malt Distillers (SMD – see page 115), and became an increasingly important player within DCL's blends, especially Johnnie Walker. It was, however, still bottled as a single malt.

The coal supply from the Brora mine was stopped in 1961 and the distillery's floor maltings ceased five years later, the same year that electricity finally arrived. The modern world was encroaching, and in 1968 SMD built a new, larger and more efficient distillery, also named **Clynelish**, closing the old site down.

Clynelish remains a significant element within the blenders' arsenal because of its peculiar character. Every distillery has its own DNA unique to that site. Clynelish has something extra. It's a rare beast – a 'waxy' spirit with candles and sealing wax on the nose and a sticky mouth-clinging quality to the palate.

The fact that the spirit stills are larger than the wash helps to add more copper and reflux, lightening the aromatics – a distinct pineapple aroma that appears near the start of the spirit run signals that things are on track. The secret to the waxiness lies elsewhere, however; it doesn't come from the stills but from the foreshots and feints receiver.

Here, if distillation is kept slow, a thick grey gunk builds up, coating the inside of the tank and surface of the liquid. Push Clynelish too hard and the waxiness goes. The precursors for this have been created earlier on: the clear wort, low gravities and long ferments. No one has pinned down how the gunk works – the fact that it does is perhaps all that's required.

If the distillery character remains a constant, the building itself has had a major refit with a switch from oil to gas that has reduced carbon emissions by a third, a spanking-new visitor centre and an all-singing, all-dancing tour. There is also an excellent bar with drams and cocktails that's open to all.

For years you could stand outside Clynelish looking down on the increasingly sad, silent old site that had a late flaring of life from 1969 to 1981. Briefly closed when the new site opened, a drought on Islay and the rebuilding of Caol Ila (see page 97) meant a shortage of peated stock for the DCL blends.

The old distillery was reopened, renamed Brora, and set to work making a fat, thick, smoky whisky.

CLYNELISH

CLYNELISH 14 YEAR OLD 46%

Waxed jacket/snuffed candle. Some pineapple and sea breezes. Discreet. The silky viscosity with fresh crumpets, butter and honey is succulent, waxy thick and clinging. Steely acids.

FRUITY/SOFT

Its waxiness was heavier, with more oils and leather. The fact that it had worm tubs rather than shell and tube condensers helped, and a different shape of still. From 1974 the peat levels fluctuated, dropping away by the 1980s until the distillery's final production in 1983.

Closed distilleries are romantic, the finite nature of the stock exciting collectors, so a distillery with as unorthodox and compelling a character as Brora was guaranteed to build a reputation. This old workhorse became a cult, the forgotten musician, the lost film. As prices rose, it steadily became unavailable even to those who had revered it.

Then, on 21 May 2021, **Brora** reopened. It has been rebuilt, repurposed and reroofed. An old-style rake and gear mash tun has been installed, using four waters, the ferments top 115 hours and the vapour from refurbished stills once again runs into worm tubs.

It's a considerable achievement for Donna Anderson and her malt distilling team to try to recreate not just one but three styles from Brora's past: smoke, oily and earthy. Will it be the same? As former Johnnie Walker master blender Jim Beveridge said at the opening, 'We have set it up, but then it will do what it will do...and it will be Brora.' A distillery is a living thing.

Above *Clynelish's stillhouse contains the tanks where its famous gunk forms.*

ROSS & CROMARTY

Proceeding southwards, we go past Golspie and Dornoch to Edderton. Hard by the railway tracks are the warehouses of **Balblair**, the second of International Beverage's northern sites. It was the location for the distillery where the cask of Malt Mill was stored and auctioned in Ken Loach's film *The Angel's Share*, a story of redemption through whisky that came as a welcome note of positivity in these times when neo-prohibitionists are ever more vociferous.

Why did Loach choose Balblair? Because it looks like what most people imagine a distillery should be: low buildings, thick granite walls, slate roofs, cupola, brick chimney. It seems unchanged, though the original site, dating from 1798, was further up the hill. It moved because of the railway arriving, opening up what was previously a remote part of the country. Barley and casks could come in, as well as coal – even here to 'the parish of the peats' – and whisky could go out. It made commercial distilling viable.

There is a nod to that former name in the tiny amount of peat in the malt, not enough to make it smoky but, as IB's master blender Stuart Harvey says, take it away and the texture changes.

A thick bed on the mash tun gives a clear wort, while the small squat stills help to give a fascinating new make whose top notes are apples and fragrant fruits, but which also has a creamy, bold body and slightly spicy end. It's also the only one of IB's distilleries not to have worms. All of this gives a complex array of characters that evolves at different times during maturation.

The casks are aged solely in eight, earth-floored dunnage warehouses. Debate ranges across the world as to the best warehouse type, but ask any old hand in Scotch and dunnage wins every time. The earth floors give a damp humid atmosphere, while the thick walls provide an even temperature year round. It promotes slowness.

For years, these different stages of maturation were framed in vintage bottlings, but in 2019 a new age statement range was introduced, while Stuart is working on a series of extra-matured releases utilizing Australian wine casks as well as Moscatel and Marsala. 'We're in no rush to release them,' he says. 'When I think it's ready, I'll tell the team.'

BALBLAIR

BALBLAIR 12 YEAR OLD 46%

The distillery's estery character to the fore; apples, greengage, pineapple sweets and bubblegum. Softer on the palate, but deliciously mouthwatering.

FRAGRANT/FLORAL

BALBLAIR 15 YEAR OLD 46%

Sherry casks are adding weight and Seville orange elements. Pineapple now cooked. As it opens, there's coffee bean, prune, cola and some earthy elements, but still has estery drive.

RICH/DEEP

Above *Balblair has quietly built a deservedly high reputation for its single malts.*

Above *Glenmorangie remains at the forefront of innovation in Scotch.*

Innovation is a vexed word in whisky. It's a spirit that is continually, often slowly, evolving. While some of that change is driven by necessity, it can also be the result of advances in science or engineering. Whisky stands still at its peril, and yet in Scotch you have an industry whose image, whose whole *raison d'être* is based around the notion of being a fixed point, a drink that is reassuring, familiar. Without innovation, however, things become stuck and die. Managing to maintain this forward momentum without it becoming little more than window dressing? Whisky can't do short-term. It needs time.

That those thoughts spring to mind when approaching **Glenmorangie** is no great surprise. This was the very first distillery to commercialize a range of finished whiskies, the first to use chocolate malt (for Signet) and an adopter of local barley (for Cadboll Estate). In recent years, the A Tale of... series pushes the 'What happens if...?' question: a whisky that tastes like cake; another using barley kilned with botanicals. Some of these are adaptations of techniques readapted; some are new. Dr Bill Lumsden is the name behind them all. Glenmorangie's distilleries director is a restless soul who always speaks of a desire to be seen to be moving whisky forward.

Glenmorangie's substantial red sandstone building (originally a brewery) on the shores of the Dornoch Firth seems an unlikely location for such modernist thinking. It only uses Scottish barley, and there are long ferments and distillation in the tallest stills in Scotland whose height prevents any heavy elements getting across. All of the 300,000 casks are matured on site. Tradition seems hard-baked in. But those stills are so tall because they were originally used to make gin and in 1887 were the first in Scotland to be heated with steam. And the water sourced from the Tarlogie Springs is hard. Glenmorangie has always been different.

The casks are predominantly American white oak, many made specifically to Bill's specifications – slow-grown on north-facing slopes in Missouri and then air-dried. Casks are only filled twice, giving signature vanilla and minty accents to the passion fruit and orange of the new make. There are also casks that have held wines from Sauternes to Burgundy and

GLENMORANGIE

★ GLENMORANGIE 10 YEAR OLD 40%

Sweet oak and passion fruit. Vanilla pod, frangipane and mango sweat. Lush and chewy texture. Canned mandarin. Gentle yet complex.

FRUITY/SOFT

★ GLENMORANGIE CADBOLL ESTATE 43%

Lifted and citric, with the amontillado cask adding mixed nuts and acidity balancing the Greek (peach) yogurt, mint, chamomile and melting chocolate. Energetic yet relaxed.

FRUITY/SOFT

elsewhere, fortified wines and many more. Balancing those bold flavours with a light, complex new make is no easy task.

Now all of that 'What if?' thinking is housed in a new glass-fronted experimental distillery called The Lighthouse. There are two Glenmorangie-style stills but with modifications allowing multiple tunes to be played. Here is the playground, the place where the needle is moved: different cereals, specialized malts, yeast types, distillation.

Whisky is also about place. The question is, can innovation damage that connection? 'We don't truly believe that the differences in taste of the many whiskies made in Scotland are specifically to do with geographic location,' says Bill. 'We believe it's more to do with the way the products are distilled and matured.' Does location really not matter? Could you make Glenmorangie in Edinburgh? If you could, why not do it? Why bother moving all those casks to the distillery? What of that water source, that climate? We're back to the tension between the fixed and the need for change. Maybe that's where creativity sits.

EASTER ROSS

The northeast has been a place for the wealthy to indulge in their interests. In 1839, Alexander Matheson founded **Dalmore** in the town of Alness next to the Cromarty Firth. He had made his fortune as a partner in the powerful Asia-based trading firm Jardine Matheson, which at that time was controlling exports of opium into China. Purchased by the Mackenzie family in 1867, it remained family owned until 1960 when one of its main customers, the Glasgow blending house Whyte & Mackay, took charge.

It has retained one of the most unusual distillation regimes in Scotland – one that rivals that of Mortlach (see page 52). The wash stills have flat tops, and are of different sizes: three 13,000 litres (3,434 US gallons) in capacity, the fourth double that. This affects reflux and gives two different characters. The four spirit stills each have what looks like a copper surgical collar on its neck but are in fact water coolers also used to increase reflux.

The spirit stills are charged when the low wines and feints receiver is full, meaning that the strength of the charge fluctuates, so yet more variation. The condensers are horizontal – of course they are. All these distillates are vatted before being casked. Quite how a consistent new make is achieved is a mystery to all bar the people running the stillhouse. They know. In 2023, as part of a complete redevelopment, the team discovered previously unknown cooling tubes inside the neck of the stills.

Dark fruits and cereal are key elements in the Dalmore character, and the widespread use of ex-bodega rather than seasoned casks in its older expressions – which reach up to 45 years old – helps to boost this inherently sweet nature. While a seasoned sherry cask will have wine in it for two years, an ex-bodega cask will have matured sherry for decades. There is less active oak but greater vinous impact. The result is a lush, sweet, concentrated malt – and one that is attuned to the Asian palate. This embrace of luxury seems somehow appropriate.

Just 3 miles (5km) along the coast is the **Invergordon** grain distillery. Though an unlikely location for such a plant – all of the company's other sites are in the Lowlands – it was the perfect solution in 1959 when a regeneration project for the area was required after the closure of Invergordon's naval base three years earlier.

Invergordon is a deep-water port, which allowed grain to be transported by sea, while sufficient barley was grown nearby to provide the malt content.

Owned by Whyte & Mackay since 1993, it has been refurbished with new stills, and has switched from corn to wheat as part of Whyte & Mackay's 'Grow Scotland' initiative that aims to source all raw materials from within 50 miles (80km) of its distilleries. It now also runs off green energy in line with the sustainability programme. A new range of age statement whiskies was released in 2023.

DORNOCH

★ DORNOCH 5 YEAR OLD CASK NO. 124 57.4%

Baked fruit, banana, nutty cheese rind, bracken and Nutella, with an underlying mix of ripe pear, pepper and earth. Rapeseed oil. Young yet complex.

FRUITY/SOFT

OTHER ROSS & CROMARTY DISTILLERIES

Diageo's **Teaninich** sits close to Dalmore. One of the firm's largest sites, it's the only one with a mash filter (see page 14) and as a result is used when experiments need to be scaled up, difficult to process malt appears and other grains such as rye are being used. The character is green and oily.

At the opposite end of the scale size-wise is **Dornoch** (see below) run by the Thompson brothers, who took over the town's tiny old fire station and installed two direct-fired Hoga stills and open-topped fermenters, then got to work using old barley strains and a range of different yeast types. Their back-to-basics approach is driven by a deep love of old styles of whisky and reinforced by an equally profound understanding of distillation. They are a beacon for many new distillers.

You can see a similar thought process at **GlenWyvis** in Dingwall, which released its first whisky in 2021. Local barley, malted in Inverness, is given a six-day ferment and aged in a dunnage warehouse.

Both the aforementioned seem miles away from Teaninich's technological approach and yet underpinning both – and Glenmorangie's Lighthouse – is this idea of what can whisky be. Dornoch and GlenWyvis's approach is far from Luddite, but forward-thinking, learning from the past and adapting for the future.

THE CITY OF INVERNESS

Inverness was once home to three distilleries, but the travails of the 1980s, when the industry was faced with a serious stock surplus, saw all (by then owned by Diageo) close: Glen Mhor and Glen Albyn in 1980, and Millburn in 1985. Leaving the capital of the Highlands and a prime tourist destination without a distillery was a major oversight.

In 2023, however, on the banks of the River Ness, the **Uile-bheist** (Gaelic for 'monster') distillery opened (see left). A 'brewstillery', it also makes beer and therefore uses a brewer's yeast for its whisky. It also utilizes a heat pump to harness heat from the water it draws from the river, as well as solar panels to generate its own electricity.

A single malt distillery will always be constrained by its size, but what if you have global expectations for a brand? Why not create one that covers multiple distilleries? That was the solution taken by Diageo with Singleton, the umbrella name for single malts from Dufftown (see page 53), Glendullan (see page 53) and **Glen Ord**. The last makes up the bulk of the production, most of which is destined for Asia.

Glen Ord is a behemoth, with two stillhouses and a capacity of 15 million litres (4 million US gallons) per annum, and thanks to the large drum maltings on site is one of the few distilleries that malts all of its own requirements. It's located on the Black Isle, whose deep (black) soils are ideal for barley and therefore historically home to not only illicit distilling but the legendary Ferintosh distilleries owned by Duncan Forbes, who supported the Hanoverian side in the 1715 Jacobite Rebellion. In gratitude, the king permitted him to distil tax-free from his own grain. By the time the indulgence was scrapped in 1784, the four Ferintosh distilleries were producing two-thirds of the (legal) whisky in Scotland.

No one knows what Forbes's whiskies tasted like, but Glen Ord's new make has remained intensely green and grassy for many years (though the hint of smoke in older expressions has gone). There is plenty of reflux in the stills, while the water in the shell and tube condensers is kept hot, thereby extending that copper conversation. It's only in the aftercooler that the vapour turns back to liquid. The hot water is looped back to the maltings, further saving energy.

INVERNESS-SHIRE

Tomatin, which lies 15 miles (24km) south of Inverness, was once the largest single malt distillery in Scotland. In 1974, its 23 stills were making 12.5 million litres (3.3 million US gallons) per annum, an enormous amount for the time – and a reckless one, given that the industry was about to head into a serious dip in fortunes. Japanese producer Takara Shuzo came to the rescue in 1986, purchasing the distillery primarily to supply bulk malt for its Japanese blends.

To some extent, history was repeating itself – Scotch is nothing if not cyclical. Built at the end of the 19th-century boom, and also to take advantage of the railway, it only ran until 1906, though it reopened three years later, gussied up by the legendary distillery architect Charles Doig.

There are only 12 stills now, 6 wash and 6 spirit, and there has been a pivot away from being primarily a supplier of fillings for blends (including its own, excellent Antiquary) to the world of single malt.

Tomatin has been a quiet player that has gradually gained greater recognition thanks to a series of remarkable extra-aged releases filled with decadent amounts of the tropical fruit character that only comes with relaxed maturation, itself a time-driven development of an intensely estery/pear eau-de-vie new make, the product of an extra-long 168-hour fermentation.

A boil ball adds reflux, while the downward kink on the lyne arm stops things getting too light, adding a slight vegetal note. It's those esters that, over time, concentrate due to evaporation, with the most volatile flying off to satisfy the angels. Reduce the active oak and allow oxygen to take charge, and the tropicality emerges.

One of the few distilleries in Scotland to have retained a cooperage, all the whisky is matured on site. It houses all the firm's offices, as well as most of the staff, giving a rare sense of this being a truly local distillery.

The arrival of a biomass boiler has reduced carbon emissions by 80%, and draff is sent to a biogas plant that feeds into the mains gas network, while trials are underway with farmers and maltings of varieties more suited to the (changing) Scottish climate. With a now established lightly peated variant, Cù Bòcan, and 'secret' trials running every summer, there are signs that the fourth age of Tomatin is underway.

GLEN ORD

★ SINGLETON OF GLEN ORD CELEBRATORY BOTTLING 51.8%

Rich. Gingerbread with thick butter, cooking spices, date and a flash of green grass. Lovely balance between sour and sweet. Has breadth and complexity.

RICH/DEEP

Some 15 miles (24km) east of Inverness is Cawdor Castle and next to it, surrounded by barley fields, is **Royal Brackla**. An old site, it was founded by Captain William Fraser who, in 1835, was granted the first Royal Warrant by William IV. A keen publicist, Fraser immediately gave Brackla the tagline 'The King's Own Whisky'.

While the prefix is still in place, Royal Brackla's fame has dimmed somewhat. On a back road, off the tourist trail with no visitor centre, and rarely seen on the shelf even as an independent bottling, its life was to provide fillings, first for DCL/Diageo and, in more recent years, its new owner John Dewar & Sons.

Now, like Tomatin, Royal Brackla is another modest player that is finally emerging. Its new make even has some similarities with its Highland companion – needle-sharp and floral, with some pineapple.

While most Royal Brackla is aged in ex-Bourbon casks, the single malt releases have all been given a period of finishing by master blender Stephanie Macleod in different types of sherry cask, adding another layer and a degree of richness on that high-toned character. The 18-year-old, finished in Palo Cortado, is particularly noteworthy where the sherry's austerity matches the brightness of the distillery character and the slight oiliness of the new make is amplified. It confounds the expectation that the older the whisky is, the heavier and more woody it becomes.

TOMATIN

TOMATIN 12 YEAR OLD 43%
Estery lift (pear, rhubarb) but also butterscotch and golden raisin. Water brings out barley and menthol. Soft and unctuous with Tomatin's energy cutting through.
FRAGRANT/FLORAL

CÙ BÒCAN SIGNATURE 46%
Very gentle, sweet toffee and oak. Kumquat. The whisper of smoke on the palate adds another texture. Some pine and a gently rounded mid-palate.
SMOKY/PEATY

ROYAL BRACKLA

ROYAL BRACKLA 12 YEAR OLD 46%
The intensity of the new make cuts through the oloroso finish, which in turn adds a sweet mulberry-like depth. Opposites can attract.
FRUITY/SOFT

ROYAL BRACKLA 18 YEAR OLD 46%
High-toned, lemongrass, cooked grapefruit, bergamot and spiced orange. Waxiness and a savoury edge. Cooked quince and pear. Lightly oxidized. Silky feel. Kiwi, pear and a steady build of racy acidity.
RICH/DEEP

Above *Tomatin has both scale and deep stocks, and is steadily becoming a significant player.*

SPEYSIDE

The Speyside region contains the highest concentration of malt distilleries in Scotland. This wedge-shaped area roughly follows the watersheds of the Spey, Lossie and Deveron rivers, although its legal boundaries are political rather than geographical. Its landscape is varied, with a flat coastal plain, deep valleys and high moorland, and its whiskies reflect this. There is no such thing as a single Speyside style – there are many.

Above *Mountain, moor, river, plain: Speyside has it all.*

ELGIN & ENVIRONS

Stretching out along the coast from Inverness is the alluvial plain of the Laich o' Moray, prime barley growing land, although records show that by the 1920s, the bulk of Scotch whisky was being made from cheaper imported barley.

Whisky may have started to be made here as a small, farm-sized enterprise, but when scaling up became a factor, resources were limited. There was resistance in Moray from farmers and some distillers to this trend. In 1927, *Ridley & Co.'s Monthly Wine and Spirit Trade Circular* ran a series of 18 articles entitled 'Practical Notes on Whisky Manufacture'. When discussing barley, it reported that, although imported barleys produced a 'more silent' whisky with less body, 'it has the advantage of maturing more quickly'. Highland barley on the other hand, farmers and some distillers claimed, had 'fatness'.

'There is no doubt, however, that good Home-grown Barley, when obtainable, produces a Whisky of greater body and of a finer bouquet and flavour and was at one time used, to the exclusion of Foreign Barley, by Highland and North Country Distillers, the best samples being then obtainable from the Morayshire district.'

There was nevertheless growing pressure from the big distillery combines to import, with the lack of grain-drying facilities in the north being cited as another contributory factor. Today there are three large maltings in this area: Bairds in Inverness and Diageo's pair of Burghead and Roseisle. The barley used is local.

Clusters of distilleries soon appear as we head deeper into Speyside, the first in the small town of Forres. By the 1980s, both of its distilleries were silent, but in 1998 **Benromach** restarted, part of the first glimmerings of the single malt category re-establishing itself. A former DCL site, it had been bought by Elgin-based independent bottler

Gordon & MacPhail (G&M – see box, page 42). Intriguingly, although all the equipment was new (bar the mill) and the stills smaller, the new make retained the same character as in the past. G&M's aim was to make a whisky that spoke of an older style, one with a richer body, more fruit and a hint of smoke – a celebration and reimagining of what had been lost. As a result, there are 10–12 phenolic parts per millon (ppm) in the malt (less in the new make), while a cloudy wort adds cereal-led depth. But this character is then layered with fruits coming from long ferments – 67–115 hours that utilize 3 yeast types (2 distiller's and an IPA yeast). It's cut at 61% ABV to get some of the phenols and weight. That fatter style returns. Organic barley has been run, as has Golden Promise – the 1960s variety that adds a thick mouthfeel.

Benromach is only aged in first-fill casks, a mix of ex-Bourbon from Speyside Cooperage (see box, page 49) and ex-Williams & Humbert sherry, and, naturally given the overall ethos, all are placed in on-site dunnage warehouses. It's another example of how some players in Scotch are looking to the past to move onwards. You can't turn the clock back, but you can be inspired by it.

BENROMACH

★ BENROMACH 15 YEAR OLD 43%

Malty. Booze-soaked raisin and a hint of bonfire smoke. A punchy 'old'-style dram. Thickly textured, Lady Grey tea, plum, cherry and soft tannins. Real substance.

RICH/DEEP

Miltonduff, near to Elgin, is part of the Chivas Brothers stable and lies close to Pluscarden Abbey. Romantics like to suggest that the early monks would have distilled. It's possible, but there is no evidence of it. All the Abbey's records were lost when it was demolished in the 17th century (it was rebuilt in the late 1940s). There was, however, distillation underway at the Milton farm (originally the Abbey's mill), which led to the building of the distillery on the site in 1824. By the end of the century it was triple distilling over 1 million litres (264,000 US gallons) a year.

It became part of Ballantine's' blended stable in the 1930s, and from 1964 to 1981 a pair of Lomond stills (see box, page 81) produced a malt called Mosstowie that very occasionally surfaces as an independent bottling.

A significant expansion in 1974 saw capacity increased to more than 5 million litres (1.3 million US gallons) per annum, with 3 pairs of stills and 18 washbacks now operating. In 2005 it became part of Chivas Brothers. Further expansion is underway at the time of writing that will double capacity, along with the introduction of Mechanical Vapour Recompression (MVR) technology (see box, page 47) to reduce carbon emissions.

MILTONDUFF

★ MILTONDUFF 12 YEAR OLD RESERVE COLLECTION SINGLE CASK EDITION 59.2%

Sweetly floral: privet and bluebell. The cask adds sawdust sweetness. Palate mixes of grass, clover honey, vanilla sponge and lemon.

FRAGRANT/FLORAL

Above *One of Chivas Brothers' 'workhorses', Miltonduff's gentle character is highly prized by blenders (and malt lovers in the know).*

SPEYSIDE

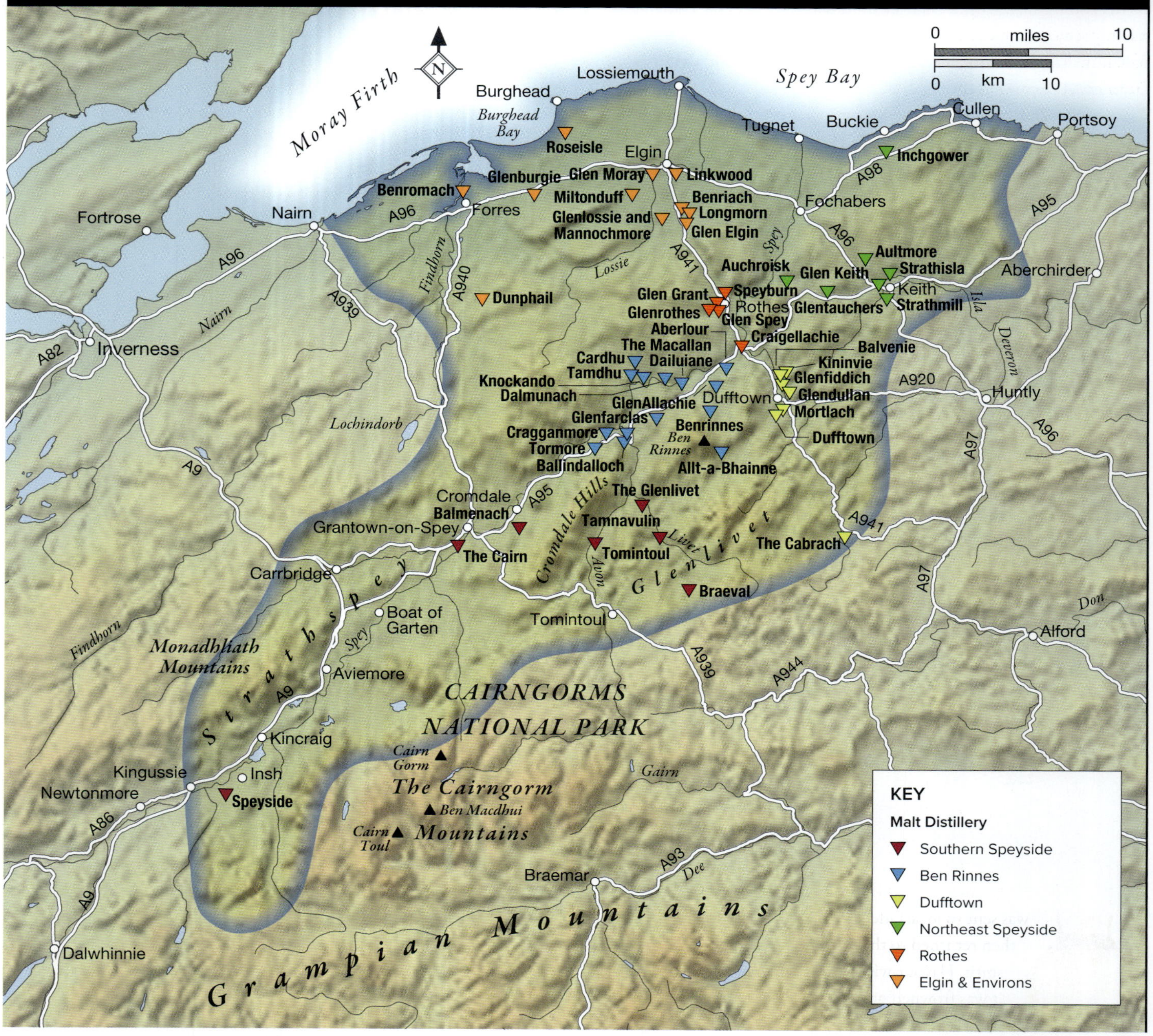

ROSEISLE & DUNPHAIL

In 2010, Diageo opened a large distillery next to its **Roseisle** maltings (see left). Wrongly believed to be an attempt to centralize production and close existing plants, it was built to allow the firm to balance the demand for some single malt styles while also maintaining the flavour and volume requirements for blending. A highly flexible site, its 14 stills make (at least) 3 styles: 'nutty', 'grassy' and 'meaty'. A biomass plant allows it to generate its own power and waste heat is then looped between the Roseisle and Burghead maltings.

Forres has regained a second distillery with the building of **Dunphail**. Combining floor-malted barley (including peated), long ferments and direct-fired stills, it's another in the back-to-the-future grouping of new distillers.

The new make has a floral, green/oily note with touches of cucumber and a crisp finish, fitting perfectly into the Ballantine's style. Normally it's only seen as an independent bottling, but Chivas (which bought Ballantine's in 2005) does release the occasional bottling under its Distillery Reserve collection.

Miltonduff's sister distillery **Glenburgie** lies 7 miles (11.3km) west. Started in 1829 as Kinflat, it produced fillings for the Old Smuggler blend and, from 1936, Ballantine's. It too had Lomond stills, making Glencraig. A new site was erected in 2004. There are now three pairs of stills making a sweet, grassy style.

ELGIN

The largest town in Speyside, Elgin is home to Gordon & MacPhail's amazing repository (see box, page 42) and two distilleries.

Glen Moray sits on the banks of the Lossie – too close at times, as the river is prone to bursting its banks. A substantial site that was built, like so many, in the late Victorian boom of the 1890s, it lay silent between 1910 and 1923 when Glenmorangie's then owner Macdonald & Muir bought it. An expansion in 1958 saw a Saladin maltings being installed, which operated for the next 20 years. Although Glenmorangie used it for some of its finishing experiments (Chenin Blanc and Chardonnay were both trialled), the firm positioned it as a budget brand, which often sold for less than standard blended Scotch. Volumes might have been healthy, but it did nothing for the whisky's reputation – these things work both ways, you see.

It was purchased by French firm La Martiniquaise (Label 5 and Cutty Sark) in 2008. The new owner has invested heavily with three phases of expansion that has seen capacity rise from 2.2 million to 8.5 million litres (580,000–2.2 million US gallons) per annum. With more washbacks, ferment times have been raised from 44 to 60 hours, allowing some more fruit and depth to the new make. There has also been an increased focus on single malt bottlings and at a more sensible price point. The finishing programme has continued, the most newsworthy coming in 2018 with a 'cider cask' finish. An empty cask of Glen Moray was sent to local cidery Thistly Cross to age its cider and then returned to the distillery to be refilled with mature spirit. Flavour-wise it was a success. Legally, however, it was frowned upon – cider casks unlike beer are not permitted to be used to age Scotch.

On the other side of the town is **Linkwood**. An older site (it started production in 1824), it has steadily grown in size and reputation. As with Clynelish (see page 31), in 1972 a new distillery was built next to the original and the plants ran in tandem until 1985.

The old site – which had worm tubs – was then used as an experimental distillery. Much of the research into copper conversation happened here. The old building was demolished in 2011 and a new stillhouse with another pair of stills was constructed. Capacity now sits at 5.5 million litres (1.5 million US gallons) per annum.

Above *Glen Moray remains a leading proponent of 'finishing' whiskies in different casks.*

Although considered as the Speyside representative of Diageo's Classic Malts – it had been part of its precursor the Ascot Malt Cellar – it was passed over in favour of Cragganmore (see page 60). Today it can be found as a 12 year old within Diageo's Flora and Fauna range and from G&M who have had a gentleman's agreement to bottle it for many years.

Again similarly to Clynelish, Linkwood's spirit stills are larger than the wash, the result of a desire to make everything as aromatically light as possible – no grass, no heavy fruit and definitely no sulphur. This doesn't mean that it's neutral. Rather, the new make is fresh and spring-like, with apple blossom and peach skin. The palate, however, has an almost

ROSEISLE

★ ROSEISLE 12 YEAR OLD 56.5%

A blend of three of the distillery's styles: grassy, fruity and nutty. Lemon puffs, burned vanilla. Fragrant, bright, green grass, elderflower cordial, crème anglaise, red berries, light praline.

FRAGRANT/FLORAL

LINKWOOD

★ LINKWOOD 12 YEAR OLD 43%

Nothing comes close to Linkwood's mix of green apple, peach skin, William pear and lime blossom. Mouthwatering and lightly perfumed. A silky, oily palate. Highly complex.

FRAGRANT/FLORAL

GLEN MORAY

GLEN MORAY CLASSIC SINGLE MALT 40%

Foam bananas, a bready back note. Simple, easy-going and light. A mix of peanut and soft fruits on the palate.

FRAGRANT/FLORAL

oily, clinging quality, adding surprising depth. You can see why blenders and true whisky aficionados adore it. Even with that extra capacity, sadly, there is no sign of it becoming a frontline malt. A secret waiting to be discovered.

SOUTH OF ELGIN

Another cluster of distilleries is found on the road from Elgin to Rothes. Although four prefer to hide from public gaze, the first can hardly be missed. **Benriach** has proudly painted its name in huge black capitals on the side of one of its warehouses.

Whisky making in this part of Speyside came into being in the late 19th century, which turned out to be the least propitious time for building distilleries. A downturn in sales saw many of the new plants closing. One of those was Benriach.

The third of the stills built in the area by John Duff, it opened in 1897 and closed two years later. It lay silent until 1965, although its malt floors were used to supply neighbouring Longmorn. In 2001, Chivas Brothers closed it once more, but this time fate smiled kindly on Benriach. Two years later the site was sold to Billy Walker (see page 56) and, over 100 years after it first opened, Benriach's story finally began.

It's one of the few distilleries still using four waters in the mashing, which they claim helps to increase the sweet extract that's needed to create the distillery's distinctly orchard/stone fruit signature. There is an underlying maltiness – more noticeable when young – and a spiciness that comes into focus as it matures. These three elements allow many tunes to be played with casks.

The fruit and spice template has been taken up and broadened by Rachel Barrie, master blender at Brown-Forman's Scottish arm (the

Above *Benriach's reopening of its floor maltings saw it join a small but select group of producers.*

GORDON & MACPHAIL & THE INDEPENDENT BOTTLERS

In 1895, James Gordon and John Alexander MacPhail started as grocers and wine merchants in South Street, Elgin. A year later, they were joined by John Urquhart, who, after the death of the two partners in 1915, took over the firm. It's still run by his great-grandchildren.

Unlike other grocers who went into blending, it was single malt that intrigued John's son George, who is, rightly, seen as the father of the modern malt whisky category. In the days when single malts were rarely bottled by distillers, those interested in them turned to independent bottlers such as Gordon & MacPhail and others like Cadenhead's.

Gordon & MacPhail became the de facto official bottlers for Ardmore, Glenburgie, Glen Grant, The Glenlivet, Glentauchers, Linkwood, Longmorn, Miltonduff, Mortlach, Scapa and Strathisla. A wide range of other distilleries formed the rest of its range.

All were filled as new make into the firm's own casks, aged in its warehouse and bottled when deemed ready. The result is a treasure house of casks, some of which are in excess of 70 years old. Think the National Library of Scotland but in liquid form. No other firm has such range in such depth.

In 2023, however, G&M announced that it was stopping filling casks. Now with two distilleries (Benromach, see page 38, and The Cairn, see page 66), it was taking control of its own destiny, a 'conscious uncoupling' if you like.

Independent bottlers are always reliant on distillers. A few fill new make, while others will buy from distilleries and some will buy from brokers. There might be a glut of casks at times when stocks are high and a shortage when they are low, necessitating planning and a lot of forward-thinking to maintain the business.

The rise in single malt has seen distillery owners holding onto their own stock. No wonder that many independent bottlers (Signatory, Adelphi, Dewar Rattray, Douglas Laing, Hunter Laing, Morrison Distillers, Thompson Bros) have become distillers themselves.

The single casks that the bottlers specialize in are all one-offs, giving insights into distillery character. It's a way to discover those distilleries that aren't frontline, the ones that work away behind the scenes and the closed ones, as well as more unusual examples of some better-known names. They are a vital element in understanding single malt Scotch.

Above *The worms at Glen Elgin are run hot.*

GLENLOSSIE & MANNOCHMORE

Diageo has another two sites close by. **Glenlossie** was the first of John Duff's distilleries, built in 1876. It has three pairs of wash and spirit stills, the latter fitted with purifiers that feed heavier alcohols back into the base of the still for redistillation. This helps to create a fascinating green oily character in the new make and a textural quality in the mature spirit, while the extra reflux promotes a floral top note.

In 1971, another distillery, **Mannochmore**, was built on the site. Though also light in character, it has more of a sweet, almost grappa-like quality as new make. Cut flowers come out as it matures and in time a softer fruit. Mannochmore was the whisky used for the 'black whisky' Loch Dhu in the 1990s and also played a part in the brief Cardhu blended malt fiasco (see page 60).

Kentucky-based giant bought Benriach in 2016). Her approach is to use fully matured Benriach from different cask types to layer different flavours. The 10- and 16-year-old expressions are blends of ex-Bourbon, sherry and virgin oak casks, while the 12-year-old uses a Bourbon/sherry base as well as port.

All of these are unpeated, but peated and triple-distilled spirit is also made. Peating actually started under Chivas in the 1970s, meaning there is mature stock to draw from, while triple distilling began in the 1990s. In addition, the floor maltings were reopened in 2013, and while not capable of supplying the distillery's requirements, the one-month season is sufficient to produce a stand-alone annual release. Benriach is making up for that lost century.

Its older sibling **Longmorn**, built by John Duff in 1894, survived the downturn and has run (virtually) continuously ever since. It's notable for being the favourite single malt within the Chivas distillery teams (though it takes a little persuasion for them to admit to that). Like most old distilleries, its life has been spent in quiet support of blends. For many years, the only way in which to try Longmorn was through independent bottlings. It built up a considerable following in Japan, perhaps due to one of the country's whisky founding fathers Masataka Taketsuru spending a short time working there during his fact-finding mission in 1921. The stills at Yoichi (see page 274) and Miyagikyo (see page 276) are modelled on Longmorn's.

Wash from a relatively short (50-hour) ferment is distilled in 8 small, onion-shaped stills. Although the fill is low to increase copper availability, their size adds weight. The coal fires that ran until 1990 added even more punch and can be seen in older bottlings.

External shell and tube condensers also mean a slightly shorter copper conversation. It's creamy and richly fruited – fruit cake and ripe banana come through in the new make. That weight enables it to start showing its fully mature character in its mid-teens.

A hefty kick of the ball over the fence would see it land in the garden of Diageo's **Glen Elgin**, yet another distillery that sits in the shadows. In these single malt-focused days, 'blending distilleries' can be wrongly assumed to be of lower quality. It's not the case. Every bit as much care is taken in them to produce quality, characterful and individual spirit. You could argue that one reason they aren't frontline single malts is because they are so good and idiosyncratic, which is the case here.

Glen Elgin runs clear wort and long ferments, but is then worm-tub condensed. Often that means sulphur in the new make, but running the stills slowly and giving them air rests by opening the still doors between runs helps to enliven the copper, and keeping the flow of water to the worms slow removes sulphur. All of this allows fruit to emerge and at Glen Elgin they are luscious – peaches, apricot and banana – while the worms add thick weight. This was also the site where Diageo trialled Pombe yeast, which produced an intensely green apple distillate. Always keep an eye on the quiet ones.

BENRIACH

BENRIACH 10 YEAR OLD 43%

Macadamia nuts, sweet pear, apple and cinnamon pie (with custard), then American oak. Very sweet: barley sugar, some sultana and golden syrup.

FRUITY/SOFT

LONGMORN

LONGMORN 23 YEAR OLD DOUBLE CASK 48%

Mix of dark and tropical fruit, shortbread, Bramley apple (with skin), Comice pear and a hint of Shaoxing wine. Light, filmy feel.

FRUITY/SOFT

GLEN ELGIN

GLEN ELGIN 12 YEAR OLD 43%

Green apricot, mash, green apple, then more malt. Medium weight. The worms here adding an almost burned note. Water adds Muscat grape.

FRAGRANT/FLORAL

ROTHES

As the Elgin road dips into Rothes, you get a quick glimpse of a distillery nestled in a deep gully. While this might seem a slightly odd location, it made sense when it was built in 1897. The road runs along a ridge, but the railway ran through the valley. Because of the latter, this was the ideal place to build **Speyburn.**

It's only when you finally turn into the courtyard that you realize the size of the site, which has doubled in capacity over the past decade, although its buildings appear unchanged. The old drum maltings have long gone – they shut down in 1967 – but there are now 19 washbacks and an energy-saving Thermal Vapour Recompression (TVR) system (see box, page 47). It might be bigger, but the distillation remains the same – and you suspect pretty much so since the start. The two wash stills run into condensers, but the four spirit stills have worms attached and it's this last touch that leads to a meaty, vegetable, sulphury new make.

SPEYBURN

SPEYBURN 10 YEAR OLD 40%

Some density thanks to the worms. Muesli, honey, raspberry, then curry leaf. Light palate but with some grunt in the middle. Mature character emerging.

FRAGRANT/FLORAL

SPEYBURN 15 YEAR OLD 46%

A hint of vegetable, Highland toffee but also dried flowers and lavender. Palate-clinging with soft tannin.

FRUITY/SOFT

There is a significant difference between distillate sulphur and sulphur from casks. The former, when intentionally made, is the result of low copper conversation, which is why worm tubs are so effective. The latter comes from casks (usually wine or ex-sherry) where sulphur has been used to stop bacterial infection during shipping and is unwanted.

Distillate sulphur might be the dominant character of the new make, but it's what lies beneath that's important. Sulphur is a marker of what comes next, not the end result. 'Speyburn has an amazing top end,' says Stuart Harvey, master blender at Speyburn's owner International Beverage. 'That's why we mature it in ex-Bourbon casks. The char helps to remove the sulphur and take those aromas out into butterscotch and toffee. That top end then comes out – which people don't expect.

'If you put a heavier new make straight into a Spanish oak sherry butt, the sulphur will remain in the final whisky. I only use them for further maturation after the sulphur has been converted.' Merlot and rum casks are also being trialled for this second maturation.

Although available as a single malt targeted particularly at the US market, low pricing damaged Speyburn's reputation. 'It was sold too cheaply,' Stuart explains, 'and we've had to carry that negative baggage.' Now, though, with the new 15- and 18-year-old winning medals, there is a new upbeat feeling to the distillery. In 2023, it finally opened its doors to the public. 'It's great to see,' says Stuart. 'I've always thought it was underrated, a hidden gem.'

Above *Nestled in its tight glen, Speyburn has been massively expanded. World domination must be next.*

Above *Speyburn's distillers now have 19 washbacks to check.*

Above *Glen Grant was one of the 19th century's pioneering single malts.*

Founded in 1840 by brothers James and John Grant, by the late 1800s **Glen Grant** had established itself as one of the first global single malts, predominantly thanks to the efforts of Major James Grant ('The Major'), big game hunter, owner of the first car in Speyside, who added a remarkable garden to the back of what was by then a substantial distillery. He also built a replica distillery, Caperdonich (see box, page 49), across the road. Key to Glen Grant's success was The Major's decision to lighten the spirit by adding purifiers to the necks of the stills.

Glen Grant's status as a malt brand continued in the 20th century when it became the biggest-selling whisky in Italy – the result of promotional work by another far-sighted individual, The Major's grandson Major Douglas Mackessack.

Glen Grant wasn't just 'a light whisky', though. It had weight on its mid-palate, the result of coal fires under the stills and partial triple distillation, which only stopped in the late 1970s – one reason why bottlings from the 1950s until that change have lasted in cask so well.

Today, owned by Campari, the ferments are relatively short, the coal has been replaced with steam, but the purifiers still work. Lightness is its signature style – fresh, delicate, whispering, yet still with a trace of that silken core. Production is up at around 4 million litres (1 million US gallons) per annum.

Located next to the town's graveyard, and close to a ruined castle, **Glenrothes** is another substantial Rothesian site. It was built in 1878 by James Stuart, whose fortune was made in the guano trade, in anticipation that Scotch's next phase would need distilleries with scale.

Along with Robert Dick of the town's Caledonian Bank and its Elgin accountant William Grant (no relation to William of Glenfiddich), and an Elgin lawyer John Cruickshank, Stuart had already bought The Macallan and built Glen Spey (see box, page 49). Now the quartet were ready to build their second site in Rothes, but their plans coincided with a Europe-wide harvest failure, coupled with a major banking crisis.

The partnership was dissolved, with Stuart staying at The Macallan and Dick and Grant continuing with the Glenrothes plans, securing loans from a rear admiral and, in an unlikely development, the minister of the local United Presbyterian church.

On 28 December 1879, it opened, complete with a newfangled heat exchanger for cooling the wort, and switchers in the washbacks, which eliminated the need to employ small boys with brooms to beat down the froth should it start to foam over. The aim from the outset was always to make something lighter that would appeal to blenders.

Glenrothes runs at an unusual tempo. The ferments are relatively short, adding maltiness, but distillation in its tall stills is slow. Maybe

GLEN GRANT

GLEN GRANT 12 YEAR OLD 43%

Clean. Icing sugar, white flowers, fresh orange juice and dry oak. Soft with low oak allowing zingy acids to show. Palate is whipped cream and green walnut.

FRAGRANT/FLORAL

★ GLEN GRANT 18 YEAR OLD RARE EDITION 43%

More American oak (vanilla, baby lotion), but still fresh green fruit. Good mid-palate complexity. Mature. Concentrated orchard fruit on the finish. Peach purée and white flowers.

FRUITY/SOFT

Above *The cathedral-like stillhouse at Glenrothes was once haunted and had to be exorcized.*

this contrast is why it needs time in cask and in glass. When young, it's the maltiness that comes across, but with years in cask it changes – deepening and broadening, becoming more generously fruited. The vanilla note that is discernable at new make stage becomes ever richer. It's a whisky that gains in elegance, as long as you are patient.

Ageing is mostly in ex-sherry casks, another indication that what might seem light actually has substance. Although the distillery was owned by Edrington, the brand was owned by the venerable London wine merchant Berry Bros. & Rudd, who released it in small batches of different vintages, each one allowing different facets of its personality to be revealed.

In 2017, Edrington brought it back in house and the year after the vintages were replaced by age statements. In 2023, however, the younger expressions were discontinued with only the 18-year-old remaining. It's a bold move, but one that shows a deeper understanding of how the distillery style evolves. Volumes will inevitably be limited, but supported by a range of one-off releases. It's a pitch for the growing premium/luxury market where Edrington already has experience.

A first-time visitor to Rothes might, naturally enough, think that the bulky building on the other side of the Glen Grant roundabout is another distillery. It's actually the town's dark grains plant. Run by the Combination of Rothes Distillers (CoRD), comprising Highland Distillers, Chivas Brothers, Inverhouse Distillers, Glen Grant Distillers, Diageo, Benriach Distillers and more recently John Dewar & Sons, the plant was established in 1904.

'Dark grains' might sound like some occult practice, but basically the plant takes distillery by-products – draff and pot ale (the non-alcoholic remnants from the spirit still) – and through evaporation and drying converts them into pellets for animal feed. In 2013, a joint venture between Helius Energy Plc and CoRD completed

GLENROTHES

GLENROTHES 18 YEAR OLD 43%

Some oxidized sherried notes. On the funkier side, sultana, vanilla pod, date loaf. Good palate weight: cereal, horchata, figgy, dried redcurrant and chocolate brownie on the finish.

RICH/DEEP

★ **GLENROTHES 25 YEAR OLD** 43%

Classically elegant old Rothes. Very soft, rich and warming. Light tannins. A hint of malt, caramel toffee, blackberry and cedar. Ever-evolving.

RICH/DEEP

a new biomass heat and power plant at the facility, with a pot ale evaporator plant. The draff, along with wood chip, is now used as fuel to produce renewable energy, and the evaporator then concentrates the pot ale into syrup – again for animal feed.

There are two other dark grains plants in Speyside, one at Glenlossie and another at Dailuaine (see box, page 61).

The former Caperdonich site (see box, page 49) is now the HQ of Forsyths Coppersmiths. The Forsyth family has been working in the trade since the 1890s and established the firm in 1933. Now the fourth generation runs the business. Besides diversifying into the oil and gas industry, Forsyths has grown to be the premier supplier of hand-hammered stills to the world's distillers. There is a long waiting list. Not only constructing and repairing stills, the firm designs, engineers and effectively builds the process for new distilleries. It's a further example of how whisky is about more than just the distillery. There are many trades contributing to its creation, including malting, coopering, vat construction and coppersmithing – a web of people, craft, skill and place.

CRAIGELLACHIE

The idea of place within whisky should be an expansive rather than reductive approach. Geology, climate and geography will all play their part in its making, but so do commerce and people. Speyside is one of those places. It had the conditions to make whisky, but what elevated it in terms of significance was transport.

TVR & MVR

Saving energy remains a crucial element in distillation – not just from a cost perspective but for environmental and sustainability reasons. As distilling consultant (and thermal engineering expert) Brendan McCarron puts it, 'Distilling is about making cold things hot and hot things cold. It also uses a lot of energy and can be inefficient.

'For example, there's a lot of water coming out of the condenser at the end of the process, but it's a medium-temperature water. That low-grade heat isn't particularly useful.

'If, however, you pull a vacuum by dropping the pressure, the water will evaporate (a process called flashing off), giving low-pressure steam. If you can then compress that, you have high-pressure steam, which can be injected back into your boiler, meaning you use less oil or gas.

'Steam is loose,' he adds. 'It's like kids running about in a big playground. Compressing it is like being the teacher coming out and shouting at everyone to behave, stand close to each other and not move.'

There are two systems now being used extensively across Scotch. Thermal Vapour Recompression (TVR) uses the Venturi effect (forcing steam through a narrow chamber) to create the vacuum and increase the pressure and temperature. Easy and cheaper to install, it still needs extra energy to function.

Mechanical Vapour Recompression (MVR) works to the same principle: the compressor pulls a vacuum, flashing off steam into a compressor and converting it into high-pressure steam. It uses less energy than TVR, as the compressors are powered by electricity, not steam, allowing all the compressed vapour (which is higher in temperature than in a TVR) to be used. The downside is that it's more expensive and considerably more bulky.

Above *When the aroma of sulphur permeates the Craigellachie stillhouse, distillers know all is well.*

Above *The Telford Bridge at Craigellachie was a major tourist attraction in Victorian times.*

ROLE OF THE RAILWAY

The modern whisky industry didn't spring into immediate rude health with the change in law in 1823 (see page 64). Many of the newly licensed distilleries in rural parts foundered because of poor communication. While it was advantageous for an illicit distiller to be hidden away, it was less practical when you had to get your now legal goods to market.

They were also competing with large urban malt and grain distilleries in the Lowlands who were more savvy in marketing, had access to the newly developing coalfields with their cheaper fuel and had an ability to import barley and grain.

Change would only become possible with the help of the railway, but progress was slow. In 1845, the Great North of Scotland Railway linked Inverness and Aberdeen along the coast and from there to the Lowlands. A spur in 1852 joined Lossiemouth with Elgin, which was subsequently joined to Rothes in 1862.

In 1863, all of the disparate lines across the region were linked to the Strathspey Railway at Craigellachie, which in turn met the main Highland Railway at Boat of Garten. Speyside was finally open and distilleries were built along the railway's length – in Keith and Dufftown as well as Aberlour, Carron (Dailuaine and Imperial), Knockando, Tamdhu, Ballindalloch (Cragganmore) and Cromdale (Balmenach). It was the saving of some, and the spur for others to build.

At the start of the 19th century, one reason for heading to Craigellachie was to gaze at the bridge over the Spey built by Thomas Telford in 1812–15. In 1863, the village's significance rose further thanks to it becoming the railway junction linking Keith, Dufftown and Elgin to the Strathspey Railway, which would carry goods to the south (see box). It not only provided ease of transport for whisky but also ease of travel for tourists keen to experience the wild country that had been popularized by the Romantic poets, Walter Scott and Queen Victoria.

Enter a young man called Alexander Edward. He had inherited Benrinnes distillery from his father, then came a move to Craigellachie, where

CRAIGELLACHIE

★ CRAIGELLACHIE 13 YEAR OLD 46%

Rounded and substantial. Cooked cabbage under heavy florals and canned pineapple. Creamy oak on the palate, juicy fruit. The worms add solidity and light earthiness. Appetizing and unconventional.

FRUITY/SOFT

SPEYSIDE COOPERAGE

Without casks, there would be no whisky. Without good casks, quality would fall. Another part of Scotch whisky's ecosystem, therefore, is the cooperage.

Just outside Craigellachie on the road to Dufftown is the Speyside Cooperage (see left). Founded by the Taylor family in 1947, it remains the largest independent cooperage in the UK (now owned by Tonnellerie François Frères), with workshops at Craigellachie and Alloa, as well as Kentucky and Ohio in the USA.

Here casks are repaired, re-charred, toasted and also built from scratch. Speyside imports oak from around the world, including mizunara from Japan, and is a partner in the Scottish oak project (see box, page 72). Open to the public, it is a must-visit.

he built substantial holiday houses, a grand railway hotel and, in 1890 in partnership with Sir Peter Mackie of White Horse (and Lagavulin) fame, a distillery.

Craigellachie occupies an interesting point in whisky's evolution – built at the start of the great blending era and also therefore on the cusp of a change in character dictated by blenders.

It was already understood that each distillery had its fingerprint in terms of flavour and texture. What the blenders wanted was three-fold: greater consistency and clearer definition, giving them the ability to create broad flavour camps from which to draw. Distilleries were different but could be grouped together. Some firms such as Johnnie Walker would then create vats of each of these groupings, which in turn would give consistency to its blends.

If Islay gave smoke and Campbeltown weight, then the Strathspey whiskies as they were still known could give fruit and any overt peatiness could be dropped. There were (and still are) variations on that theme. The fruits in some Speyside drams move into floral, while others are deep, rich and robust. Craigellachie manages to bridge these two camps. It's muscular, but also in possession of a pineapple aroma.

As a new make it's sulphury, the result of worm tubs giving out Orwellian guffs of boiled cabbage. This is amplified by sprinkling sulphur onto the malt at kilning. Needless to say, it takes time for this to shift into something more acceptable.

The 13-year-old was partly chosen because all of the Craig's age statements are prime numbers, but also because that's how long it takes for that sulphur to be absorbed, dispersed and converted. The muscle remains (helped by bottling at 46% ABV with no chill-filtering and no colour addition), but out now come fruits, and heavy florals. Thick in feel, it's one of those drams that, while not conventionally beautiful, is still compelling.

OTHER ROTHES DISTILLERIES

Rothes (see right) is also home to **Glen Spey**, another of the late 19th-century grouping and also founded by James Stuart. In 1887, he sold it to gin maker W&A Gilbey, which was then expanding into Scotch. It makes a light style (almond and popcorn).

Rothes's other still **Caperdonich** (or Glen Grant No. 2, as it was rather unromantically known) sat over the road from the original. The new make used to be pumped in a pipe that ran above the road to Glen Grant's filling store. Single casks do appear, but stock is dwindling.

Above *Glenfiddich's traditional warehouses contain decades-worth of slowly maturing stock.*

DUFFTOWN

Dufftown is an example of the result of the improvements in agriculture in the 18th and 19th centuries. Rather than the Clearances seen in the north and Hebrides, here there was a wide relocation of population as farms became larger. New towns (New Keith, Grantown-on-Spey, Aberlour, Rothes) were built, each with ancillary industries – weaving, wheelwrighting, blacksmithing, tanning – and distilling.

There is a limestone seam running close to Dufftown that provided fertilizer for the new farms. It gave the young William Grant his first job, though he left to learn the trade of distilling at Mortlach. In 1886, he made the dramatic move to build the town's second still. With his wife and family helping, **Glenfiddich** was raised and on Christmas Day 1887 the first spirit ran. William Grant's great-great-grandchildren still run a company that now has global reach.

Speyside is the perfect place to observe the evolution of the modern industry from farm sites to illicit bothies; the early licensed sites, those caught up in the railway boom, then the late Victorian sites as blends took over. The most recent development has been the return to the 'self whiskies' (aka single malts) that made the area initially famous.

Although single malts were available from some distillers and through independent bottlers like Gordon & MacPhail (G&M – see box, page 42), the first modern single malt brand was Glenfiddich.

Launched in 1961 (and into export two years later), it has become the world's largest-selling single malt brand. That needs volume and that's what you get. A massive recent expansion has seen the building of a new stillhouse. Today there are 43 stills in operation, with space for more. Despite this, the two mash tuns remain traditional, the washbacks are wooden, the stills small and the cut points high. Change without change.

Volume can be a double-edged sword. Scale can mean dilution, but there's something else at work here. Charles Gordon (William's great-grandson) who with his brother Sandy ran the firm from the 1950s believed in self-sufficiency, so there's a coppersmith and cooperage on site as well as a bottling line. (Not all Glenfiddich is bottled on site, but any that isn't is

GLENFIDDICH

GLENFIDDICH 12 YEAR OLD 40%

The world's bestselling single malt. Fresh apple with a hint of lemon, light grass and a hint of malt on the back. Soft feel but with great length.

FRAGRANT/FLORAL

★ **GLENFIDDICH 15 YEAR OLD** 40%

Dates, caramel, raisin and a concentrated red apple element. Flirts with savouriness. Balanced and layered. Black Forest gateau yet still fresh. Excellent balance.

RICH/DEEP

still reduced with water from the distillery's Robbie Dhu spring.) There is a community within the community.

The single malts are all married in large vats before bottling to add harmony, while the 15-year-old comes from a solera vat. The same ratio of mature whisky from first-fill Bourbon, refill and sherry is emptied into the vat and the same amount withdrawn. As the vat is never emptied, the depth and complexity of the whisky is increased. All of this is expensive and it adds space, but it gives character. It's what the family wants. What could be a loss of place is actually reinforced.

Glenfiddich starts apple-fresh with a drying touch of malt as new make. Gentle oak amplifies this at 12 years old, but it's long-lived. In time the apple ripens, becomes cooked and sweetened by oak sugars, while a chocolate element develops. It gains in grace.

BALVENIE

★ BALVENIE DOUBLE WOOD 12 YEAR OLD 40%

Honey on white bread, malt, sugar puffs, muesli with sultana. Incredibly soft in the mouth, then sherry adding sweet plum and cherry chocolate. Subtle.

FRUITY/SOFT

BALVENIE 16 YEAR OLD FRENCH OAK 47.6%

More oak, slightly vinous. Soft fruits, some nectarine, coffee and cardamom, new shoes. Firm tannins, attractive bitterness, clove, chocolate, just-burned cask.

RICH/DEEP

In 1892, William Grant built **Balvenie** next to Glenfiddich. It's a sister, but with its own distinct personality – the peculiarities of specific location at work again. Balvenie has its own malt floor, and barley grown on the family's farm is also distilled separately each year. The stills are fatter, the spirit richer. There is some cereal in the new make, but from the outset it's sweeter and more honeyed than Glenfiddich. As the first distillery to experiment with finishing, it's an avenue that has continued to be explored. The original Double Wood 12-year-old (finished in ex-sherry) was joined by a 14-year-old (finished in rum casks), a 17-year-old Double Wood and the 21-year-old Port Wood.

This range has been expanded to include a 15-year-old Madeira cask, a 16-year-old French oak and an 18-year-old PX cask finish. More may yet appear. The key to finishing is balancing the distillery character with the high activity given by the second cask. It can often end up with the tail wagging the dog. A deep understanding of the nuances of the distillery character as it matures is required to match with the finishing casks used. Having a cooperage team of 20 helps.

That idea of the community within the community was made manifest by the launch of the Balvenie Stories series, which celebrates many of the craftspeople who have worked at the distillery. In 2023, David C Stewart MBE, who had nurtured Balvenie as its malt master for 60 years handed over the reins to Kelsey McKechnie.

Above *Balvenie's floor maltings remain in use.*

WORMS

No burrowing invertebrates are involved in the production of Scotch whisky. The Scottish Whisky Association wouldn't allow it. A 'worm' gets its name from the old term for serpent, which makes sense when you see how this sectioned copper pipe is coiled inside its water-filled tub (see below).

Worms are one of the oldest methods of condensing. When the vapour passes from the lyne arm into the worm tub, the drop in temperature causes the alcohol vapour to turn back into liquid. The worm tub itself is being constantly refilled with fresh, cold water.

There is less active copper in a worm than in a shell and tube condenser, meaning that they will tend to produce a heavier and often sulphury new make. Worm tub distilleries can, however, be worked against type. Slowing the flow means the water will warm up and prolong the vapour stage and that copper conversation. Glen Elgin, Royal Lochnagar and Oban all operate in this way.

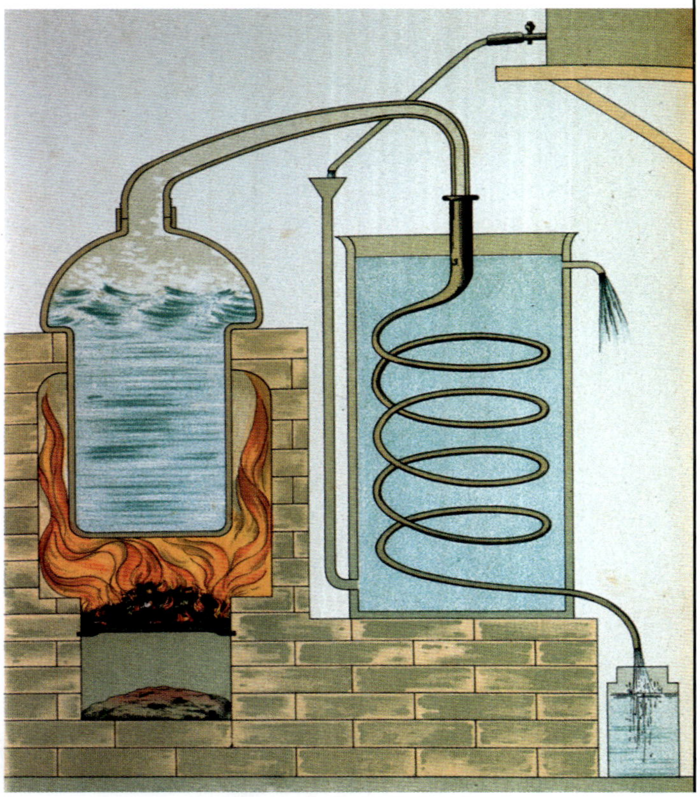

A third distillery, **Kininvie**, joined the Dufftown family in 1990 mainly to relieve pressure on Glenfiddich, which up until then was supplying fillings for the Grant's blends. It uses a separate mash tun and washbacks within the Balvenie tun room. Nine stills make a light, floral, estery new make, again different to its sisters. It has also been used for experimental distilling – the trials for Grant's Tullamore single pot still were done here.

The old settlement of Mortlach was absorbed into Dufftown in 1817. An important medieval abbey, it was also the site of a major battle in 1010 between the Scots and Vikings, one of the first victories that ended the latter's control.

It was also the location of the town's first legal distillery, which took out its licence in 1824. Whisky drinking was already rife in the area, much to the chagrin of the local minister who in the *Old Statistical Account of Scotland* (1876) reported that '...the drinking of whisky instead of good ale is a miserable change, and so likewise is the very general use of tea. These put together have been exceedingly hurtful both to health and morals.'

Maybe his irritation worked, as **Mortlach** remained the town's only distillery for 66 years. Its date of origin also gives an indication of its style. Many of the area's oldest stills produce a heavy style and this is an irredeemably old-style whisky. There are worms (see box), an embrace of meaty sulphur, weight and power. Others try a similar tack, but nothing quite comes close to the charred roast beef and gaminess of that new make. Mortlach is feral, with the occasional whiff of badger sett, turned earth, autumn woods.

This is achieved through a bewilderingly complex distillation system. Mortlach has six stills, three wash and three spirit. All are different shapes. All are run hard to cut down on any copper conversation. One of those wash and spirit pairs works as per normal, its new make kept to one side. The other two wash stills work as a pair, and 80% of the run is collected as 'strong feints' and forms the charge for No 2 spirit still. The remaining 20% 'weak feints' is redistilled in a tiny still called 'The Wee Witchie', which runs three times to build strength before any cut is made. (This system of 'dud runs' to build weight was also used by Glen Grant until the 1970s.) This is where Mortlach's meatiness comes from. The three new makes are then vatted together and aged. It's not double distillation, it's not triple, it's 2.81-times distilled.

So much is involved in Mortlach's weird stillhouse that the role of its worms is overlooked. To add yet more chaos, the tubs are all different sizes, meaning more variation. The water is also kept as cold as possible, and some is even sprayed onto the top of the lyne arm as it starts to dip into the water. All of this keeps that interaction between vapour and copper to a minimum.

Does this detail matter? If you're a geek you lap it up, but what really matters is that the people making it understand it and that the character remains the same. Knowing what effect it has is, let's face it, more important than breaking it all down.

Here's another distillery that needs time and also one that's a blender's dream. That heft adds a resonance to any blend, which is where most of the spirit ends up, but finally it's also being promoted as a single malt. Is it a new Macallan, or will it forever remain an enigma?

MORTLACH

★ MORTLACH 12 YEAR OLD 43.4%

Meaty but with thick fruits, leather, cigar ash, rowan jelly, brambles. Gamey in the mouth, sandalwood, petrichor, but with some sweet persimmon.

RICH/DEEP

GLENDULLAN

SINGLETON OF GLENDULLAN 12 YEAR OLD 40%

Delicate fruit is there alongside toasted barley, white pudding, wet autumn leaves and plump, late summer berries, cherry and cinnamon.

FRUITY/SOFT

OTHER DUFFTOWN DISTILLERIES

Dufftown is also home to Diageo's other two Singleton sites. **Dufftown** (see left) was a Bell's distillery and made whisky to that house's spec: short ferments, fast distillation, nutty/spicy variations. In recent years fermentation has been lengthened, the wort runs clear and a grassy new make is made.

Glendullan started life in 1897 when it was built by the Aberdeen blender and broker William Williams and makes a gentle fruited style – there's some blueberry in mature character. The town is also home to the shells of Parkmore, Convalmore and Pittyvaich. Hubris is never far away in Scotch.

Heading southeast from Dufftown you reach the high plateau of **The Cabrach**. Here, as part of a project to bring people back into the area, a small distillery has been built in an old farmsteading. The aim is to have one pair of stills making a relatively modern style, and a small one producing something more akin to what The Cabrach made when it was one of the area's centres of illicit distillation.

None of Dufftown's whiskies tastes the same. Place cannot be pushed into a box like that. It flexes because it's about a reaction to conditions and circumstance. Whisky is about place and place is also about people.

Above *The black mould on a distillery building (here at Glendullan) is a good sign of whisky maturing.*

THE BEN RINNES AREA

If you want an example of one way in which the modern whisky industry has evolved, then head over the Spey – en route to Scotland's whisky mountain, Ben Rinnes – to **The Macallan**. This old distillery, which took its licence in 1823, has gone harder and further into the luxury market than any other, winning new friends and also irritating many old followers by doing so. Maybe that's the price you pay for being a pioneer.

The move has been underway since 1979, when the first glimmerings of interest in single malt were starting and The Macallan, which until then had been a fairly anonymous malt used in numerous blends, began to assert its individuality.

A clever advertising campaign started. The 200 hand-drawn ads that ran for 16 years told stories, spun tall tales and pricked pomposity. By targeting opinion formers, they ran in broadsheets and *Private Eye*. The Macallan took cues from wine: it was 'château aged' and called itself 'the Cognac of single malt'. It was also happy to speak of how it was made: small, direct-fired stills; a heavy, slightly sulphury/earthy new make; Golden Promise barley; a mix of yeasts; and ex-sherry casks. In this way, it set itself off to one side of an industry that was becoming more conventional in its choice of new barley strains, distiller's yeast and American oak.

It worked. Sales rose, a cult formed, the distillery expanded. By 1975, there were 21 stills. Single malt became a category. In 1996, The Macallan was bought by Highland Distillers.

Above *The Macallan's new distillery took its architectural inspiration from the new wave of wineries.*

SHERRY & WHISKY

The links between Scotland and sherry date back to the 16th century. By the 19th century, most of the production from Jerez was destined for the UK. In John Strang's account of *Glasgow and Its Clubs* (1857), he recounts that 'the liquor in common use was sherry, presented in mutchlin stoups...The quantity allowed was...almost incredible.' [A 'mutchlin' was just under a pint.]

The modern whisky industry had taken full advantage of this bounty. From the 1820s, maturation of Scotch became a deliberate decision and distillers needed casks. They turned to what was available from wine merchants and shippers: rum, wine and sherry.

The sherry boom wasn't to last. By the 1870s, a decline in the wine's popularity saw a shortage of casks coming at the precise moment when the whisky industry was expanding. This led to blending houses – particularly Glasgow-based W.P. Lowrie – to import staves of new American oak and season them with 'sherry' in Scotland to imitate the now more scarce 'shipping cask'.

In 1986, by law all sherry had to be bottled in Jerez, so distillers began to source their casks in Jerez (or other regions such as Huelva or Montilla), again made from new wood, then seasoned for a period. American oak began to be replaced by Spanish oak, giving more tannin, colour and extract.

A further dip in sherry's fortunes also saw a large number of ex-solera casks being exported. Sherry is an oxidized wine, meaning that producers don't want active wood. The old solera casks were rich in wine, but not active. They were good for one, maybe two fills, but were then just inert containers.

Today the demand for ex-sherry casks is on the rise once more and there is greater legal control over their production. The wine used, however, though legally sherry (oloroso remains the most popular), is not the same quality that would be employed for bottled product. It tends to be used for seasoning twice and then sent for distillation or vinegar.

There is now a move by some distillers to work more closely with bodegas and use quality sherry and a wider range than just oloroso or PX. With whisky growing globally, and other spirits also wanting 'sherry casks', there is severe pressure on stock. Factor in the decline in the area under vine and a 60 per cent fall in harvest in the past three years, and there is trouble brewing. Only with closer cooperation between winemakers and distillers will the growing crisis be averted.

Above *Unusually, The Macallan's stills now sit in a circle.*

By the early 2000s, its then director Ken Grier was asking the question: if single malt whisky is truly premium and a signifier for success in the same way as wearing an expensive watch, or a bespoke suit, why shouldn't it also be seen as a luxury product? It's a fair question. If people are willing to pay high prices for an 'experience', then why deny them the pleasure?

The subsequent question was: in doing so, did The Macallan stop being a single malt and become 'The Macallan'? And then if it did, does that even matter? Different rules apply in the luxury world. Sales projections indicated that the old distillery couldn't keep pace with projections. The decision that was then taken was dramatic.

It takes balls of steel to close down one distillery and rebuild another on the same site but to a different configuration. But that's what The Macallan did. The old Macallan was utilitarian; beautifully, brutally functional. There were pipes and copper, and fires (though steam was introduced in 2010). The transition, however, required an appropriately luxurious distillery. The new £140 million distillery, which opened in 2018, was part art gallery, part winery, the inspiration for which came from Bodegas Ysios in Rioja, northern Spain.

Now the mash tun holds 17 tonnes (18.7 US tons), the still shapes are the same, but now there are 36 (12 wash and 24 spirit), which are steam-heated. The orientation has changed as well. Now they sit in three circular 'pods' of 12 (8 spirit and 4 wash) with washbacks in the centre. These have to be fitted with external cooling because of the ambient heat.

The constant is the ex-sherry casks from Spanish (*Quercus robur*) and American (*Quercus alba*) oak made to the distillery's specifications at Tevasa Cooperage in Jerez. So much wood is now required (each cask only has two fills) that in 2023 The Macallan bought 50% of Bodegas Grupo Estévez in Jerez, parent company of the legendary Valdespino sherries. The evolution continues.

THE MACALLAN

THE MACALLAN 12 YEAR OLD DOUBLE CASK 40%

Racy. Marzipan, dried citrus peels, currant leaf. Palate has prune juice, light malt and Madeira cake and tea-dunked sweet biscuit. Struck match and a little meatiness.

FRUITY/SOFT

THE MACALLAN 18 YEAR OLD SHERRY OAK CASK 43%

Deep and slightly oily. Assam tea, stewed fruit, date, tamarind and malt. The palate has rounded tannin, fig roll and slight astringency. Medium weight.

RICH/DEEP

SOUTH FROM ABERLOUR

Many of Speyside's distilleries exist because of the railway, but many more first appeared because of the area's geography. Fertile soil allowed barley to be grown, and there were plentiful supplies of peat (the region's whiskies only stopped being lightly smoked relatively recently). Then there's the rock, and in particular its sentinel mountain Ben Rinnes, a triangular granite outcrop of the Cairngorms. It gathers clouds and brings forth water, which floods the bogs creating conditions for peat, and runs into the burns and springs that supply seven distilleries.

One of those is **Aberlour**. James Gordon who, as Provost of Elgin, was one of the instigators of extending the rail links, built the first distillery in the village in 1826. That first distillery foundered, and it wasn't until 1879 that Aberlour's new distillery was built, closer to the railway.

With a narrow entrance, the distillery runs alongside the Lour Burn that rises on the mountain. A strong export brand, particularly in France, Aberlour is doubling capacity to almost 8 million litres (2.1 million US gallons) per annum, and the combination of a bio plant and Mechanical Vapour Recompression (MVR) technology (see box, page 47) means it hopes to be carbon neutral by 2026.

There's one pair of fat-bottomed, onion-shaped stills and a style that flexes between different ratios of ex-sherry and ex-Bourbon. A blackcurrant note emerges in some and a toffee-like element picks up in time. How the distillery character works with sherry is seen most fully in the 100% sherry cask, cask-strength A'bunadh. As this is released in batches, the variations make the exercise more interesting. It's a surprisingly ferocious and bold expression from a distillery that seems to typify modesty.

Just over 1 mile (2km) uphill is **GlenAllachie**. A modern site, built in 1967 to provide fillings for the Mackinlay blends, it was latterly part of the Chivas Brothers stable. Then in 2017 it was put on the market. As it happens, Billy Walker's former partners in the Benriach Group had sold that firm (comprising Benriach, Glendronach and Glenglassaugh) to Brown-Forman. GlenAllachie was too good an opportunity to miss. He had already

ABERLOUR

ABERLOUR A'BUNADH BATCH 74 60%

Muscular. Coffee grounds, bitter orange peel, old leather, hard toffee, briar fruits. Some heat, but the core is sweet, balancing the tannins.

RICH/DEEP

GLENALLACHIE

GLENALLACHIE 12 YEAR OLD 46%

Deeply, sweetly fruited with a perfumed note. Nocino and a little bright citrus. Palate is dried fruits, damson jam and violet. A bitter twang to the finish gives balance.

RICH/DEEP

★ MEIKLE TÒIR THE ORIGINAL 5 YEAR OLD 50%

Woodsmoke, but simultaneously sweet. Tarragon, verjus and a rum-like phenolic note. The palate is dry grass, lemon flower, waxes and fragrant smoke.

SMOKY/PEATY

Above *Under new ownership: GlenAllachie has been reimagined by Billy Walker and team.*

Above *The flames still blaze under Glenfarclas's massive stills, adding weight and power to the new make.*

built one unknown distillery into a name. Could he do it again? 'Well, I'd a feeling that there was unfinished business, so when I was given the heads-up, I thought, why not?'

You might think that would mean examine the stock, maybe tweak a few things and raise the profile. Instead, there has been a complete recalibration. At a time when distilleries across Scotland are expanding, some might say recklessly, GlenAllachie is making less than it used to. Deliberately so. What is a 4.2-million-litre (1 million-US-gallon) per annum size is now making 900,000 litres (238,000 US gallons) per annum.

'We could never make 4.2 million litres,' Billy says. 'Our footprint couldn't cope with that. We'd need so much more warehousing. Also, we're not supplying fillings.' The fact that he has extended ferments from 52 to 140–180 hours also constrains production. 'If this plant was working to capacity, ferments would have to be 52–58 hours,' he says. 'If you went back 100 years, I reckon they'd be as long as we're doing now. They allow us to get the yeast to go full term, engage friendly bacteria and give the wash oxygen, making it very benign. If you distil after 52 hours, it's still fermenting.' He has also lowered the strength of the wash and tweaked the cuts to make things 'more muscular'.

BENRINNES

★ BENRINNES 15 YEAR OLD 43%

Robust. Antique shop, old-growth forest, beef jerky and cacao. The meatiness becomes more savoury on the palate, overlaying a sweet heart.

RICH/DEEP

The advantage, as with Benriach, was that GlenAllachie was a blank slate in terms of the consumer. Chivas had bottled it occasionally, there had been the odd independent bottling, but as Billy says, 'Now we're painting our own picture. GlenAllachie was never given the chance to showcase itself. Now is its chance.'

Heading further up the hill and you reach **Benrinnes**. Another of Diageo's meaty sites, for many years it ran partial triple distillation similar to Mortlach except without the dud runs. Today, distillation is more straightforward, but the worms still turn out a rich and substantial new make that shines in ex-sherry casks.

We are in the sherry triangle. As well as Macallan's adherence to the cask, there's Aberlour's A'bunadh, Billy Walker's recalibration of GlenAllachie, Benrinnes' meat and now, as we head around the mountain, there are another two proponents of the style.

Glenfarclas has been owned by the Grant family (no relation to the Glenfiddich Grants) since 1865 when they bought it for £511,19s. Under their stewardship it has become a touchstone for bold, older-style, sherried single malt. Much of that depth comes from the family's belief in heating its six stills with live flame. In 1981, as part of a modernization process, steam was brought in, but the guts went out of the new make and, though expensive, the flame came back as did the fire in the spirit's belly.

Above *Although the River Spey gives its name to the region, none of its distilleries use it as their water source.*

'The direct fire gives us the old toffee and caramel flavours that people crave in older whiskies,' says manager Callum Fraser. There are long ferments as well, hitting 106 hours and those stills are large, meaning there are fruits and reflux going on as well as the power given by the fires.

It's another one that builds slowly, partly down to the new make but also due to the family's adherence to ex-sherry casks. By law, all Bourbon casks must be charred. Even after they have been emptied and transported to Scotland, this charcoal layer acts as a filtering mechanism, removing aggression and sulphur from the maturing spirit.

Ex-sherry casks are not only larger (500 litres/132 US gallons against 200 or 250 litres/53 or 66 US gallons), but have only been toasted. Lower spirit to wood ratio and less charcoal means maturation will progress more slowly. Glenfarclas really only starts its climb to peak maturity after 15 years. There's another reason for this that's location-driven. This is a cold part of Speyside and all of the distillery's casks are matured in 42 earth-floored dunnage warehouses. The average evaporation rate in Scotland is 2% a year – here the angels only get 1%. Maybe all of this results in a different conception of time. The family is proud to say that it has survived 22 recessions. It plays the long game, laying down stock when others might be pulling back. The result is a distillery with the deepest reservoir of old stock in Scotch. The core range includes a 40-year-old expression and older ages are common.

'The location might not give specific flavours, but it does influence our mindset, and understanding of tradition,' says Callum. 'That means it is vital.'

WEST OF THE SPEY

Across the river is **Tamdhu**. Like Imperial (see page 61), this distillery was built in that late Victorian period of wild overenthusiasm, as sales of blends rose globally, necessitating greater capacity. Blenders, seeing demand rising, began to buy distilleries to secure supply while other consortiums also began to speculate. It was a return to the distillery-building mania of the 1820s.

GLENFARCLAS

★ GLENFARCLAS 15 YEAR OLD 46%

Roasted cereal, sweet raisin, heather root, dry bracken (it's of its place). A sense of concentration but not cask-dominated. Deep, off-dry and slow.

RICH/DEEP

★ GLENFARCLAS 35 YEAR OLD 43%

A mix of Russet apple, dried peach, propolis, wax polish, mead and waxiness, but still the distillery's earthy density and mouth-filling quality. Complex and elegant.

FRUITY/SOFT

In Tamdhu's case it was established by a group headed by William Grant, a director of Highland Distillers (which became Edrington) who we've already met in Rothes. Handily enough, he was on the board of the Elgin bank that built the distillery in 1897 for the equivalent cost of £20 million. Described as being 'the most modern of distilleries', it was 'filled with the latest engineering and labour-saving devices'.

Like many sites it had a rocky time, closing from 1927 and remaining silent for two decades. Soon after its reopening, the old floor maltings were replaced with 10 Saladin boxes, each capable of holding 22 tonnes (24.3 US tons) of malt. It malted all its own requirements, as well as supplying malt for Glenrothes and the unpeated requirement for Highland Park. Deemed surplus to requirements by Edrington in 2011, it was quickly snapped up by Ian Macleod.

Under its previous owner, Tamdhu had already embarked on a 100% sherry-maturation regime and today all of its whisky is matured in ex-oloroso casks made from American and Spanish oak. There's a tiny amount of peat in the malt, clear wort and a fairly short (59-hour) ferment, giving a fruity new make. Tamdhu is medium weight and supple; it glides alongside sherry and European oak, softening tannins rather than being gripped by them, its sweetness acting as a foil to the drier tones from the wood.

'The slight challenge with Tamdhu is that there's not enough stock,' says Gordon Dundas, Ian Macleod's brand development and advocacy director, 'so at the moment we have to manage any holes in stock. It's very much a long-term project.'

As a blending malt – in Edrington days it was home to the (underrated) Dunhill blend – there's also the matter of bringing a 'new' distillery to people's attention. 'Every year we take it out on the road and the response is amazing. It's liquid on lips,' says Gordon. 'In ten years we've taken this distillery and given it its own voice. It will take time, but we know that it's a luxury product, but not at a luxury price.' As prices head ever upwards, you can still get great sherried whisky for a fair price. Smart move.

It's possible that the first distillery in the area to take out one of the licences prompted by the passing of the 1823 Excise Act (see page 64) was **Cardhu** (Gaelic for 'black rock'). Originally a farm called Cardow, it was leased in 1811 to John and Helen Cumming. This was the height of the illicit whisky trade and Cardow operated as a site of moonshining and an early warning station for distillers in the Glenlivet area (Helen would hang out a flag to show that gaugers were on their way), but also, amazingly, as a billet for the Excise. It's believed that Helen was the whisky maker – as would have been the case with many farm distilleries across Scotland in less fraught times.

Helen and John's son Lewis took over the business but died in 1872. His wife Elizabeth then took up the reins and it is she who is rightly

TAMDHU

★ **TAMDHU 15 YEAR OLD** 46%

Chewy even on the nose. Baked fruit, light malt, plum pudding, resin, then in the palate treacle, cardamom and supple tannins.

RICH/DEEP

★ **TAMDHU 18 YEAR OLD** 46.8%

Mature. Dark figs, more treacle toffee, chocolate-covered black cherries and walnut. More grip. Savoury, sloe, mulberry and the earthy sweetness of beech mulch.

RICH/DEEP

Above *Under new owner, Ian Macleod, Tamdhu has established itself as a leading sherried single malt.*

Above *Tormore is both a striking distillery and one that is eccentrically laid out.*

credited with building (the now renamed) Cardhu's fame. In 1884–5, she built a new distillery on nearby land then, in 1886, sold its old stills to William Grant who was setting up Glenfiddich. She also established a strong working relationship with John Walker & Sons.

You wonder whether the new distillery also ushered in a new style. When whisky's first historian Alfred Barnard visited the old site, he described the make as being 'of the thickest and richest description', and though he added it was 'admirably adapted for blending purposes', the following years saw a lightening of character as whisky and soda replaced the older, heavier toddies. Cardhu remains a key malt in the Walker blends portfolio.

Cardhu seems permanently embroiled in whisky history. When in 2003 both Walker and Cardhu sales were booming, something had to give. The bright idea was to make Cardhu a blend of three different malts. The only problem was that 'Cardhu' was the name of a single distillery. The industry shouted foul. A bad day at black rock indeed. The resulting debate resulted in new labelling rules and definitions, such as 'blended malt'. Cardhu remained a single and Diageo created Singleton.

Now Cardhu has a new visitor centre and is one of Walker's 'Four Corners of Scotland' strategy. It makes a grassy spirit, slightly less intense than Glen Ord, with a citric note.

HEADING SOUTH

Very few of the oldest Speyside sites, the ones hidden away on farms to escape the attentions of the Excise, survived. It would be those that were closest to transport links – in towns, beside roads or in **Cragganmore**'s case beside the railway – that stood a better chance of prospering. In fact, when Cragganmore was built in 1869, it was the first newly commissioned distillery to be constructed since 1840. It was the brainchild of John Smith, who had plenty of experience as a distiller. He persuaded the landowner Sir George Macpherson-Grant to lease him land next to the proposed line where he could build a distillery – the first to do so.

Rated 'A1' by blenders, it has stuck to Smith's methods. The stills he designed remain the same cunning shape. They have to. Without those tall wash stills with steeply angled lyne arms or the small, flat-topped spirit stills with their offset long lyne arms, the character would be lost. The wash builds weight, while the flat tops means that a fascinating type of reflux is going on, with only congeners with a specific weight being carried through to the worms. The conversation that's stopped in the wash is extended in the spirit, then the worms add density. Complex flavour-making machines. Sulphur is in the new make, but Cragganmore's complexity comes out in maturation. There are dark and orchard fruits, tobacco and a tiny touch of leathery meatiness at the back. Refined and subtle, quite why it hasn't become a cult remains a mystery.

Tormore doesn't look like a distillery. It's a strange modernist/neoclassical mashup in the middle of moorland built – as Oliver Chilton, head blender of its new owner Elixir Distillers, puts it – 'by four eccentrics'. One was Lewis Rosenstiel, the former bootlegger who owned Schenley Distillers. In 1959, he and press baron Brendan Bracken hired the 79-year-old architect Sir Albert Richardson to build the distillery. Somehow Tory MP Bracken got author Neil M Gunn (a lifelong Scottish nationalist) to find the site.

If none of that seems to make sense, then neither does the distillery, one of whose three tun rooms is inside a clock tower. At a time when distillers were building functional boxes, it's a chaotic mess.

CARDHU

CARDHU 12 YEAR OLD 40%

Wet grass, chocolate orange and a mix of oxidized notes and old cellar (earth floor, damp wood). Some struck match. Crisp and cereal-driven in the mouth.

MALTY/NUTTY

It's a big site – capable of making 4 million litres (1 million US gallons) per annum, which went straight for fillings. What Oliver and the distillery team led by manager Polly Logan are doing is finding out how the place works. The more they uncover, the more bizarre it becomes. The short ferments are 50 hours, the longs are 100. Each of the eight stills has a purifier, but they've discovered a rectifying plate in the wash stills and possibly the same in the neck of the spirit stills. 'There were copper chains inside the stills at one point,' he says, 'and at another there were water-cooled copper pipes inside the neck.'

Production is being kept at roughly the same level. The shorts and longs are being lengthened, while there's a change in yeast. The team will look at the purifiers and see how reflux can be extended. The aim is to build more fruit above the firm nuttiness of the new make. 'The future is about harnessing its weird eccentricity,' Oliver says, 'and also bringing back the culture.' Richardson built an arc of houses on both sides, some of which are now being bought back by Elixir. The workforce has already trebled. As Oliver puts it, 'We're investing in the plant, but also the community.'

OTHER BEN RINNES DISTILLERIES

If you stand at the end of the Ben Rinnes road and look north you'll see a plume of steam rising from a valley. It's **Dailuaine** (or to be precise its dark grains plant). Once the largest malt distillery in Scotland, it has been a workhorse for multiple blends and has made meaty, fruity and grassy spirit.

Dailuaine's one of those distilleries which seemingly has always been able to cope with change and the vagaries of the blending market. When, in the 1890s, less smoke was demanded, the pitch on its kiln roof was heightened and the first Doig cupola installed. Whatever character has been demanded though, Dailuaine has remained sweet with a soft mid-palate. Amenable, always.

Its nearest neighbour's tale was one of tragedy. **Imperial** was built next to Carron station in 1897 by Thomas Mackenzie, the year of Queen Victoria's Diamond Jubilee. The maltings even sported a gigantic cast-iron crown. Two years later the market crashed and Imperial was mothballed, only to reopen in 1919, then close again in the 1920s and lie silent until the 1950s when another whisky boom was underway. In 1985, it shut once more, then ran between 1991 and 1998. In 2013, the buildings were demolished. Empires crumble. Independent bottlings of its gentle cream soda style whisky appear but are increasingly rare. In 2014 a new distillery arose on the same site. **Dalmunach** (see above) is owned by Chivas Bros. It makes a light fruity style. The site is still haunted by its predecessor. Behind Dalmunach is a large grassy knoll. Under the turf are the imperial remains. Imperial however exists in spectral form. Behind the new distillery is a large grassy mound. Under the turf are the Imperial remains.

Another from Chivas's Speyside stable sits at the oppsite end of the Ben Rinnes road. Dating from the 1960s, **Allt-a-Bhainne** (it means 'the milk burn' in Gaelic) makes a style which is also at the opposite end of the flavour spectrum. Estery and light, there is also occasionally some peated made which is reminiscent of a garden bonfire.

In 2014 the MacPherson-Grant family who still live in Ballindalloch Castle opened a distillery in an old steading next to their 9-hole golf course. **Ballindalloch** is run on pretty traditional lines – barley from the estate, stills leading into worm tubs. Its first bottling appeared in 2023.

NORTHEAST SPEYSIDE

Located in the fishing port of Buckie on the Moray Firth, **Inchgower** was built in 1871 and was owned by the town council for a brief period in the 1930s. It then became part of the Bell's stable (Dufftown, Pittyvaich, Blair Athol). All operated on similar lines – quick mashing, short ferments and a nutty-spicy character. Inchgower, however, plays its own variation on this theme.

Mashing here is slower, which cuts back on nuttiness, while by increasing the temperature of the second water, spiciness is accentuated. Ferment times are also lengthened. Somewhere within this lies the answer to why Inchgower can be remarkably saline and even at times slightly waxy. Is it the proximity to the sea? Maybe, but not all coastal whiskies are 'salty'. It remains a mystery. Thankfully.

Built in 1896 to take advantage of the blend boom, **Aultmore** has played a central role within John Dewar & Sons since 1923. Indeed, the deal-breaker when Dewar's was bought by Bacardi from United Distillers & Vintners (UDV) was over who would hold onto Aultmore. It's another forgotten jewel. A modern plant, it stands strangely isolated on a bluff looking south towards Keith, the sea wind at its back. Its slightly anonymous interior is more than compensated for in the new make coming from its small stills with their downward-facing lyne arms. Its penetrating, grassy intensity with subtle depth gives energy and drive to a blend – think of it as Dewar's Linkwood. It became part of the firm's malt portfolio in 2014.

A significant expansion started in 2023 that will see capacity double to 6 million litres (1.6 million US gallons) per annum, helping to balance the requirements of Aultmore as a growing single malt brand and the needs of its signature character within the Dewar's and William Lawson's blends. Long ignored and forgotten, maybe Aultmore's time has finally come.

There is a degree of confusion over which is the oldest (operational) legal distillery in Scotland. The significance of the 1823 Excise Act and the subsequent creation of the modern

AULTMORE

AULTMORE 12 YEAR OLD 46%
Sweet start, white chocolate, hazelnut with a little sweet cereal. The mid-palate thickness comes as a pleasant surprise.
FRAGRANT/FLORAL

★ AULTMORE 18 YEAR OLD 46%
Firms up a little. A typical gentle lifted dried flower/ yellow fruit nose, but the action is on the palate where there's substance and a clinging quality.
FRUITY/SOFT

Above *The town of Keith was one of the first to be 'improved' in the 18th century, ushering in the start of legal distilling.*

Above *Wood, water and copper; Strathisla seems unchanged by time.*

Above *Strathisla's stills reach to the rafters.*

Scotch whisky industry are regularly taken as the starting point for whisky itself. This ignores the fact that there were legal distilleries on Islay before that, and a northern distillery (Royal Brackla) was legally producing in 1815. The oldest, however, is on Speyside. **Strathisla** received its licence to distil in 1786.

Originally known as Milton (Mill-town rather than in homage to the blind poet), it's located on the outskirts of Keith. The town itself is made of three distinct parts: Old Town (700 CE), New Keith, which dates from 1750, and, in 1817, Fife Keith.

New Keith was built by the Earl of Findlater to settle people moved from the land during the Improvement movement. The 'lotts' they were given were sufficient to raise crops and cattle and set up workshops. New industries started up along the banks of the River Isla – woollen and flax mills as well as those milling grains.

Built on the site of a medieval monastery, there is some suggestion that brewing would have taken place at Milton. Distilling was simply the next logical step. Laws dictating that only one still was permitted per parish (see page 64) and Milton's location close to the town centre meant that an illicit operation was extremely unlikely. A licence made sense.

Fast forward to the 1950s, and Milton was the first Scotch whisky distillery bought by Canadian whisky mogul Sam Bronfman (see page 249), who snapped it up for £71,000. Bronfman had bought the ailing Chivas Regal brand in 1949 and was in the process of relaunching it as a 12-year-old.

STRATHISLA

STRATHISLA 2007 15 YEAR OLD
(EXCLUSIVE TO THE WHISKY EXCHANGE) 60.3%

Light nut, vanilla, peppermint, hazelnut ice cream and straw. The palate mixes the dry grass with apple syrup. Balanced.

FRUITY/SOFT

Strathisla, as he renamed it, was the first piece in the puzzle that would see Bronfman's Seagram owning nine malt distilleries and one grain distillery in Scotland.

Today, Strathisla runs a 5.1-tonne (5.6-US-ton) mash, the wort fermented in one of 10 wooden washbacks for 56 hours. The stillhouse has a low roof and the necks of the stills reach into the rafters. There's a slight optical illusion here, as they are in fact flat-bottomed with short necks. This means that while Strathisla tries to push towards light, there is a heaviness to its new make – even a touch of sulphur. All of this gives guts and helps in terms of long maturation. It's the lighter elements that rise first in the mature whisky, the weight and fruits biding their time until the late teens when it begins to put on extra muscle. Another underrated malt.

In 1957, with Chivas Regal sales increasing in the USA and Passport and 100 Pipers also on the rise, Bronfman commissioned a new distillery on the other side of the river behind Strathisla. **Glen Keith**'s tall, slender-necked stills (which are all different sizes) were the first in Scotland to run off gas. It began producing triple-distilled new make and alternated that with double distilled, until its mothballing in 1999.

It was always Chivas's experimental site. A wheat mash was distilled here, as well as peated variants mostly produced in the orthodox way by kilning, though one experiment saw smoke being passed through peaty water on a purpose-built plant on the Isle of Lewis. The water was then concentrated and used for mashing. It was reopened in 2013. It's rarely seen and the peated expressions, bottled as Glenisla, are rarer still.

Keith's milling heritage brought another distillery to life. **Strathmill** was originally a meal mill converting to legal distilling (as Glenisla) in 1892. It became part of the W&A Gilbey portfolio (with Glen Spey and Knockando), becoming a player in the J&B brand. In 1968, as part of an expansion, purifiers were added to the stills. These give a light olive oil-like character, adding an unctuous texture to the new make. Quick maturing, it can be found as a member of Diageo's Flora and Fauna range.

WEST & SOUTH OF KEITH

To the west of the town on the road to Rothes are the substantial granite buildings of **Glentauchers**, originally a distillery built to supply the Black & White blends (it also briefly made a 100% malt distillate in a column still). Mothballed in 1985, it passed into the hands of Allied Distillers and then Chivas Brothers, who reopened it in 1989. It's still a significant player in the blending world and is also the only wholly manually operated site in the Chivas Brothers' stable and therefore used as a training site for new operators. A balanced distillation gives a fruity/floral style that's most often seen in the ranges of independent bottlers.

Auchroisk, which lies to the south of Keith, was built in 1972 by J&B. Here, a heavy character is produced by fast mashing, quick fermentation and, in the wash stills, a rapid boiling regime that almost cooks the solids and allows controlled carryover of some solids. This almost singed character (shared with Blair Athol – see page 76) is most obvious at new make but recedes with maturation, being replaced by a balanced honeyed cereal sweetness.

It was bottled as single malt in 1986 and was the first to carry the prefix 'Singleton'. It was also one of the first 'finished' whiskies, aged initially in ex-Bourbon before being decanted into ex-sherry.

SOUTHERN SPEYSIDE

The landscape changes dramatically in the Speyside's southern part. Roads narrow and become more winding, hills rise more steeply, heather encroaches on farmland, villages replace towns and then there are only isolated farms. Here is space. That all made this area, Glen Livet, prime moonshining and smuggling country in the late 18th and early 19th centuries.

A chaotic series of laws intended to curb distillation was imposed. In the 1784 Wash Act, the 'Highland Line' was drawn imposing a different set of legal requirements on distilleries north of the line, specifying larger stills and only permitting one to operate in each parish. It effectively criminalized the farmer-distillers who had been using their own crops to make spirit that they had used as barter – and hospitality – for centuries. Distilling was an important extra income, especially in the poorer and more isolated lands in the south. Moonshining was the inevitable result.

In 1823, it all changed when a landowner-led reaction to the rise in smuggling helped create the Excise Act. This remarkable piece of legislation is rightly seen as the moment when the modern Scotch industry began.

It was also far-reaching in its rules, which covered everything from building to pipes, washbacks, stills and taxation. Every facet of distillation was included up to and including the housing for each new distillery's resident Excise officer.

Above *The Glenlivet's isolated location is an indication of its roots in illicit production.*

Above *Tamnavulin has spent all of its years producing light spirit for blends.*

It's often seen as the moment when the moonshiners were given (or forced to) take out licences and go legal. It wasn't as clear-cut as that. While some of the new licences were given out to reformed moonshiners, the majority were granted to hand-picked, loyal tenant farmers who could afford to build one of these new distilleries.

Quite where George Smith of Glen Livet sits in this remains unclear. Was he a moonshiner or a good tenant doing his landlord's bidding? As that landlord was the Duke of Gordon, one of the figureheads of the new legislation, there was some pressure being applied. Smith took out one of the new licences in 1824, but by 1827 he was on the verge of closing because of the expense. He was bailed out because his whisky had a reputation and also because the enterprise couldn't be seen to fail.

The fact that Smith's Glenlivet became the only distillery in the area until the 1960s speaks volumes for its quality, and the stubbornness of its owner. Ironic, therefore, that the fame of Glenlivet was such that distilleries as far away as the coast added the suffix to their name. A court case over the right to use the name ended with a compromise that Smith's distillery could call itself **The Glenlivet**, while others could use Glenlivet with a hyphen (as in Macallan-Glenlivet).

Today, The Glenlivet is (potentially) the largest single malt distillery in Scotland with an estimated capacity of 20 million litres (5.3 million US gallons) per annum. It has grown and grown and grown again. Now the mash is 14 tonnes (15.4 US tons) and there are 28 stills (14 each wash and spirit).

The style is light, fresh, and spring-like, all green apple and flowers. Everything is done to try and keep cereal and heaviness at bay, making this technically a difficult distillery character to achieve. Different cask types add spice (French oak) and richness (ex-sherry) to older expressions of the spirit.

Its nearest neighbour is the Whyte & Mackay-owned **Tamnavulin**, which was built (complete with Saladin maltings) by Invergordon Distillers to take advantage of the North America-led blended whisky boom of the 1960s. Its role as a supplier of fillings (it can make up to 4 million litres/1 million US gallons per annum) has not changed significantly since. The new make is light, clean and dry. Until 2016, it was rarely seen as a single malt, but in 2016 it began to sidle onto that stage. In recent years, wine cask finishes have led the way.

The River Livet is joined by the Crombie Water, opposite Tombae, a mile (1.6km) south of Tamnavulin. In front of you is The Bochel, a

THE GLENLIVET

THE GLENLIVET 12 YEAR OLD DOUBLE OAK 40%

Good distillery character: apples and flowers, elderflower, sweet malt. Flimsy. A whisky sorbet.

FRAGRANT/FLORAL

Above *Braeval's first mash took place before the distillery roof was in place.*

hill that acts like the cork on a bottle, the body of which is the Braes of Glenlivet. This is high but fertile land, remote but able to support people. The perfect place, in other words, for illicit bothies to be established and the whisky to be smuggled out over the Ladder Hills to the southern markets.

In the 18th century, this remoteness resulted in the Braes being home to a secret seminary at a time when the Catholic faith was being suppressed in Scotland. The Scalan seminary remained in operation until 1799.

In 1973, the Braes became the somewhat unlikely location for another Seagram distillery, **Braeval**, whose first mash took place before the roof had been put on. Its half-dozen stills with slim necks and upward incline to the lyne arm make a light style.

Answering the 1960s blend boom lay behind the building in 1965 of **Tomintoul**, which lies 5 miles (8km) southwest of The Glenlivet. It has been part of Angus Dundee since 2000. Its oldest expressions show a lush tropical fruit character, though at times it has flirted with a drier, more cereal-accented make.

On the way to Grantown-on-Spey, in the village of Cromdale, sits in one of the area's oldest sites, **Balmenach**. Another of the 1823 set, this is definitely built on the site of an illicit still run by one James MacGregor and has retained older methods of production with worm tubs giving a meaty new make, a character that carries through to the mature spirit, which mixes autumn fruits, petrichor and chestnut honey. It's such an important player for International Beverage's (IB's) blends that no official bottling has appeared...yet.

Grantown is now home to Speyside's newest distillery, **The Cairn**, owned by Gordon & MacPhail, which opened in 2023.

The final member of the region lies next to the river in the Cairngorms. **Speyside** was the brainchild of George Christie, though it would take 30 years from conception to it finally making spirit in 1990. In 2013, the distillery and the firm was bought by Harvey's of Edinburgh.

Above *Gordon & MacPhail's The Cairn is modern in design but run on traditional lines.*

DEFINING 'SPEYSIDE'

And so, beside the A9, we leave this complicated area. What is Speyside? That's hard to pin down. The drawing of boundaries works in two ways: it allows us to comprehend a specific area, but by doing so it can also constrict. It's known and unknown, in full sight and hidden. Complete, self-contained, layered and complex, but misunderstood. Those geographical (or in this case political) limits apply not just to space but comprehension. We search for patterns, for some unified character, only to see them fade to mist. Speyside can be light and floral or fruity, or grassy. It can be rich or meaty. 'Speyside' is a multifaceted response to landscape, circumstance, creativity and commercial necessity, and each distillery needs to be appreciated on its own merits.

There is another question, however. Until recently, Speyside's production has been fixed to the success or otherwise of blends. Its distilleries' characters have been formed to suit that model. Making a single malt and creating a single malt brand are two very different things. There's more to it than just putting the liquid in a nice bottle.

Single malt is also about identity. It's about why the distillery is there; it's about who makes it as much as what it makes. It's human, it has provenance, a resonance that goes beyond 'process' that has to be captured. It's about finding your roots. The best do that and some do it at scale.

How will Speyside respond to 21st-century Scotch with its greater concentration on location? It will still need its workhorses, the unsung heroes making quality whisky, because you still need a lot of liquid for Walker, Chivas, etc., but as single malt develops, so different demands will need to be met.

Above *Speyside is, thankfully, hard to define as a single entity.*

KEY
Malt Distillery
Eastern Highlands
Central Highlands
Southern Highlands
Grain & Malt Distillery
Southern Highlands
0 miles 30
0 km 30
Cape Wrath
Thurso
Wick
Flow Country
The Minch
Loch Shin
North West Highlands
Brora
Ullapool
Tain
Moray Firth
Loch Maree
Glenglassaugh
Banff
Fraserburgh
Elgin
Macduff
anCnoc/ Knockdhu
Dingwall
Nairn
Portree
Isle of Skye
Inverness
Peterhead
The GlenDronach
Huntly
Dufftown
Spey
Deveron
Ardmore
Ellon
BrewDog Distilling
Kennethmont
Glen Garioch
Inverurie
Findhorn
Loch Ness
Aviemore
Fort Augustus
Don
Aberdeen
Mallaig
Royal Lochnagar
Ballater
Dee
Burn O'Bennie
Banchory
Eigg
Dalwhinnie
Grampian Mountains
Stonehaven
N. Esk
Fort William
Ben Nevis
Fettercairn
Laurencekirk
Glencadam
Brechin
Pitlochry
Edradour
Blair Athol
Montrose
S. Esk
Arbikie
Aberfeldy
Ben Lawers
Loch Tay
Kythe
Isle of Mull
Coupar Angus
Arbroath
Dundee
Methven
Glenturret
Perth
Crieff
Tay
Strathearn
NORTH SEA
Firth of Lorne
Loch Awe
Tullibardine
Auchterarder
Deanston
Loch Lomond
Stirling
Forth
Dunfermline
Jura
Firth of Forth
Killearn
Glengoyne
Alexandria
Falkirk
EDINBURGH
Clyde
Glasgow
Bute
Islay
Southern Uplands
Kintyre
Arran
Firth of Clyde
Ayr
Campbeltown
N

THE HIGHLANDS

If Speyside is hard to corral, then the Highlands are an even looser bag of possibilities. The same principles apply: take each distillery on its own terms and discover its story, character and why it's in that location and what impact that might have. We start on the northeast coast near the Speyside border, head through Aberdeenshire and down the east coast, before swinging into the middle and finally drifting towards Glasgow.

EASTERN HIGHLANDS

The starting point is **Knockdhu** (bottled as anCnoc) and the final member of International Beverage's (IB's) portfolio. Like its sister stills, tradition is reflected in the retention of a worm tub on the wash stills (unusually, both stills run into a single tank).

Knockdhu, however, occupies an interesting halfway point between such older ways and more modern approaches. The mash tun is semi-lauter, but as of 2023 the spirit stills have had shell and tube condensers fitted. The shift will move the new make away from the cabbagey notes of the past and accentuate the lighter and more estery notes. 'Knockdhu has the biggest top end of all our stills,' says IB's master blender Stuart Harvey, 'a fruity floral fragrant citrus element, but when people try it they're surprised by big mouthfeel – and that's coming from the worms. It's a really interesting one to play tunes with.'

It's also where his brewing background is, finally, being given full rein. Prior to joining Inver House (as IB was then known), Stuart worked at London's Mortlake brewery before moving to Webster's in Halifax. Now he has a chance to go full circle. anCnoc is being given second maturation in Imperial stout casks, while he and his team are also now starting to explore different yeasts.

'In whisky, yeast is seen as the workhorse. A distiller's yeast is used to give you yield, but then you think about the yeasts that are about flavour and how you can play

ANCNOC

ANCNOC 12 YEAR OLD 40%

Yellow fruit gums and a touch of freesia, fresh mint and just a hint of residual sulphur adding weight. The palate shows concentrated fruits emerging. Highly approachable.

FRUITY/SOFT

★ ANCNOC 15 YEAR OLD PEATED 46%

The smoke is on the green, mezcal-like side of things. Minty in time. Water brings out a bittersweet, layered quality with more date and prune.

SMOKY/PEATY

Above *In such a large region, there's little surprise that the Highlands produces a wide range of different whiskies.*

THE DEVERON

Keep heading along the coast and you pass Banff, home to Scotland's unluckiest distillery, which caught fire three times, exploded once and was bombed during World War II. **Macduff** across the river Deveron was more lucky. Now owned by John Dewar (and bottled as Deveron), it's a major player in the William Lawson blend.

It's an odd site, which starts off as classically nutty-spicy (fast mashing, short ferments), but things become strange in distillation. The two wash and three spirit stills have upward-slanting lyne arms with an odd right-angled kink to them. The condensers are horizontal and are run hot. In other words, things start heavy and end light and grassy, which sometimes is pushed out as far as having the aromas of high-grade cannabis. Not that I would know what that's like.

Above *Back to the future: direct fire has been reinstalled at Glen Garioch.*

tunes with those. It's an area I'm excited about, as is having the chance to play with speciality malts and see what difference they could make at different sites.'

The trials are under discussion, but if they go ahead, the smart money is on Knockdhu being where the trials will take place. It is, after all, already stretching things out a little by making a peated variant, a nod to what would have been the original style from 1894.

When you come across a distillery that's hidden away in a valley nowhere near a main road or railway, you can immediately suspect how distillation began there. Such is the case with **The GlenDronach**. Its founder, James Allardice, got a licence in 1826, but there's plenty of evidence that he had previous experience in whisky making.

There is a reassuring solidity about the place, giving the impression that not only has little changed since Allardice's day but that it hasn't needed to. The mash tun is rake and plough (a good way to get clear wort if you're patient), the washbacks made of larch. The stills are substantial with, now cold, fire pits underneath – this was one of the last distilleries in Scotland whose stills were coal-fired.

Direct fire adds a burliness to new make and any taste of GlenDronach from new make on reveals its latent power. This also gives an indication of where the spirit is happiest maturing. That muscularity is needed to cope with the tannins of European oak. Lighter styles can be swamped and any complexity is lost as astringency takes hold. GlenDronach shrugs, says 'bring it on' and relaxes into sherry casks' embrace. Its underlying sweetness is key not just in terms of giving balance but also to its success in Asian markets where this combination of dried fruits and sweetness appeals.

There has been £30 million invested by Brown-Forman since it took over in 2016, some of it on upgrading the visitor centre, and at the time of writing, work is starting on adapting the old maltings as a new production unit, which will double production. There's certainly commitment here.

All seemed set fair for **Glenglassaugh** when it was built at the fishing port of Portsoy in 1874. Highland Distillers bought it in 1892, but the downturn saw it close in 1907 and remain silent for half a century. With the good times back, it was revived, providing its make for Highland's blends such as The Famous Grouse and (at that point) Cutty Sark. It was, however, a tricky one to blend – some whiskies are just awkward characters – and when the next dip in Scotch's fortunes came in 1986, it shut once more. Billy Walker bought it in 2008 and it became part of Brown-Forman in 2016.

The 12-year-old gives a good idea of blender Rachel Barrie's approach of layering different fully matured casks. A peated variant, Portsoy, uses sherry, Bourbon and port, while Sandend has a distinctly maritime edge thanks to the use of ex-manzanilla casks, helping it further reflect its place.

GLENDRONACH

GLENDRONACH 12 YEAR OLD 43%

Deep and earthy. Liquorice, tobacco tin, polished wood, cocoa nib and some malt extract. The palate mixes dunnage floor with black fruits. Softer with water.

RICH/DEEP

GLENGLASSAUGH

GLENGLASSAUGH 12 YEAR OLD 45%

Nutty. Coconut (gorse flower), then olive oil, crisp malt and almond paste. Cask-forward but some berry fruit, the grip of sherry, black and red wine gums. Chewy.

FRUITY/SOFT

★ GLENGLASSAUGH SANDEND 50.5%

Seemingly light on the nose. Barley, dry sand, minerality, then mimosa and Chanel No 5. Manzanilla cask adds saltiness and green olive to the distillery's floral character. Of its place.

FRAGRANT/FLORAL

THE GARIOCH

Pronounced *Geerie*, the Garioch is a 150sq-mile (390sq-km) area of Aberdeenshire stretching along the watercourses of the Don and Urie. The earth is thick and black, the mountains low and long, like sleeping figures resting on the rich farmland. As with many agricultural regions it tends to be a place that most travellers pass through. The whisky pilgrim certainly does so, their eyes fixed on Speyside, but the Garioch is worth pausing in.

Some towns have a church at their centre; Old Meldrum has a distillery. **Glen Garioch**, which was licensed in 1798, has however been a hard one to pin down with regular switches of style. When part of Distillers Company Limited (DCL)/Scottish Malt Distillers (SMD) between 1937 and 1968, it made a heavily peated malt.

This continued for a period under Morrison Bowmore, before the peat was dropped and a high-toned new make introduced. Then it switched again, still with no smoke, but allowing a tallow-like quality to be pushed through.

Suntory Global Spirits changed things yet again, investing £6 million to restore the large floor maltings, which now provide 5% of the malt for each mash.

GLEN GARIOCH

GLEN GARIOCH 12 YEAR OLD 48%

A malt/white pudding note to the nose, with raspberry jam and light oak. There's robust earthy weight here along with poached pear and a hint of white chocolate on the finish.

RICH/DEEP

Most intriguingly, direct fire has been brought in for the wash still, a direct result of Suntory's research at Yamazaki and Hakushu (see pages 285 and 280), where flame has been used to add depth to the new make. The higher temperatures achieved by direct fire cause a Maillard-like reaction between yeast cells and the copper. Remarkably, at Glen Garioch it's more energy-efficient than the old system.

You can also see the Suntory Global Spirits hand in the use of ultra-clear wort, the mix of distiller's and brewer's yeasts and the lengthening of ferments. All make the distillery less 'efficient', but that can often mean more flavour. Now the make has an oily depth, while the fruits have been punched up. It's big-boned and will need time in cask but, hey, it's been around for over 200 years; it's used to that. Whisky is cyclical in commercial terms. It's also cyclical in thinking. The past returns.

If Glen Garioch is rooted in its town, then **Ardmore**, Suntory Global Spirits' second still in the Garioch, seems slightly out of place. It's a sprawling site, red brick in a place of granite, obvious while others hide. It looks urban, but there's no city within 30 miles (48km). Why here? That goes back to its founder, Adam Teacher, who in 1898, seeing blend sales increasing, wanted to secure supply. While staying nearby at Leith Hall, he found this site, conveniently located next to the railway line.

It's Victorian in the extreme: heavy machinery, greased cogs and pulleys, steam and pumps, the product of an era when engineering could solve every problem. Clyde-built in Aberdeenshire, there is something forthright about the heft of those machines and the scale of the operation, from mash tun to the stillhouse with its long row of eight stills.

It was the last in Scotland to use coal fires. When they came out, any potential lightening of character was averted by adding some kinks in the

Above *Ardmore was one of the last distilleries to be coal fired.*

steam coil to mimic the hot spots you get with direct fire. The lyne arms point down, adding to the weight, but the very long ferments add fragrance.

Ardmore pulls in opposite directions. It's smoked, but also light. As the latter elements come at the start of the spirit run and the peat at the end, getting both in balance is a real achievement.

The barley as well as the peat are sourced locally and, as part of Suntory Global Spirits' £400,000 Peatland Water Sanctuary project (see box, page 103) that aims to restore twice as much peat as it uses to create its Scotch whiskies, 13 hectares (32 acres) of peat moss have been restored in the neighbouring Gartly Forest.

Ardmore's been a hidden player, but has built a dedicated following thanks predominantly to independent bottlings. In it you can see the fuliginous signature that typified old Glasgow blends such as Teacher's. Maybe living in a smoky industrial city resulted in the local palate wanting soot in its drams as well as in the air. There's also the lightness and freshness of the Garioch. A product of places.

SOUTH OF ABERDEEN

Heading south some 30 miles (48km) from Aberdeen, **Fettercairn** sits between the fault line of the Highlands and the thick-soiled coastal plain known as the Mearns. It's been distilling quietly for almost 200 years. At first glance, the distillery seems to fit that idea of an old place. The mash tun is still rake and plough, the washbacks are wooden and the stills are in an odd configuration placed seemingly haphazardly as the distillery expanded.

Then you look at the necks of the spirit stills, each of which has what looks like a slipped halo around its upper neck with steam rising off it. The halo is a cooling ring that's running cold water down the sides to create more reflux.

BRECHIN TOWN DISTILLERY

There are relatively few distilleries remaining on this eastern seaboard. The great cull of the 1980s bade farewell to **Glenugie**, **Glen Esk** (aka **Hillside**), **Lochside** and **Glenury-Royal**. The town of Brechin had two. While **North Port** (aka **Brechin**) has gone, **Glencadam** remains. Owned by Angus Dundee, it makes a fragrant floral style. Despite its apparent lightness, the aroma of the fruits ripening and 'thickening' makes it become more buttery as it matures.

SCOTTISH OAK

The idea of using Scottish oak for maturing whisky was dismissed for generations. Depending on who you asked, the reason was either there wasn't any or there wasn't enough, or the quality was bad. None of these is true.

Oaks have been growing in Scotland for 9,000 years, both sessile oak (*Quercus petraea*) in poorer acidic soils, and common oak (*Quercus robur*) in the rich glacial soils. Oak became the dominant broadleaf tree, a sacred being, the emblem of hospitality. It soon became a resource, used for housing, ships, furniture and pit props. Forests were felled to feed industry and make way for houses. By 1914, Britain was importing 90% of its timber.

Now, thanks to Gregg Glass, master blender at Whyte & Mackay, all of that is changing. He started an industry-wide initiative aiming to re-forge links between forester, sawmill, cooperage and distiller. New Scottish oak casks are being built at Speyside Cooperage and being used by Fettercairn and GlenAllachie, while new plantings such as the one next to Fettercairn are being established. In 100 years some will be used to make casks.

Scottish oak seems to contribute dark fruitiness, a honeyed element, coffee and low levels of vanillin. It will be rarer – and more expensive – than the oak from France, America or Central Europe, but as well as its own flavour it adds to the importance of provenance and helps to revitalize rural industries and crafts, while being sustainable and regenerative. Everything that 21st-century whisky needs to be.

Above *The Mearns south of Aberdeen is both a barley-growing and whisky-making area.*

Above *One of a number of farm distilleries, Arbikie is one of 21st-century Scotch's pioneers.*

BREWDOG DISTILLING

The final Aberdeenshire distillery is contained within BrewDog's enormous Ellon complex. Originally called Lone Wolf, it's now been brought more into the fold (or pack) as **BrewDog Distilling**.

Two styles of new make are made: single malt and a rye. As there's no designation (yet) for rye whisky in Scotland, this has to be called 'grain whisky'. The advantages of being part of a brewery runs through the entire process. There is a mash filter (see page 14) as well as a lauter tun, which extends the possibilities in terms of mash bills and grains. The mash filter can easily cope with the sticky 100% rye mash for the grain. The wort for the malt spirit is boiled after leaving the lauter (as in brewing) to kill any lactobacillus. 'We let the yeast be the star,' says distiller Steven Kersley.

The yeasts are blends (the rye only uses one) and fermented in temperature-controlled washbacks for five days. 'It's a bold, long chain ester-forward wash,' as Steven puts it.

Both spirits are double distilled. The wash still and one of the spirit stills have triple bubbles in their necks, massively increasing copper availability. The other spirit is plain. Once again, it's about creating as many flavour possibilities as possible.

Without the ring on, the new make is malty and cereal-accented. Turn the water on and you get fruity. It was a simple and clever invention from the 1950s – unlike the stainless-steel condensers that were installed in 1995 but thankfully scrapped in 2009.

Now owner Whyte & Mackay is helping to reveal Fettercairn's quality. It's all about the fruits, which in cask quickly move to the tropical. In some casks, this is pushed into a herbal quality akin to Yellow Chartreuse.

The author Lewis Grassic Gibbon was born here. His best-known work, *A Scots Quair*, is a trilogy telling of Scotland's progression, starting in the Mearns in an era of small farms and self-sufficiency, to market town, and then to city. At the end of the final book, its heroine Chris Guthrie returns to the Mearns, a circle completed.

There are parallels with Fettercairn, its journey from rural to urban, malt to blend and now back home and reconnecting with its roots. These days, the distillery sources its barley through direct contracts with local farmers, and in another initiative, Whyte & Mackay has planted 13,000 oak saplings next to the distillery (see box, opposite).

The idea of self-sufficiency is the ethos behind **Arbikie**, 20 miles (32km) south, next to Lunan Bay. The Stirling family has been farming here for four generations. In 2014, brothers John, Iain and David decided to open a distillery. If you grow it, why not make it? The approach, John Stirling says, is a modern interpretation of when most of Scotland's distilleries were farm-based, utilizing their own cereals.

The whisky making is somewhat more modern. 'One of the key components was to fully understand the relationships between our natural environment, our soil and our farming techniques, and how these all determine what we can grow and the qualities of the resultant harvested grains,' his brother Iain adds.

Distilling also took a different tack. Single malt is maturing, but Arbikie's first release was a rye (made from malted and unmalted rye plus some malted barley), the first to be commercialized in Scotland.

Unusually, the wash and distillation is 'grain in' (no filtering in the mash tun). Part of the wash is distilled in pots with a downward lyne arm; the other portion goes through the distillery's column. The former adds waxiness, citrus and some of rye's herbal/caraway element. The column ups the spicy element. Ripe fruits run through both.

'We think it opens the door for much more challenging and diverse Scotch to be produced,' says John. 'Switching from imported maize [corn] or wheat to home-grown alternatives such as rye or oats will not only mean we are making truly Scottish spirit but can also improve farm biodiversity, increase crop resilience and lower inputs.'

The brothers' push to sustainability continues with plans to make Arbikie the world's first green hydrogen-powered distillery using renewable wind power to create hydrogen that will power the site. Joined-up thinking.

FETTERCAIRN

FETTERCAIRN 12 YEAR OLD 40%

Vibrant fruits leading with pineapple, light peppermint, peach, milk chocolate. Weight on the palate. Mango, coconut, light oak and nutmeg.

FRUITY/SOFT

★ FETTERCAIRN 16 YEAR OLD 46.4%

Lush tropical fruits. Barbecued pineapple, baked fig, golden syrup, pink grapefruit. The fruits continue in the mouth. Sultana, carrot cake and ginger. Coffee bean on the finish.

FRUITY/SOFT

Above *Dalwhinnie is the coldest settlement in the UK and used to have a bunkhouse for travellers stranded in the snow.*

CENTRAL HIGHLANDS

Deeside, 40 miles (64km) due west of Aberdeen, is home to two distilleries, one old, **Royal Lochnagar**, and one considerably newer (see page 75). This was also a centre of substantial illicit production in the 18th and 19th centuries, centred in the foothills of Lochnagar, Braemar and Glen Muick. It was often a family affair with both men and women making and smuggling the *cratur*. One such family was the Stewarts who lived at Tilfogar on the Abergeldie estate.

In 1823, James Robertson took out a licence and built a distillery that he called Lochnagar. It was burned down three years later by his former moonshining associates. He moved the site closer to the Dee and started again, selling it to the Farquharson family who employed Charles Stewart, the Tilfogar moonshiner, as distiller. After this too caught fire, in 1845 John Begg rebuilt the place into what it is today.

This was the site of the first recorded distillery tour. In 1848, Begg invited his new neighbours Queen Victoria and Prince Albert to come and see the place. Impressed with Begg's 'works', the Queen awarded it the 'Royal' prefix.

It's a fascinating site – Diageo's smallest – that has an open-topped mash tun, using four waters, and a stillhouse with small stills and worms. Everything about Lochnagar points to heavy, but it runs against type. There's clear wort, and a mix of long and short ferments, while the doors to the stills are opened after each charge to refresh the copper. The fill level is low and, most importantly, the temperature of the water in the worms is kept warm by only being slowly replenished. Everything in distillation here is being done to maximize the copper conversation.

By doing so, the character of the whisky shifts from what you would expect – sulphur/meaty – to a dry grass/meadow hay. The worms still play their part, adding weight and succulence to the mid-palate. It's mostly filled into ex-sherry casks and can cope with that often challenging environment.

Worm tubs also play a significant role at **Dalwhinnie**. Technically speaking, the distillery falls within the Speyside boundary, but Diageo

THE REBRANDING OF SCOTLAND

The 18th century was when the idea of Scotland started to change. The previous image, of the Highlands especially, was of a rebellious, heathen place where people spoke a different language and were of a different religion. By the end of the century, however, its epic scenery fitted the Romantic notion of 'the sublime' – mist, mountains, lonely places, legends. It could be admired, rebranded. It was safe. Sir Walter Scott took charge of the task. In his time the most popular novelist in the world, his tales were of an idealized past.

The most significant shift in Scotland's image came when Queen Victoria and Prince Albert bought Balmoral. That seal of approval completed the rebrand. The Highlands became a theatre set, a game reserve, a place for mass tourism. As people gazed at empty landscapes, few asked the question, why were they empty?

Whisky became an integral part of this. It was a sensible move. If you were trying to sell your product in England, then further afield, you needed to convey 'Scotland' in shorthand. Labels sported images of castles, mountains, stags, sporting pursuits, soldiers, tweeded gentlemen and old Highlanders. Scotland became Scott-land, then Scotch-land. All of this appealed to the gentleman drinker and the expat. Without this, Scotch would have failed, but what was lost in the process?

Above *Edradour's origins as a farm distillery can still be clearly seen.*

prefers to call it Highland. There's certainly high land on either side – the Cairngorms Massif to the east, the Monadhliath Mountains to the west.

High and isolated, it was a meeting point for drove roads that transported cattle from the west and north to the markets in Perth and Falkirk. Illicit whisky was, allegedly, tied beneath the bellies of the hairy beasts. No gauger would want to get in the way of those horns.

It seems a strange place to build, however, until you realize that the Highland Railway runs behind the distillery or, when it was built, the front. As with all from the late Victorian era, it was constructed in order to take advantage of what seemed the endless rise of blended Scotch. In 1926, it became part of Distillers Company Limited (DCL)/Scottish Malt Distillers (SMD) and was licensed to Buchanan's, which, one assumes, had used it for its lighter-toned blends such as Black & White.

Dalwhinnie makes great play of being the highest (technically the joint-highest with Braeval) in Scotland as well as being the coldest settlement in Britain. Both of these impact on its flavour. Cold, snowmelt water running into the worms makes for a very sulphury new make that's a mix of pea and ham soup, fermenting sauerkraut and car exhaust, none of which you want in your mature whisky.

ROYAL LOCHNAGAR

★ ROYAL LOCHNAGAR 12 YEAR OLD 40%

Fruits come through (blueberry muffin) along with lightly herbal meadow grass. There is subtle power in the delivery. Lovely.

FRUITY/SOFT

In 1986 those worms were replaced with shell and tube condensers. The sulphur disappeared, and the new make went grassy. Back went the worms.

It's only after 15 years that the sulphurous carapace cracks and out comes a mature whisky filled with honey and night-scented flowers. Refill American oak takes things into crème brûlée. It works with chocolate – the distillery visitor centre does a roaring trade in hot chocolate. The fact that one of its signature serves is straight from the freezer seems highly appropriate.

PITLOCHRY TO ABERFELDY

Edradour had the distinction of being Scotland's smallest distillery for many years – and was one of its oldest, being established in 1825 as Glenforres by a consortium of local farmers. The combination of that grouping plus the fact that there was a mill nearby is suggestive that illicit stuff had been made in this tight glen north of Pitlochry for many years previously.

Until relatively recently, a visitor to Edradour would have thought that little had changed in the intervening years. The mash tun is open-topped and works with a rake and plough system; the wort is cooled not in a modern heat exchanger but in a Morton's Refrigerator – a 19th-

BANCHORY TOWN DISTILLERY

Burn O'Bennie started life in 2017 as Twin River but moved to its new site in the Deeside town of Banchory in 2020. Allying itself closely with brewing, it produces a high-strength wash. The use of speciality malts (pale, crystal and chocolate) also widens the flavour spectrum.

CHILL-FILTERING

Mature whisky contains numerous elements that give it flavour and texture. Among these are fatty acids, which can come out of solution when water is added to the whisky (or it's stored at cold temperatures), causing it to go hazy. To stop this – and to stop customers complaining that the whisky's gone off – before bottling distillers will drop the temperature of the liquid and pass it through a filtration system that removes the fatty acids. Result: clear whisky.

The problem with this is that these fatty acids are, as the name suggests, fat. They add mouthfeel and, some would argue, also carry flavour alongside. Remove them and the whisky might be clear, but it's thinner, something that is more apparent in those that are bolder and thicker to begin with. Increasingly, single malts are being bottled without chill-filtering. Check the label.

Above *Aberfeldy's honeyed character somehow reflects its bucolic setting.*

century design (a stainless-steel one has now been installed). The stills are tiny and squat and their lyne arms dip into equally minuscule worm tubs. Not a museum but a time capsule.

It became part of Campbell Distillers (now Chivas Brothers) in 1982, having previously been owned by blending house William Whitely since 1933. It was then bought by renowned independent bottler Signatory Vintage in 2002, giving it a brand to build as well as warehousing for its extensive selection of casks.

Nearby Pitlochry is a tourist magnet and tens of thousands of people came to Edradour, but the reality was that people on bus tours don't spend money. Post-pandemic, the decision was taken to close the visitor side of the business and concentrate on distilling.

A second distillery was built on the site in 2018. It's set up in exactly the same way as the original. The style is sweet, oily and rich, with the bulk of the new make being aged in ex-sherry. A peated version, Ballechin, was launched in 2006.

John Dewar hailed from Shenavail, which is close to Dull, though he was anything but, and his son Tommy even less so. Within a single generation, Tommy had built the Dewar's blends into a world brand. The first great whisky traveller, mythologizer and embracer of advertising, Tommy was one of a cohort of extraordinary Victorian entrepreneurs who turned Scotch from a drink of one small country into a globally renowned phenomenon.

The success of blends required liquid and lots of it. Blenders, the people who had taken others' product and made a drink that was greater than the sum of its parts, also became distillers in order to secure supply. They bought and they built. Tommy Dewar and his brother John Alexander did both these things.

The one they built was at **Aberfeldy**, 2 miles (3.2km) from their father's birthplace. Not just home, but also conveniently close to Perth, which was already establishing itself as a blending centre to rival Glasgow and Edinburgh.

Perth's days as the home of blends has gone, but Aberfeldy's distillery remains. In its Distillers Company Limited (DCL)/Scottish Malt Distillers (SMD) days, its whisky was waxy. Today the wax has, more or less, gone and the aim is grassy and fruity with a cereal back note, though it's hard to eliminate these things completely. Honey rather than wax comes into play.

That languorous quality helps it to stretch out during maturation, developing, deepening, becoming more complex as it ages, mostly in refill and re-charred casks, allowing that distillery character to show. This base can then be overlaid with ex-fortified wine casks, or wine. A light touch is all that's needed.

Aberfeldy has a discreet weight, even more so now that the strength has been upped to 46% and the spirit is no longer chill-filtered (see box) and any adjustment with caramel stopped.

OTHER CENTRAL HIGHLANDS DISTILLERIES

Blair Athol is, at the time of writing, the most visited distillery in Scotland, a fact that makes the fact that it's only rarely seen as a single malt somewhat surprising. Being located in the tourist trap of Pitlochry helps – as does the excellent visitor centre. You can only hope that they all buy a bottle. Another of the old Bell's distilleries, it takes the nutty-spicy template to a different extreme than Inchgower. Here, the wash still is run sufficiently hard to allow some particulate matter to carry over, adding a deeper nuttiness to the new make. Malt extract and elegant fruits are the end result, with the latter emerging over time.

Thirty miles (48km) southwest is **Kythe**. The brainchild of Angus MacRaild and Jonny McMillan, it is another back-to-basics site using locally grown, floor-malted heritage barley varieties, two-week ferments with brewer's 'trub' and a wood-fired wash still. Inefficient? Yes, but as Angus says, 'Efficiency is the enemy of character.'

ABERFELDY

★ **ABERFELDY 21 YEAR OLD** 46%

Slightly understated nose, but the palate has compressed, syrupy dark fruits, heather honey and beeswax, then some Madeira-driven blueberry jam.

FRUITY/SOFT

SOUTHERN HIGHLANDS

Heading 20 miles (32km) south from Aberfeldy, you arrive in the old spa town of Crieff, home to **Glenturret** and a strange, winding saga of how a farm distillery started, closed, was forgotten, re-emerged as a blending whisky, became more famous for animals than its drams and then was reborn as part of a luxury group. Whisky rarely moves in straight lines, but this distillery's story is more convoluted than most.

It possibly started off as The Hosh (confusingly, there were two distilleries of the same name in Crieff, one of which closed in 1850), one among many farm distilleries across Scotland that got, or were forced to take, licences.

Most foundered – there's a lesson there – as owners didn't grasp the need for a hobby to become a business, that quality had to be maintained and, as the industry developed, volume was also important. Whisky making is also expensive. Renamed Glenturret in 1875, it closed in 1923 and was demolished. The site was subsequently rebuilt in 1959.

Under Highland Distillers' ownership, it became the home of Perth blend The Famous Grouse (the city is only 20 miles/32km away) complete with a giant fibreglass bird. It was a well-intentioned and effective way of telling the story of blends (Aberfeldy still does the same for Dewar's), but the distillery became more famous for the big bird, and the fact that Towser (the distillery cat) was a Guinness world record-holder mouser, than its single malt.

Then, out of nowhere, it was bought by Lalique. Ken Grier, ex-Macallan, had a lot to do with it, as he had worked with the legendary glass-maker on Macallan bottlings. He hired Macallan's former whisky maker Bob Dalgarno into the same role at Glenturret. If a whisky maker suddenly appears in an 'unknown' distillery, it's because they know the stock, and the potential. Glenturret was a gem that needed to be polished.

It's small-scale (only 220,000 litres/58,000 US gallons per annum) with Douglas fir washbacks and two stills that make a citrus-sweet and fruity new make. Since the takeover, there have been trials with different barleys. It's old style without being old.

The new releases have been revelatory. All are in sherry and the sweetness of fruit comes

GLENTURRET

★ GLENTURRET TRIPLE WOOD 43%

Some sherried funk. Leather sofa, roasted almond walnut and Olde English marmalade, then a light touch of vanilla and *vin santo*. Scented and complex.

RICH/DEEP

★ GLENTURRET 12 YEAR OLD 46.4%

The glossiness you can get with ex-sherry cask. Hot tiles, roasted coffee bean and dried flowers. Cask-driven, but manages to be light, not ponderous.

RICH/DEEP

Above *Glenturret is the only distillery in the world with a two-star Michelin restaurant attached.*

OTHER DISTILLERIES WEST OF PERTH

The village of **Strathearn**, 7 miles (11.3km) west of Perth, became home to what was then Scotland's smallest distillery in 2013. It was the brainchild of Tony Reeman-Clark who approached whisky making in an expansive fashion, which brought him into regular dispute with the Scotch Whisky Association. He also established the Scottish Craft Distillers Association, but then promptly resigned! His principles are now in line with much of the thinking around the world: different woods and yeast strains, and a return to older lower-yielding barley varieties. Strathearn was bought by independent bottler Douglas Laing in 2019. The first release from its new owner appeared in 2024. Ferments push 150 hours, low-yielding barley is still used and the two stills make 57,000 litres (15,000 US gallons) a year.

To make whisky, you need water. A lot of it. Finding a plentiful supply, cold and untainted by dead sheep, is therefore the first thing a distiller will look for when finding a site. Some places are more blessed than others in having that. One is Blackford, which sits on the old Glasgow to Perth road (and railway line). It's home to Highland Spring mineral water and over its history has played host to many breweries – there were three and a maltings in the 19th century. One of those, Gleneagles (the famous hotel is close by), was converted into **Tullibardine** in 1949, the first new build in Scotland for almost 50 years. It's been a discreetly steady producer ever since, ownership switching between blending houses. It's now owned by Picard, which owns Highland Queen and Muirhead's, both big sellers in the French market.

through, a slight oiliness softening the tannins. It's being recognized as a classy dram. It always was. It's just that no one knew it. There is also a two-star Michelin restaurant. Kitsch to luxury. It can be done.

Deanston, it must be said, doesn't look like a distillery. Neither should it really. It was, after all, originally an 18th-century mill, which at one time boasted the largest water wheel in Europe and was home to the development of the Spinning Jenny. The mill was here because of the water. The River Teith was harnessed to provide power and today 20 million litres (5.3 million US gallons) an hour pass through the distillery turbines, meaning that it's not only self-sufficient in power but sells the excess back to the National Grid. While other distilleries (rightly) trumpet their green credentials, Deanston remains strangely silent about being the first.

The mill ceased production in 1964 and the site was turned into a distillery, its whisky aimed at the blended boom in North America. It's now part of CVH Spirits.

It has a huge mash tun capable of holding 12 tonnes (13.2 US tons), the wort is low gravity – for fruitiness – the fermentation time is set at 85 hours and the upward angle on the lyne arm helps to increase reflux and that fruitiness.

It's one of only two distilleries to make a waxy spirit, initially at the behest of Diageo when Clynelish began to struggle to produce enough for its own single malt and fillings for Johnnie Walker. Organic whisky is made each year, though the biggest success in recent years has been its Virgin Oak bottling (see tasting note). This first appeared in 2010, but thanks to former distillery's director Brendan McCarron it became a cult during Covid as the base for Old Fashioned cocktails. It is another of the quiet distilleries that is, deservedly, picking up a new following.

DEANSTON

★ DEANSTON VIRGIN OAK 46.3%

Controlling virgin oak is not easy in Scotch, but the distillery's waxiness and weight is the ideal foil. Honeycomb, caramelized peach, coconut and a hint of pine. A wee gem.

FRUITY/SOFT

Above *Deanston's water turbine made it Scotland's first green distillery – before the term was even coined.*

Above *Under the volcanic plug of Dumgoyne, Glengoyne is one of Scotland's most frequently visited distilleries.*

THE GLENGOYNE BURN

If Speyside is defined by its administrative boundaries, then you might imagine that the Highlands are more clearly divided from the Lowlands. There is after all a geological fault that runs in a southwest-to-northeast diagonal passing through the north end of Arran all the way to Stonehaven on the Aberdeenshire coast. This means that Deanston, Tullibardine, Glenturret and the Mearns distilleries are all geographically speaking Lowland sites. The legal boundary, whisky-wise, takes a different route, following the notorious Highland taxation line from 1784 (see page 64). **Glengoyne**, sitting under Dumgoyne, the volcanic plug that's the dot of the exclamation mark of the Campsie Fells, is in the geographical Lowlands, but in whisky's Highlands. Go figure.

Its origins as a farm distillery are still clearly seen in its layout. Little seems to have changed physically since it was converted from a series of sheilings and barns in 1833. Owned by Glasgow blender Lang Brothers, it was part of Highland Distillers until 2003 when it was bought by Ian Macleod.

That proximity to Glasgow has made it one of Scotland's most-visited distilleries and one of the best to visit for anyone new to whisky. Everything is in one space, allowing you to see the flow from mash tun to washbacks to the three stills (one wash, two spirit).

It seems simple when put like that, but there is much more going on. The distillery is powered by renewable energy (wind and solar power). The wort is clear, the ferments are a mix of long and very long and the distillation is slow. Really slow. The stills are small, so to get the intense fruits and sweet hay character in the new make, time is needed for the vapour to flirt with the copper, rising, condensing, sliding back down, rising again, continually separating as it goes. There are even boil balls to increase the copper conversation. But you can never fully escape their small stature – they will always add weight. It's that which allows Glengoyne's fruits to steadily develop over time in ex-sherry.

Walk out of the still and through the wetlands that the distillery's established, a series of a dozen pools with 14,500 plants from 20 varieties that filter waste from the process. The now clean water then joins the burn. The waste gathered in the pools is harvested and as biofuel powers 350 homes.

Does it matter what side of the line it is on? No.

Follow the Glengoyne Burn and it reaches Loch Lomond, which conveniently enough is also the final

GLENGOYNE

★ GLENGOYNE 12 YEAR OLD 43%

The warming (and nicely sweaty) side of sherry. Orange, wet wool, butter. A complex palate with white pepper, coriander seed mixed with sweet dried fruit. Balanced.

FRUITY/SOFT

GLENGOYNE 15 YEAR OLD 43%

More oxidized and fruity. Brazil nut, roast almond, green fig, graphite, candied pear. Dried apple, barley and chilli flakes. Creamy finish.

RICH/DEEP

Above *Loch Lomond's multiple still types make it the most versatile distillery in Scotland.*

LOMOND STILLS

There are numerous models of hybrid (or reflux) stills around the world, including those at Loch Lomond. The first in Scotland was created by Duncan Thomas at Littlemill distillery (now demolished but also owned by Loch Lomond) in the 1950s as a way to make a triple-distilled style with two passes, running alongside a standard pot still. This hybrid ran until 1994 and served as the model for Loch Lomond's straight-necked stills.

In the 1950s, Hiram Walker installed a hybrid pot still designed by engineer Alastair Cunningham at Inverleven, the malt plant within the Dumbarton grain distillery. This had three plates that could be adjusted to increase or decrease reflux. The plates could be water-cooled or left dry, which gave further control.

They were called 'Lomond' stills, but bore no resemblance to the Loch Lomond stills. They were subsequently installed at Hiram Walker's other sites: Miltonduff, Glenburgie and Scapa. Running wash through the plates resulted in them becoming coated with dead yeast and barley residue, which reduced their efficiency and so they were all decommissioned.

Scapa still uses its Lomond still but without plates, and the old Inverleven still became Bruichladdich's 'Ugly Betty' gin still.

More recently, Inchdairnie (see page 107) has installed a hybrid still (see left) that it calls a 'Lomond Hills' still – it's named after the range of hills to the north of the distillery. There are six bubble-cap trays in the short, stubby neck, which can be flooded with condensed distillate, increasing reflux. The more that's reintroduced, the lighter the spirit will be.

stop on this trip through the Highlands. The **Loch Lomond** distillery is in an industrial park in Alexandria. Not romantic, but practical. Getting on with its own thing. The site was developed in 1966, and now there are two distilleries and 28 warehouses holding 400,000 casks.

It was established to produce whisky destined for value-for-money brands, a high-volume, low-margin business. If that's the model, then why not have everything under one roof? Why be beholden to others? So it makes grain and malt, and has a cooperage and a bottling line.

Blenders also need a wide variety in terms of flavour. Loch Lomond took that principle, plus its own principles of self-sufficiency, and created the most innovative distillery in Scotland. At a time when there was excitement about how innovative Japanese distillers were with their multiple streams and in-house blended single malts, Loch Lomond was already doing the same. They just didn't tell anybody about it.

Until now that is. Under new management, Loch Lomond, which also owns Glen Scotia (see page 85) has become more single malt focused and started to share its secrets. At a time when Scotch distilleries are, rightly, talking about new ways of approaching whisky, Loch Lomond smiles quietly, shrugs and says 'Yes, we've been doing that for decades.'

Fermentation times are steadily creeping up? Loch Lomond has always run a minimum of 92 hours, to increase lactobacillus and esterification. Distillers exploring yeast? They use wine yeasts to help boost fruits for their longest ferments.

Different types of stills? Done that as well. There are standard swan-neck pots, but also straight-necked stills with a pot still base and a column neck that has rectifying plates: one in the wash, and seventeen in the spirit, which also has a cooling ring. In other words, it's all about reflux.

'The characteristics of a swan-neck pot defines the profile of the spirit,' says master blender Michael Henry. 'The straight-necked stills, however, have flexibility and were created to produce a triple-distilled style in double distillation.' By running different cut points (high-strength at 85% ABV, low at 68% ABV) and levels of peating on the straight-necked stills, Michael collects seven different flavour streams. Then there's the new make from the swan-neck stills to add in. Each release of single malt is a blend of different distillates.

The malt side also has a Coffey still (see page 17) that distils a 100% malted barley wash (which was common practice in the 19th and early 20th centuries). The distillery asked to have a separate definition for this style but was turned down. So it's 'grain', but not grain as we know it. That's made at the site's grain distillery.

Now that the doors have been opened – and a better wood policy is in place – finally we can see what has been happening: 21st-century Scotch started in the 20th century.

LOCH LOMOND

★ LOCH LOMOND PEATED SINGLE GRAIN (100% BARLEY) 46%

Very juicy fruits (canned fruit salad), dusty florals, an old-fashioned sweet shop. Creamy texture. Remains juicy and long.

FRUITY/SOFT

LOCH LOMOND PEATED SINGLE GRAIN (100% BARLEY) 46%

A mix of spearmint, apricot, ripe melon and light smoke. Smoke tinged with lilac and violet. A rum-like quality. It works.

SMOKY/PEATY

LOCH LOMOND ORIGINAL 40%

Gently fruity with fresh wholemeal bread, orange oil and peppermint. Sugar puffs, then pineapple jam. Has substance. Caramelized black fruits. Sweet and thick.

FRUITY/SOFT

ISLE OF MULL

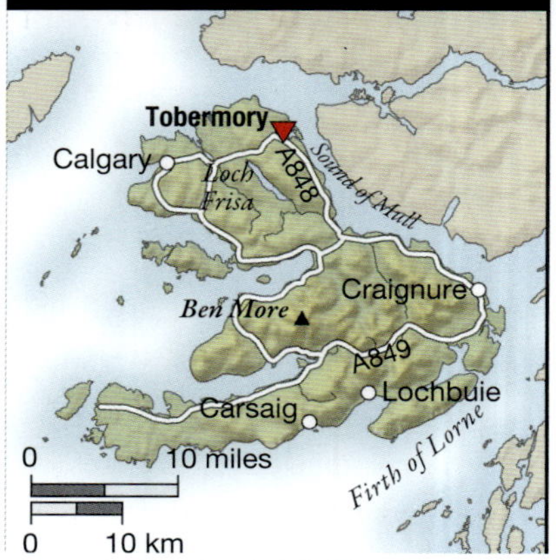

ISLE OF SKYE

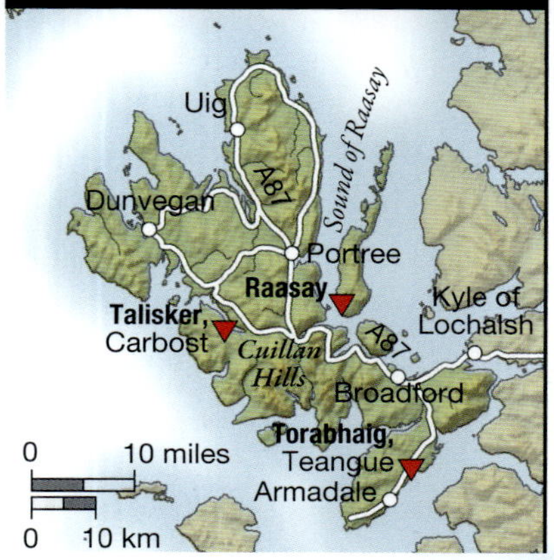

KEY

Malt Distillery

- Hebrides
- Arran
- The West
- Campbeltown

JURA

ARRAN

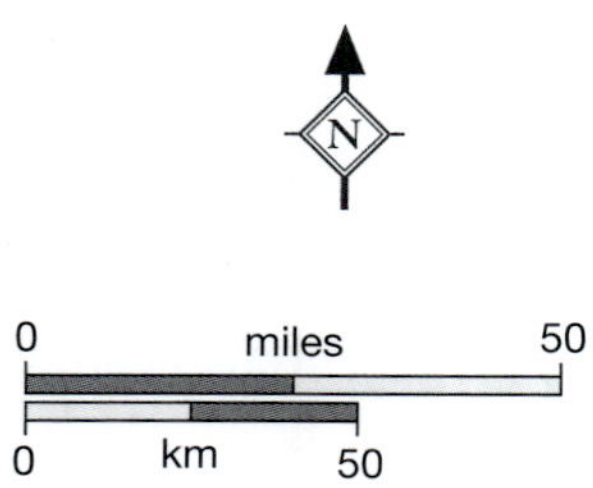

THE WEST & HEBRIDES

The arrival of the new distilleries in the west of Scotland and on the Hebridean islands has acted like a pebble being dropped into a pond. These dedicated, dynamic enterprises have brought employment, created careers and regenerated their communities. The human side of whisky is real, and it's more apparent in this part of the country than in any other of its regions.

CAMPBELTOWN

There is a sense of immutability when you walk up the lane between the Chinese restaurant and the Evangelical Church and enter **Springbank**'s courtyard. You know that inside things won't have changed. The mash tun will be open-topped rake and plough, the washbacks made of boat-skin larch, the three stills will work in their idiosyncratic way; there will be malt germinating in the malt barns, and bottles will be jingling down the line. Reassuring. Whisky as it has always been, but is that enough in a world where the new is worshipped?

Three different whiskies are made here. Hazelburn is unpeated and triple distilled, while Longrow is heavily peated and double distilled (in the wash and spirit stills).

Springbank is lightly peated and is distilled two and a half times. It starts in the direct-fired wash still and runs into a shell and tube condenser. The first 20% of the low wines go into the next spirit still charge; the remaining 80% go into the intermediate still, which is steam-heated but has a worm tub. The distillate (called 'feints') is mixed with the weaker low wines and heads and tails from the previous run of the spirit still to make the rest of the spirit still charge. This is run as normal and collected as spirit at 71% ABV.

It's complex, but so is the new make. Here's light smoke, fruit, citrus, a smear of light oils, some black olive and a touch of brine. Like some sort of whisky magpie, it seems to collect elements from all flavour camps and compress them into one distillate. All whisky is here.

Is this new or old? Interestingly, the Springbank method mirrors exactly the accounts from trade essays in the 1920s outlining how whisky was made in that era, the same essays that new distillers are poring over to find 'new' ways to make Scotch.

The cycle turns. By not remaining unchanged, Springbank now leads.

Above *Springbank has become a cult single malt, revered around the world.*

Above *Springbank's three stills are used in different configurations to make its three styles of new make.*

Production levels have long been...cautious. As sales director David Allen explains, this was down to the fact that Hedley Wright, the last of the family who owned the distillery since 1827 and who died aged 93 in 2023, 'had lived through so many booms and busts that as long as we had enough to sell in order to cover wages, then that was fine'.

In 2018, however, they doubled production. Prior to that, in true 1920s style, they ran malting and distilling separately. The team would malt for two months and then move to the distillery to distil. Now both run concurrently. That upped production to 260,000 litres (68,700 US gallons) per annum, which isn't a lot, especially when you are now a global phenomenon.

'Up until Covid, Springbank was a hand-sell,' David continues. 'It felt almost pioneering to be selling it and seeing people become engaged. Then in lockdown someone pressed the boost button.' It might have been online tastings, or people having more time to research or treating themselves to something different. The result was an astronomical rise in sales. Great news, but with a downside. Bottles went on allocation, sold by ballot, many snapped up by bot-armed flippers and immediately put onto the secondary market at massive markup. Long-term fans were priced out.

'It was a hard thing to see,' David says. 'We didn't have any more stock, but made the concerted effort to go out to market and explain, and pour people nice drams. It's gone a long way to building bridges. We're trying to ensure that whisky is getting into the right hands.'

The cult, though, has had a benefit. Springbank's visitor experience has been upgraded and people are

SPRINGBANK

★ SPRINGBANK 15 YEAR OLD 46%

Lanolin, sooty smoke, olive brine, deck oil and tangerine. Wood adds light grip. Palate is a mix of embers, tapenade and tropical fruits. All enveloping.

FRUITY/SOFT

★ SPRINGBANK 21 YEAR OLD 46%

Springbank puts on a smoking jacket and lights a cigar from a humidor. Serrano ham, oils and smoke, dried flowers. Fully integrated and hugely complex. Touch of rancio.

RICH/DEEP

★ LONGROW PEATED 46%

Eruptive smoke, some roasted barley, oils, oak spices and chypre. There's good balance on the palate, however, with balancing preserved lemon, mustard seed and a hint of salt.

SMOKY/PEATY

★ LONGROW 21 YEAR OLD 40.4%

On the meaty side. Barbecue, cordite, spent bonfire and some pumpkin. Fruit cordial, oils and smoked mussels. Light grip before raspberries on the finish.

SMOKY/PEATY

HAZELBURN 10 YEAR OLD 46%

Slightly green with a touch of lemon sponge, custard, cream, cultured butter. Has some fat on its bones. Sweet with clean acidity.

FRAGRANT/FLORAL

coming to the Wee Toon. This is a community distillery. 'The town's incredibly important to us,' says David. We employ 108 people and a good part of that is on the hospitality side. Physically we're in the heart of Campbeltown and metaphorically this is its beating heart. Us being here and being busy gives other business owners the confidence to come and open up.

'Numbers are small by other distillery standards, but if we get 10,000 people a year, that's double the population of the town and they will spend money. Campbeltown's a destination. Young people are staying here because there is work and a better work-life balance.'

All that you hear at Springbank – local barley, long ferments, direct fire, worm tubs, not chasing after yield, multiple distillation streams, a bottling line, a distillery regenerating a community – is the model for many newer sites. The model is 200 years old. Old but not old-fashioned.

'The fact that people have been trying to emulate us is a form of flattery,' says David. 'It's one thing to say we like what you do, but the practicality is very difficult and not many can carry it forward. We're a very inefficient distillery that employs far too many people!

'To make a small amount of whisky is fine when you have 200 years behind you and are seen as an one of the world's iconic whiskies, but if you are a new guy with investors looking for a return, you don't have the luxury of time. If you're caught between a distillery and the business to make a distillery viable, you have to have that business head. The Springbank way of doing things probably wouldn't work.'

GLENGYLE

KILKERRAN 12 YEAR OLD 46%

Phenolic. Lightly mossy peat and paraffin wax, but the centre is pure light white fruits with a citric edge.

SMOKY/PEATY

KILKERRAN 16 YEAR OLD 46%

A hint of sootiness, barley and some chocolate milk. Still seems young. More cereal and a light bonfire note on the palate. Drying.

SMOKY/PEATY

Next door is **Glengyle** (bottled as Kilkerran), which Springbank's owners J&A Mitchell bought and reopened in 2000 (it started production four years later). The Springbank crew runs it for three months a year making lightly peated (and some heavily peated). Changes are planned. A drum maltings will free up three months of malting for Springbank, increasing its production levels further, while a dedicated production team will start. 'The future?' asks David. 'It's about making J&A Mitchell a nice company to work for. We're in the healthiest position we've ever been.'

DALINTOBER

Around the bay in Dalintober sits the white-painted **Glen Scotia**. Although it has run steadily since its opening in 1832 – avoiding the Great Crash, which wiped out over 20 distilleries in the town in the 1920s – it has had a chequered career as a producer of fillings for blends. The collapse of the 1980s seemed to be the last straw and it closed in 1984.

It was at that point that Hedley Wright stopped any fillings contracts and started selling only single malt in order for the distillery to take control of its own destiny. Glen Scotia, however, even when it reopened in 1990 under new owners, stuck with the old formula. Loch Lomond Distillers bought it in 1999 and under that firm's new owners it has finally emerged fully formed.

It's a traditional set-up. Low, medium and heavily peated are mashed in a 100-year-old rake and plough mash tun, the wort fermented for an average of 129 hours, then distilled in dumpy stills. It makes a fat style, filled with that distinctive Campbeltown mix of oil and brine that coats the lush fruits – one for people who want some physicality in their dram. Production capacity doubled in 2024. It's reborn, as is Campbeltown. 'Glen Scotia is part of something that's bigger than the distillery,' says manager Iain McAlister, 'from both a historical and provenance point of view. Campbeltown! The former Victorian whisky capital

GLEN SCOTIA

GLEN SCOTIA VICTORIANA CASK STRENGTH 54.2%

Caramelized nuts, old citrus peels and green fig jam. Thick and lightly oily on the palate (vellum and linseed), moustache wax and peaches.

FRUITY/SOFT

★ **GLEN SCOTIA 25 YEAR OLD** 48.8%

Supple and the most overtly 'west coast' in style. Vaseline, old books, light smoke, fragrant, old paint but with water a balancing estery lift. Violet, then white pepper.

FRUITY/SOFT

Above *Manager Iain McAlister's belief in Glen Scotia is now deservedly paying off as it builds a new reputation.*

Above *When Arran was built in 1994, it was seen as a crazy idea… How things have changed.*

of the world, whiskyopolis, spiritsville…all those names point to that massive influence of the Wee Toon on the whisky industry. We've had good times, then hard times, and now a whisky renaissance.'

MACHRIHANISH

At the time of writing, Isle of Raasay distillers is building a farm-to-bottle distillery, **Machrihanish**, about 4 miles (6.5km) inland of Campbeltown. The intention is to grow a range of older strains of barley. All of the new make will be aged in ex-sherry casks.

ISLE OF ARRAN

It's funny to think back to 1994 when the opening of **Arran** was viewed as a slightly quixotic enterprise. The industry, still shaken by the crash of the 1980s, gingerly picking itself up, seeing stock levels remain high, production levels low and single malt still only accounting for 2% of total sales.

Then you tasted the new make. It was dynamic, citric, lightly fruity. It somehow spoke of optimism. Arran is now almost 30 years old. The new make hasn't changed. Those three decades weren't always plain sailing. It takes time for a distillery to establish itself, for stocks to be built up to a brand being established. Even with the single malt boom that kicked off in the early 2000s, it took Arran 10 years to be financially stable – an important lesson there for new distillers. Whisky takes time. It needs patience, understanding, investors and quality.

Now, Arran's distillery in the bay at Lochranza under the peak of Goat Fell is not only one of the most visited in Scotland but a player on the global scene.

In 2023, Stewart Bowman (former brewer and the man who reopened Brora – see page 32) took over from James MacTaggart (ex-Bowmore), who in turn had succeeded Gordon Mitchell (who had set up Cooley). That's some distilling pedigree.

Fruitiness is key here. The wort from a 2.5-tonne (2.8-US-ton) mash sits on one of the 10 washbacks for 75 hours to build up an orchard fruit character, then distilled in one of the 2 sets of stills and mostly filled into first-fill Bourbon, though Arran in sherry is a remarkable experience, as it has the fruity depth that carries well, while the citrus adds an almost

ARRAN

ARRAN 10 YEAR OLD 46%

A warming, rewarding mix of orange zest, malt, nutmeg and canned apricot. The palate is almost oily (orange oils) with some honeycomb chocolate on the end.

FRUITY/SOFT

★ ARRAN 'THE BODEGA' 55.8%

Bouquet garni, cut grass, olive, dried mixed peels (lemon) and the oxidized nutty notes of sherry. Berry fruits. Not your average sherry bomb.

RICH/DEEP

bittersweet tang. Stewart calls it 'engaging, accessible and complex. A whisky that everyone can enjoy, a delicious and honest malt whisky.' It's more than that, though.

Lochranza, geographically, is in the Highlands, while **Lagg**, the firm's second distillery in the south of the island, is in the Lowlands.

Lagg arrived in 2019 and like its sister emerged fully formed. Larger than Lochranza, it only makes peated spirit, starting with a cloudy wort with a standard peating level in the malt of 50 parts per million (ppm). The ferments reach 86 hours average. The wash still is short and bulbous, and both it and the spirit still have a steeply downward rake to their lyne arms. Distillation is fast. The aim here is power, with some grassiness, but the long ferments bed in the sweet elements that are vital for balancing any peated spirit.

'When James MacTaggart and I set out to create the spirit, we wanted something that could be described as uniquely Arran, but not just a copy of the peated spirit at our sister distillery,' says manager Graham Omand. 'The philosophy was if the spirit was delicious enough straight out the stills, the maturation would be a bonus.'

THE WEST

The fishing and ferry port of **Oban** has had a distillery since 1794 when the Stevenson brothers built it – and then went on to build the town around it. Wedged into the town centre with a cliff behind it, there's a pair of plain stills and worms that are run in a similar way to Royal Lochnagar: low fill, air rests, warm worms to make a lighter and in this case fruity new make. It does make you wonder what 'heavy' Oban must have tasted like. What is sometimes picked up as salinity is, for me at least, more of an intense gingery citric spice like togarashi pepper.

The 14-year-old built a following in the USA, although, somewhat cruelly, this was claimed to be the result of the name being easy to pronounce (though o-BAN is actually incorrect). I'd prefer to think it's because it's a direct, hugely enjoyable dram that also works well with ginger ale.

LAGG

★ LAGG KILMORY 46%

Sweet peat, the phenols lifted. Oak moss, some acetone, subtle oils. Direct. This is young but the sweetness balances the smoke well. Dynamic.

SMOKY/PEATY

LAGG CORRIECRAVIE 55%

Sherried, oily and layered. Fireplace, pickled walnut and, oddly, jelly babies, then charred dark fruit. Though sweet, already moving to the savoury side.

SMOKY/PEATY

OBAN

OBAN LITTLE BAY SMALL CASK 43%

More rounded and intense than the 14 year old. Crystallized ginger, some dark chocolate, Lucozade, popping candy and a crackle of (very) sweet spices.

FRUITY/SOFT

Above *The Lagg distillery on Arran only makes peated single malt.*

It was joined in 2015 by Little Bay, which is matured in refill American oak hogsheads, European oak sherry casks and refill casks with new ends before being married in ex-Bourbon barrels.

It's 40 miles (64km) north to Fort William and what was for many years the only other distillery on the west coast. **Ben Nevis** was established in 1825 by 'Long' John Macdonald. By the end of the 19th century, under his son Peter, the 'Dew of Ben Nevis' was a successful single malt brand, showing that blends didn't have it all their own way. Sadly, it wasn't to last and Ben Nevis produced only occasionally until 1940 when it was bought by a former Canadian bootlegger Joseph Hobbs.

After the war, Hobbs added a Coffey still and blended the new make malt and grain prior to maturing. A bottling from one of these 'Blended At Birth' casks appeared as a 40-year-old in 2002. Since 1989, Ben Nevis has been owned by Japanese distiller Nikka, to whom it has supplied fillings. It's only in recent years that its single malt has received the attention it deserves. The use of brewer's yeast, short ferments and squat stills with downward-sloping lyne arms gives a thick, oily new make – the very style that many new malt lovers lust after.

Above *Ben Nevis's oily character is in line with current tastes.*

THE COAST

When looking at a whisky map of Scotland, one of the key questions isn't where the distilleries are but where they aren't and then why? In somewhere like Speyside, the answer seems clear – they aren't in some places because of a lack of transport links. Islay had no distilleries inland because all the raw materials came in by boat.

Then look at the west coast and islands. With the exception of Islay there were slim pickings. Why? No barley? No heritage? Poor transport links? Barley could be grown, though admittedly in tougher conditions, but much of the rest of Scotland trucks it in from maltings. Getting it to the west is, however, difficult, as the roads are narrow and winding. But it isn't impossible.

Look again. The reason that the west's distilleries were in Oban and Fort William was because they were the only towns of any size – and at

Above *Nc'nean on Morvern is Scotland's only 100 per cent organic distillery.*

Above *The barley used by Nc'nean is exclusively organic.*

railheads. The rest of the coast and the islands beyond were empty, or rather emptied. The impact of the Clearances and the lack of subsequent investment have left a long shadow.

This is a stunningly beautiful part of the world, but it isn't an easy place to build a career or a business. What if you ignore all of this and build a distillery in this seemingly unpromising environment? And if you build it, will they come?

To the west of Fort William are three peninsulas: Morvern, Ardnamurchan and Moidart. None are particularly easy to get to. All are sparsely populated. Two have distilleries that are among the most exciting in Scotland. Both at their heart are about regenerating a community.

Deciding to build a distillery on an estate at the end of the road on Morvern was always going to be challenge. Deciding to then only use organic barley and also bottle on site was adding more layers of potential complications. That, however, is what Annabel Thomas has done at **Nc'nean**. Situated on the Drimnin Estate in a repurposed steading, Nc'nean's basic approach is organic barley, clear worts, long fermentations (68–114 hours), 2 styles of yeast, slow distillation and a high cut point for its 'young' new make. Each year, yeast experiments are run for potential special releases.

'We're focused on the spirit as a philosophy, because we find it most interesting,' Annabel says. 'The organic isn't about vanity or just a selling point. The contribution that organic barley makes to the texture and mouthfeel of our whisky is key.'

It all comes back to the new make. The 'young' new make is fresh and vibrant, while the 'old' is heavier, with lower cut points and intended for longer-term maturation. Everything is focused on flavour.

Does the west seep in somehow? 'Place is important as both inspiration and a guiding philosophy' says Annabel, 'be that our respect for nature, born out of being in such a remote area, or our mission to create skilled, long-term jobs in a small community.'

Sustainability is integral to both Nc'Nean and its nearest neighbour **Ardnamurchan**. Each is powered by a wood-chip boiler that is fuelled by trees from their respective estates. At Ardnamurchan, extra power comes from a small hydro plant in the river next to the distillery.

If Nc'nean has gone for a fresh light style, when Ardnamurchan's founder Alex Bruce was seeking

NC'NEAN

NC'NEAN ORGANIC 46%

Grassiness to the fore. Slightly dry with cut flowers, lemongrass, light barley. Upfront fruit, ginger and galangal along with buttered white bread. Appetizing and rather lovely.

FRAGRANT/FLORAL

ARDNAMURCHAN

★ **ARDNAMURCHAN** 46.8%

Sense of weight. Clam juice, sweet leather and hessian. Honeyed, quince, light smoke and hot oyster shells. Oily, overripe pear, membrillo, orange bitters, honeysuckle and mineral finish.

SMOKY/PEATY

Above *Once a year, Torabhaig's distillers are given free rein to make any type of new make they like.*

PPM

A 'ppm' figure is a measure of phenolic compounds in 'parts per million'. These figures are bandied around by some distillers, writers and whisky lovers as if within them lies the secret of all smoky whiskies.

The problem is that the ppm figure quoted refers to the malted barley, and most of the phenols are lost throughout the process. Some are left in the draff (mainland cows famously refused to eat smoky Islay draff) and in distillation. Phenols have a high boiling point, meaning that they come across late in the spirit run. Most of them end up in the feints, not the final spirit. Some phenols aren't even detectable by the nose.

The ppm in the spirit is a more accurate measure, but as Torabhaig's research illustrates, it is more to do with which branch of the phenolic family is captured. Peat isn't peat, smoke isn't smoke.

inspiration, he looked to the old, dirty, oily whiskies from Campbeltown and Fort William, drams that spoke of the west. The distillery looks over a sea loch; the peninsula itself is a massive ancient caldera. Salt and eruptions. Some whiskies just do reflect where they are from.

Unpeated and peated new make are both made from barley that is grown on the Bruce family farm in Fife. These are then aged separately before being blended together. Everything is aged on site, predominantly in dunnage warehouses in a mix of ex-Old Forester Bourbon casks as well as oloroso and PX. There is heft and oils but also a slinky, lightly smoked elegance that belies its youthfulness. It seems to have come from a different time and yet, like Nc'nean, be completely contemporary.

Both have helped to regenerate their communities by bringing in employment and attracting tourists, the distillery becoming more than whisky and the reason that a community doesn't just exist but prospers. It's a rewarding and inspiring return to the old ways.

HEBRIDES

The nature of a distillery is changing. In the past, they were at the mercy of the blend market, their success tied to its performance. But 21st-century whisky thinks of things differently. Now, with single malt becoming a category in its own right, new rules apply. In remote locations, what is still an expensive enterprise to start up is doubly so, but there is a freedom to make the whisky that you want.

ISLE OF SKYE

It always seemed odd that Skye had only one distillery. Plenty of reasons were given: it was too remote, there was too much iron in the water, there wasn't enough water. The last is true. Skye's impermeable rock means that the plentiful rains that fall here bounce off the mountains and run straight to the sea. Skye might have been too difficult in the old days. Not any more.

Torabhaig proves that it can be done, albeit that it took four years to convert a listed (and listing) 18th-century steading into a distillery. 'We couldn't change anything,' says Bruce Perry, global brand manager at owner Mossburn Distillers. 'The height of the roof, its pitch, nothing could be changed. All the equipment had to fit inside the same space.' It means that you slalom between washbacks as you make your way to the stills.

The length of the build had one advantage, as it gave the Mossburn team, headed by Neil Mathieson and Kenny Gray, time to work out what they wanted to make. 'Then, when the distillery was running, the first nine months were spent changing one thing at a time. It had to be smoky, but subtle, nuanced.' Everything was examined, not just barley but the kilning techniques of different maltings, which yeasts, length of ferment, cut points and flow rates. 'We used the island as a muse,' Bruce continues. 'It's peaty, it's wild, and no matter what the weather, it is also beautiful. How do we get all of that into a bottle?'

TORABHAIG

★ **TORABHAIG ALLT GLEANN BATCH STRENGTH** 61.1%

Intensely briny, fragrant smoke, sweet seaweed and water mint. Fresh. Palate is *sel gris*, burned bog oak, gunflint, chalky. Intense. Long. Huge potential.

SMOKY/PEATY

How indeed. The nature of the peatiness was integral. Cresol gives tar, guaiacol gives more fragrance and lift. The aim was to find ways to lower the former and raise the other. Bruce calls it 'elegantly rugged'. It's more refined than that – there's salinity, there's sweetness. Smoke but just not as you might have thought of it before.

'Neil was always going to make the art, and a lot of this is science, so over the two years we moved the science to meet the art.'

The first nine months were being used as the foundation for the series of six 'chapters', top-dressed with the now fixed recipe, before a permanent 10-year-old arrives in 2028. Each year two distillers, all local and none with prior experience, are given free rein to make their own whisky. Any level of peat, speciality malts, different yeast, ferments, cut points, casks – all are there to experiment with. These 'Journeymen' releases were planned to start in 2024.

And it is profoundly a Sgitheanach distillery. Mossburn's owner Frederik Paulsen chose the site because of his interest in preserving vanishing indigenous cultures. It was originally owned by Sir Peter Noble, who built Sabhal Mòr Ostaig Gaelic University nearby. Land, language and culture linked. 'There is a Hebridean movement,' says Bruce. 'I'd include Ardnamurchan and Nc'nean in that. We all have very different interpretations, but that is positive. It's like different artists engaging with the same landscape. All different, but all valid.'

How has **Talisker** taken this, given that it's been producing in the north of the island since 1825? It has welcomed its neighbour. Anything that helps the island is a good thing. It can afford to be comfortable, as the reason it has survived at the head of its sea loch for almost 200 years is quality. It might have started as a Clearance distillery – its founders the MacAskill brothers were assiduous in their emptying of the land – but subsequent owners navigated the problems over location and crafted an elemental whisky that captures the idea of where it's from.

The distillery is a geek's dream. Medium-peated malt (from Glen Ord), clear wort and moderate-length ferments are all fairly standard, but then the stillhouse baffles and bamboozles.

The two wash stills are huge. There's a U-bend in the lyne arm with a purifier pipe taking reflux back into the belly, before the pipe rises up again, then hooks into worm tubs. The stills are run fast, so there's less copper but more reflux. The result is a sulphury new make with smoke, oils and pear-like fruits.

TALISKER

TALISKER STORM 45.8%

Warming smoke (just-lit fire), light cereal. Roast pumpkin and blackberries. Hint of sulphur. Palate is soft with peppery cashews and sweet pear. A squall. Balanced.

SMOKY/PEATY

Above *Talisker's mix of fruit, smoke, oils, salt and pepper has made it a global favourite.*

It takes a lot of water to run worms, and every year Talisker would run dry, even with the multitude of springs on the hill above. Now, seawater is pumped from the loch into a heat exchanger, cooling the warmed water from the bottom of the worms, which can then be recycled.

It's more than that, though. Talisker can be gentle and fruity, it can smell of burning heather or seaweed, it sometimes goes salty. It will always have some pepper in the mix. It's moody and can be rumbustious, and it can be leaner and oilier, especially when in its second or third decade, as weather-beaten, it finally shows its complexity.

ISLE OF RAASAY

A 20-minute ferry ride from Sconser on Skye's east coast lands you on the Isle of Raasay, home to another of the Hebridean new wave of distilleries. It's a two-minute drive or ten-minute walk up the hill to the site.

Raasay has no (legal) whisky history. Like Skye, it was brutally cleared. In 1846, using the money he was given as 'compensation' of the loss of slaves in his plantations in British Guiana, George Rainy bought the island, moved its people from the fertile south to the harder, thinner soil of the north and then built a wall to keep them there.

Over the years there has been a steady drift of people away from the island. By the time the distillery opened in 2017, Raasay was home to an ageing population of 150 people. Now the distillery employs 30 people, with an average age of under 30.

Co-founder and master distiller Alasdair Day's whisky journey started when he began to look into his great-grandfather's time spent as a grocer and blender in the Borders. Finding his blend recipe book, Alasdair decided to recreate the Tweeddale blend, which led to musing about building a distillery in the Borders. Then he met entrepreneur Bill Dobbie who persuaded him that Raasay was a better bet.

Alasdair refers to **Isle of Raasay** as 'a distillery built by a blender', meaning that if blending is about working with flavour options, then a distillery should be able to offer as many as possible. 'I wanted the whisky to be Hebridean,' he says. 'For me, that meant light peat notes and dark fruits.

'And it had to be on this spot. We're sitting on Jurassic sandstone in a field of gneiss, meaning it was the only place we could get water. The geology dictated it be built next to the field of the white cow.'

RAASAY

THE HEARACH FIRST RELEASE BATCH 1 46%

Gentle with hot sand, wild flowers. A thread of smoke. Lilac blossom, dry grasses and yellow fruits. Lightly nutty. Subtle with good depth for a youngster.

FRAGRANT/FLORAL

★ RAASAY SINGLE MALT 46.4%

A melange of fried banana, plums and damson, stewed rhubarb, cinnamon, bacon, gentle bonfire smoke and a light touch of oregano. Pepper on the finish.

FRUITY/SOFT

Above *Isle of Raasay was set up to produce multiple styles of new make.*

Above *In common with the new Hebridean distilleries, Isle of Harris Distillery was established to help boost the local community.*

OTHER OUTER HEBRIDEAN DISTILLERIES

Marko Tayburn started his **Abhainn Dearg** (or Red River) distillery in Carnish on Lewis in 2008. There are no highly engineered Frilli stills here. He built the stills himself, modelled on an old illicit still he found, their lyne arms looking like elongated stork bills. All the barley – they now use Moonshine – comes from the Mackenzie brothers of Melbost Farms and is floor malted at the distillery.

More Hebridean distilleries are emerging. **North Uist Distillery** started in 2019 as a gin producer with Downpour gin, but has started to make whisky at the Nunton steadings on Benbecula. Fascinatingly, it will only use bere barley, the sole strain grown on the island. The island also has another distillery, **Benbecula**, also using bere. At the time of writing the distillery on **Barra** had just started to produce whisky.

Peated and unpeated are made. Six temperature-controlled washbacks run either four- or five-day ferments, with the unpeated held at 32°C (90°F) to get those dark fruits. The stills, made by Italian coppersmith Frilli, have a cooling jacket on the lyne arm that can be engaged to divert any reflux back into the still to build viscosity. Equally, they can be left off.

The spirit still has a six-plate bubble cap section that, again, can be engaged or not. 'It means we have six different streams without changing yeast or barley,' he says. 'I was fascinated by what was happening in Japan, US and elsewhere. Japan in particular resonated with me, especially the way they look at different streams. And I knew we could do it.

'The industry has been slow to develop single malt,' he adds. 'Distillery characters were developed for blends, and they couldn't explore outside that. We didn't have to make a "Raasay style" because there wasn't one until we arrived.'

The two new makes are then aged separately in three cask types: ex-rye for peppery spice, chinquapin oak giving maple syrup and ex-Bordeaux red wine for dark fruits and structure. Then the six-cask mix is blended. There's complexity and depth at young age. Other oaks and cask types are being explored as well as bere from the Borders, and other varieties being trialled, even some grown on Raasay. The island has come alive. It's possible to live and have a career here. That, ultimately, is the most important thing.

He laughs. 'And all I thought I was going to do was recreate a blend!'

ISLE OF HARRIS

The idea of a social distillery sits at the heart of **Isle of Harris**. With the island's population halving over 50 years, musicologist and philanthropist Anderson 'Burr' Bakewell saw a distillery as being a way to stem the outward flow of young people, giving them a reason to stay, increase employment and provide tourists with another reason to come. The distillery had not only to work commercially but for the profits made to be reinvested in the island. Opening in 2015, it now employs 40 permanent staff.

Harris is beautiful, its west coast fringed with stunning white sand beaches, on its east coast knuckles of gneiss clench through peat and thin soil. In Clearance times, the people were moved from the east where they grew crops to the west where little could be cultivated, bar kelp, which in these more enlightened times gives another link to the distillery.

Whisky needs to be matured in oak casks for a minimum of three years to be called Scotch. That maturation ties up a considerable amount of capital. Gin, however, can be made today and bottled tomorrow and brings much-needed cash into the business. It's a logical step for many new start-ups.

The key, though, is to find a way to stand out in a cluttered market. How, here, to make the gin say 'Harris'? A bottle like looking into sea glass lying in swirling water, and using sugar kelp as a botanical to add a marine touch. Link it to place.

As the gin started selling, so the whisky was being laid down. There's a 1.2-tonne (1.3-US-ton) mash and long ferments – the longest hitting 120 hours. The barley is lightly peated and distillation, in Frilli stills, relaxed. The whisky, named The Hearach (Gaelic for 'a person from Harris') and blended by Shona Macleod and Harry Wood, was launched in 2023.

It's hard to escape the links between it and the tweed for which the island is famous, the feeling of the weaving of flavours, of the strength given by the warp – the cask – and the complex patterns of the weft – the flavours.

As with all the new Hebridean whiskies, the aim has always been to make the place speak without being contrived. In The Hearach, it lies in how the peat speaks. It's quiet, the smell of peat from a chimney on a slightly damp autumn night. There's also the brightness of hot sand and the lift of flowers on the machair.

If you're wanting to capture Harris, then it has to contain its contrasts – the ever-changing weather, the obdurate and the shifting dunes. For generations, Harris has been caught in the tides of incomers with big ideas and no plan. Now there's a homemade solution.

ISLE OF MULL

If you can still smell the wet paint on most of the Hebridean distilleries, the walls of **Tobermory** have been recoated many times since the distillery came into being in 1798. Originally called Ledaig, it was started by John Sinclair. It appears to have stayed open – no easy task in those times – before succumbing to wider economic forces in 1837. It would be 40 years before Mull's sole (legal) distillery would restart. Even then success seemed to elude it, with a series of different owners buying it and then quickly passing it over. Part of Distillers Company Limited (DCL) from 1916 to 1930, it was shuttered once again and remained that way for another four decades.

Opening a distillery in the 1970s seemed like a sound idea. With blend sales rising consistently since the late 1950s, Scotch seemed bulletproof. In 1972, Tobermory welcomed a strange alliance between a Liverpool shipping company, sherry producer Pedro Domecq, and 'Panamanian interests'. It only lasted three years before a property firm based in Cleckheaton, Yorkshire bought it with the intention of converting some of the buildings into holiday accommodation. Others were used as a facility for storing cheese. It closed again in 1982 for 11 years, then was bought by Burn Stewart Distillers (now CVH Spirits). Finally, in 2017 a major refurbishment took place and stability was established.

It makes an oily spirit from stills that have an odd S-bend in their lyne arms. The peated

JURA

JURA 12 YEAR OLD 40%

Dry biscuity, toasted porridge oats and a hint of tweed. Tiny nudge of smoke. A firm base, mixed nuts. Smoke brings out sweetness. Light.

MALTY/NUTTY

JURA 18 YEAR OLD 44%

Winey, lots of ripe cherry, with some nutty oxidation, then thick custard, strawberry jam with a green back note. Malty core, black grape skin and light tannin.

FRUITY/SOFT

ISLE OF TIREE

Tiree is a fascinating case study of how the impact of Clearance and bungled economic decision-making by landowners impacted on Hebridean whisky. *Tir-iodh* means 'land of corn' and the island was noted for its barley crops. By the 18th century it was exporting cattle, salt beef and 300 gallons (360 US gallons) of whisky a year from 30 stills.

At the end of the century, John Campbell 5th Duke of Argyll instigated ambitious plans to 'improve' Tiree, but the promised investment never materialized. Instead of a productive agriculture and distilling (there were now two larger distilleries), he instead turned to the short-term kelp boom as a way of making money. Kelp was harvested and burned and exported as fertilizer during Napoleonic times, but when the wars ended, so did the kelp trade. Tiree was now overpopulated, its soil degraded, ironically because all of its kelp was exported.

In 1801, 147 islanders were found guilty of illicit distilling and one in every 10 was evicted. The Duke then ordered all barley not used by his legal distilleries at Kirkapool and Scarinish be sent to him in Inveraray. The two distilleries closed in 1818. By 1846, forced emigration began.

Now Tiree has a distillery again (see left). Unpeated barley is mashed in an oak mash tun, distilled in two direct-fire Hoga alembics and aged in a mix of casks (mostly quarters). Rye is also being distilled. 'Rye was a big part of cereal production in Tiree,' says co-owner/distiller Alain Campbell. 'We also enjoy the flavour, so we've mashed 70/30 rye/malt. It was tough to mash – several stuck mashes and lower yield. Worth it though!' The first whisky is scheduled to be released in 2024.

Above *One island, one village, one road, one distillery. Welcome to Jura.*

variant, Ledaig, is building a cult following among those who like their peat as tarry as possible. Think smoky dirty burger and you would be on the right lines.

ISLE OF JURA

The travails of building a distillery on an island with a small population was shared by **Jura**. Islay's neighbour has a population of 200 people, one road and one village. While illicit whisky was made, legal production started in 1810 at the Small Isles distillery in Craighouse. It produced for the rest of the century before being swept up in the Edwardian cull of the 1900s. It was expensive to run, and there was too much whisky being made.

Nothing happened until 1963 when in a premonition of the modern Hebridean approach, two landowners, Robin Fletcher and Tony Riley-Smith, decided to start whisky making once more as a way to stop further population decline. It has been part of Whyte & Mackay since 1993 and at the time of writing is the largest-selling single malt in the UK.

The stills are very large, fermentation is short – the new make was always light, cereal-accented – and there was always the sense of the hard rock beneath the bracken in the mature spirit. A reformulation has taken place in recent years. Up until 2017, two styles were made: unpeated and a month on heavy peat. The latter is still made, but now there is light peat in the new make, which has added sweetness and extra texture and depth.

THE HEARTBEAT OF A COMMUNITY

The 'Build It And They Will Come' ethos runs throughout this region. A distillery needs people to make the spirit, but also staff for the visitor centre. People arrive and need food, so there is a café on site, or maybe a new one opens up nearby. Because they are in remote areas, the visitors want to stay, so there's an increase in nights spent in hotels and bed and breakfasts. People move in permanently and start families. The school is saved and the doctor's surgery stays.

The size of the workforce in these new distilleries is notable. Automation means that a giant plant can be run by one person at a computer. Not here, because in these places the distillery isn't a remote place but the heartbeat of a community. It's interesting looking at this new paradigm and thinking back to Springbank, not just for its approach to whisky making and its whisky style but being at the centre of the town.

There is also an element here of taking back control. 'Who possesses this landscape? The man who bought it, or I who am possessed by it?' wrote the poet Norman MacCaig. Those who are possessed by it are now taking charge. Hebridean whiskies taste of where they come from because they are made by the people who live there.

ISLAY

Islay is easily defined. It's an island, therefore there are clear borders. Consequently, it would be easy to assume that there are flavour boundaries as well. How often have you heard, or read, that all Islay's whiskies are smoky? That assumption is immediately confounded as you start a clockwise tour of the island. The first distillery you come to on its coast is also its largest, and it makes unpeated spirit.

EAST COAST

Bunnahabhain was built in 1881 as a joint venture by two Glasgow blenders, William Robertson of Robertson & Baxter and Greenlees Brothers. Though its position in the far northeast corner of the island might seem remote, it was built on the shores of the Sound of Islay, meaning that supplies could easily be brought in by puffer. The partnership not only built the distillery but also a village. Six years later the distillery merged with Glenrothes to form Highland Distillers (now Edrington). Its function was fixed from the start – the provider of malt in considerable quantity for blends such as The Famous Grouse, Cutty Sark and later Black Bottle.

It wouldn't be until the late 1980s that Bunnahabhain was launched as a single malt under the tagline 'The Unpronounceable Malt'. It became part of Burn Stewart Distillers (now CVH Spirits) in 2003. Its reputation has steadily and quietly grown.

A three-year upgrade programme has left it considerably smarter, but some things have remained constant. The Highland Distillers template of sherry cask maturation has been retained. You might expect that its large stills make a delicate spirit, but the new make has a deep, slightly malty character that adds in a gingery note as it matures. Its core range starts at 12 years, but at 18 years old it finally blooms – the ginger moving into gingerbread, the malt adding slight nuttiness,

BUNNAHABHAIN

BUNNAHABHAIN 12 YEAR OLD 46.3%

Oatcake and bran, wholemeal flour, then a steady build of cream sherry, almond and Mejdool date. Nutty and fruity. Light touch of ginger, Walnut Whip (cream chocolate nut). Great energy.

RICH/DEEP

Above *Deer on Jura wonder where the smell of peat smoke is coming from.*

ISLAY

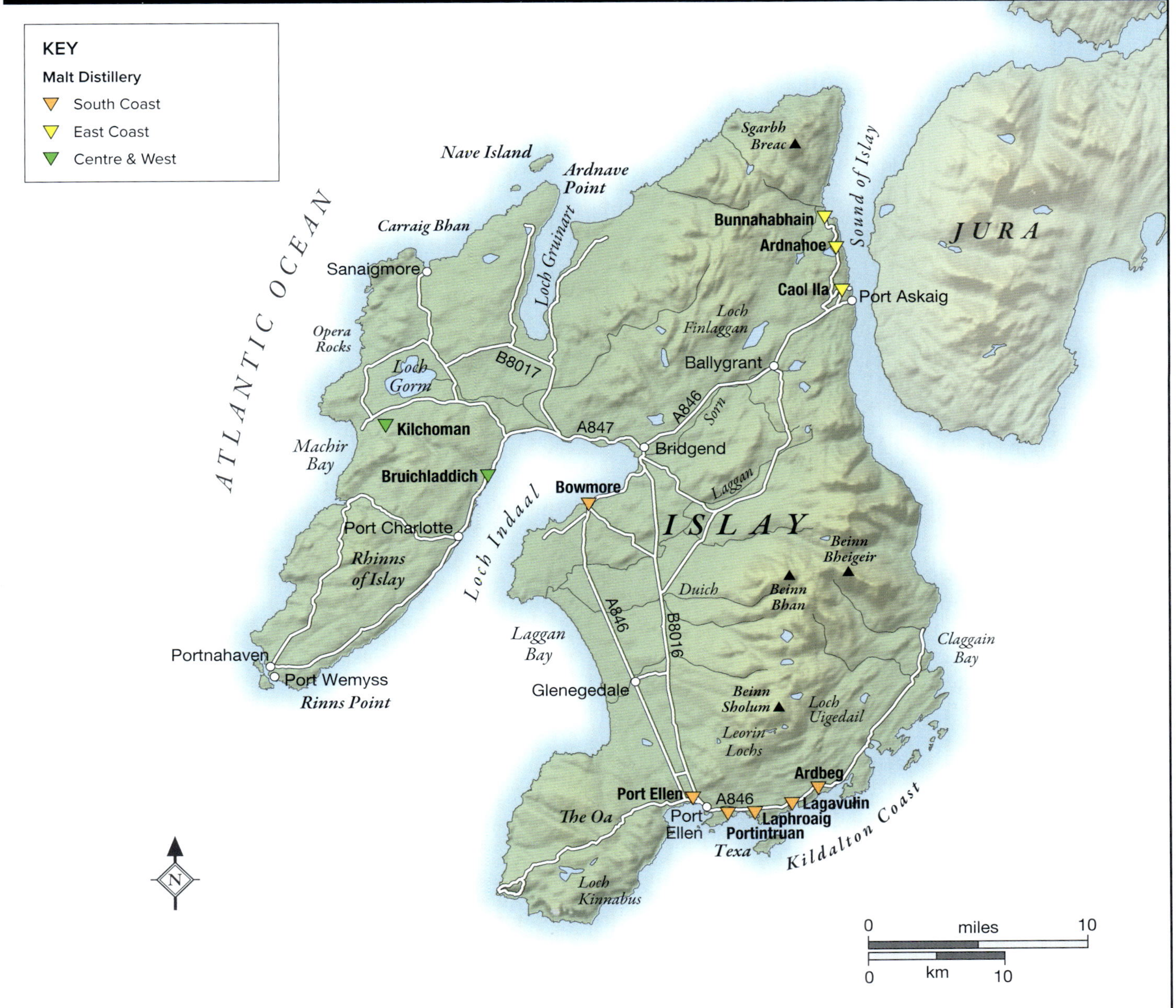

the dried fruits from the sherry softly binding it all together. The feeling is of a complete whisky.

Although heavily peated spirit was run, Bunnahabhain's success has lain in its determination to stick to its guns and make what it makes. It offers a glimpse of a different Islay and, without smoke, might well have a wider audience. A sleeping giant awakening.

CAOL ILA

★ CAOL ILA 12 YEAR OLD 43%

Briny depth, light kombu, but also sweet. Juniper, grass and lemon peel, then brine and crab creel. Well balanced yet complex.

SMOKY/PEATY

Nestled into a small bay near the Port Askaig ferry terminal, **Caol Ila** (or Sound of Islay) is Islay's second-largest producer. Built in 1846, it has been part of Diageo (originally Distillers Company Limited/DCL) since 1927 and is a significant player in the firm's blends.

It also confounds expectations by making unpeated once a year. In fact, it's flexible enough to produce a number of different characters for Diageo's blenders. In 2011, it had a major upgrade with a new lauter tun, and two additional washbacks to increase volume. Like Talisker, it uses seawater cooling.

It is, however, as peated whisky that its fame lies. The malted barley is peated to the same level as Lagavulin, but the character produced is completely different. Caol Ila's wort is clearer, the fermentations are longer, its six stills are considerably larger – more reflux – and the cut points are higher. It's another example of why the phenolic parts per million (ppm) figure is no guide to flavour (see box, page 90).

Above *Ardnahoe is another new distillery that has adapted 'old' methods – in its case, worm tubs.*

What emerges is slightly oily, but lifted, grassy with a piney juniper-like note. The smoke is constant but is shore-like, sometimes reminiscent of lobster or crab shell. Low wood brings out a slightly chilly element – like a breeze coming off the water. The key in any whisky is balance. In smoky whiskies that balance is achieved by sweetness, and Caol Ila has a sweet heart. Although seen as an official bottling, it also appears in virtually all independent bottlers' lists. What is remarkable is that every bottling is consistent in its delivery of that signature style.

Situated above Bunnahabhain with spectacular views over the Sound, **Ardnahoe** began producing in 2018. It was the culmination of a decades-old dream of Stewart Laing (of independent bottler Hunter Laing) to have a distillery on Islay. It uses Islay peat-smoked barley and ferments in four wooden washbacks.

It's at distillation stage that things become particularly interesting. The stills are large for the size of the distillery, adding reflux, with extremely long, gently sloping lyne arms again continuing some copper conversation. That is brought to an abrupt halt by the use of worm tubs – Ardnahoe is currently the only distillery on Islay to use this method of condensing. This adds a solidity to the sweet smoky new make.

SOUTH COAST

Islay's southerly Kildalton coast is a place of skerries and strange tides. Seals sing on its rocks, deer swim in the water. A mix of the pagan and the secular. Here is Fairy Hill, where the Queen of the Sidhe was said to give the women of the isles the secret of knowledge in liquid form. Close by is the Kildalton Cross, one of the few Celtic high crosses still in existence. It's not too much of a stretch to think of the area's three whiskies as having both of these dynamics at work – the earth-bound old religion and the angelic spirits.

No more than 3 miles (5km) separates them, yet each is distinct, taking their peated base in different directions. All are old sites, conceivably built next to or near older illicit operations. **Ardbeg**, for example, was founded in 1815 by the Macdougall family. Like a number of Islay distilleries, as well as providing smoky fillings for blends, it was bottled as a single malt from early days. It prospered in those times and a village grew up around the still ultimately comprising houses, a school for 100 pupils, greenhouses and a bowling green.

As blends became the dominant force in Scotch, so distilleries were wholly tied to their success or otherwise. The 1930s were tough, nothing was made in the war years, then there was an uptick in the 1960s and 1970s.

All of Scotland's distilleries were hit hard by the 1980s downturn, but Islay was particularly badly hit. Blend sales were collapsing, therefore less peaty whisky was required.

ARDBEG

ARDBEG 10 YEAR OLD 46%

Immediate sooty smoke. Camphor/naphthalene, chimney smoke and some salinity, then sage flower, thyme, smoked salmon, cereal. Lime, apple, pimenton and an oily mid-palate.

SMOKY/PEATY

★ ARDBEG UIGEADAIL 54.2%

Bitumen, dark chocolate, mint and wormwood. Layers of leather, sweet dark fruits, boggy and dense – slightly decayed. Twiggy tea, shiitake. Oily.

SMOKY/PEATY

The choice was made and Ardbeg was mothballed in 1981, running only occasionally from 1989 until 1996 when the shutters came down again. A year later it was bought by Glenmorangie and its second life began.

What its previous owners had failed to recognize was that single malt was taking off and that peaty whiskies were proving popular with this new drinker. Independent bottlings were building the Ardbeg cult. Glenmorangie got it.

It still draws its water from two lochs, Uigeadail and Airigh Nam Beist. Heavily peated barley from Port Ellen is ground, mashed and then fermented for a relatively short 56 hours. There's a purifier pipe on the spirit still that recirculates the refluxed heavy elements. Top notes are released, but an extra thickening texture is also built in. This is a sweet spirit with sooty smoke and lime accents.

The bulk is matured in ex-Bourbon casks, but sherry and heavy peat are a remarkable partnership, seen in Uigeadail (see tasting note), while French oak is used for Corryvreckan.

A new stillhouse was opened in 2021, doubling capacity. 'I arrived just as it was commissioned,' says manager Colin Gordon. 'I'm walking around the distillery thinking, OK, first new stillhouse for 100 years, new copper, doubling production middle of a pandemic. No bother!' Now they are using 50 tonnes (55 US tons) of barley to mash 26 times a week and making 2.5 million litres (660,430 US gallons) per annum.

The question with any peated distillery is how to create variations on what could be an overly dominant theme and reveal Ardbeg's multiple complexities. Glenmorangie's distillery director Dr Bill Lumsden and team (including Colin) have looked at three-week-long ferments (Fermutation), black malt (Ardcore) and upping peat levels for Supernova, while Heavy Vapours was made by sealing the purifier to see what happens (this was common practice during Allied days).

'It must have been deeply upsetting to grow up in a thriving place and it's sitting there destroyed,' says Colin. 'Now there's life in the place again. It's been a lot of work, but now it feels like a community of people who care about the place.'

A mile (1.6km) down the road is **Lagavulin**. From the roadside it has the air of a Scottish baronial castle and its malt – especially at 16 years old (see tasting note, page 100) – has a certain haughtiness to it. In its tight bay sit the darkened ruins of Dunyvaig Castle, the sea fortress of the Lords of the Isles. In 1816, a distillery was built here – conceivably the amalgamation of ten illicit sites.

Smoke pervades everything. One veteran manager, on Islay for the first time, was convinced the place was on fire when he arrived. It's there

Above *Steam (not smoke) rising from Ardbeg's washbacks.*

Above *Sooty smoke and a hint of the sea are signatures for Ardbeg.*

Above *Islay's citadel of smokiness, Lagavulin.*

at the mill (dusty, nutty), it lingers in the mash tun (sweetness) and drifts around the tun room (smoked apple). It morphs and shifts building in potency until in the stillhouse, emanating from very plain spirit stills, it reaches its zenith – complex, smoke, seaweed, pungent bog, gentian, violets. At its heart, again, is sweetness.

The 8-year-old (see tasting note) is the balance to the 16's loftiness. Here the energy and dynamism is in full flow as the nose is pulled towards the water. On every New Year's Day, there is a ritual where people jump off Lagavulin's pier into the bay. That's what this is like. The 16 is being sat inside by the fire with a blanket around you drinking Lapsang Souchong.

Another distillery once stood on the site. Malt Mill was started in 1908 by Lagavulin's then owner Sir Peter Mackie to replicate Laphroaig (for which he had lost the agency). Although it was an exact replica, it always made a different spirit. An account from the 1920s speaks of bundles of heather being added to its mash tun and peat braziers being lit in the washbacks to dry them. It made a 'heavier' style, and ran until 1962.

The Kildalton trip is completed by **Laphroaig.** Started by brothers Alexander and Donald Johnston, it remained in the same firm until the 1960s. In 1908, Ian Hunter (Donald Johnston's great-grandson) switched agency from Peter Mackie's White Horse and began to bottle it as a single malt as well as providing fillings, a practice carried on by his secretary Bessie Williamson to whom he left the distillery in 1954. She ran the distillery until 1974. It's now owned by Suntory Global Spirits.

One of three (so far) Islay distilleries with its own floor maltings, 20% of its own malt is added to each mash. After a two-day steep, it's allowed to germinate and then put into the kiln. Unlike most kilns where the barley sits on the perforated kiln floor, here it's suspended on a grid 4.9 metres (16 feet) above the floor, which means it takes longer to dry out. The fire is also kept at a low temperature, adding smoke but little heat. As phenols are oils, they cling to the surface of the grain, but only when it's wet. It's effectively cold smoked. This prolonged process helps to drive up guaiacol, giving a heavier tarry and medicinal character.

LAGAVULIN

LAGAVULIN 8 YEAR OLD 48%

Peat oils, bog myrtle, marine smoke, fresh seaweed, salt water/rockpools. Intense. It's the stillhouse sea breezes on a cold day.

SMOKY/PEATY

★ LAGAVULIN 16 YEAR OLD 43%

Rich, caramelized fruit. Lapsang souchong tea, bay leaf, light salt and incense. All-enveloping smoke. Soft with pipe tobacco, mulberry.

SMOKY/PEATY

LAPHROAIG

LAPHROAIG 10 YEAR OLD 40%

Light tar and creosote. Hint of iodine, but the smoke is slightly shy. Mussels, green apple, vanilla, pine sap. The palate is flinty.

SMOKY/PEATY

★ LAPHROAIG LORE 48%

More classical mix of oak moss, seaweed and coal tar. Suitably medicinal with some mothballs, Swarfega, calamus and wintergreen. The fruits are thickly concentrated fruits with a green herbal finish.

SMOKY/PEATY

Above *Laphroaig's self-made malted barley has an unusual character.*

After a 70-hour ferment, it's distilled. A 40-minute foreshot run means there are fewer estery notes in the new make, while a deeper cut means that those heavier phenolics are captured. One spirit still is twice the size of the others, giving another flavour to add to the mix. A major expansion is planned at the time of writing.

The bulk of the spirit is in ex-Maker's Mark barrels adding vanilla, but there is a strong following for Laphroaig aged in ex-sherry where the depth of the oak and dried fruits meld with those earthy, tarry, seaweedy elements.

Laphroaig has had great success in marketing itself as being a love or hate single malt, talking up its apparently polarizing character. In fact, it's signature 10-year-old (see tasting note) is highly approachable.

There is more to Laphroaig than the seaweed and iodine, the tar and the medicine – there's also sweetness from those American oak casks and a green mossy character that softens the impact.

PORT ELLEN BAY

The south coast is, at the time of writing, about to be host to another two distilleries. **Portintruan** just outside Port Ellen, owned by Elixir Distillers, is a good example of how a distiller on an island with well-established brands (most of which are smoky) can find a point of difference – Annandale's Professor David Thomson's 'white space' (see page 119).

'We'll be drawing from our knowledge of Scotch and "old" Islay drams as well as new developments internationally,' says Elixir's head blender Oliver Chilton. 'There are lots of aspects of traditional production that you want to take on board, but there's also science. It's not finding a USP as much as joining a culture and helping it develop and evolve.' Portintruan will, however, be different in many ways. Run by Ileach (a native of Islay) Georgie Crawford, it will consist of two distilleries: one making its new make, the other an experimental side where other spirits including rum can be produced.

There are 14 houses being built for the workforce – important on the west coast where a housing shortage is a major issue. It will be deeply embedded in the local community. 'Anyone looking to build a distillery on the island really needs to think about what positive impact they can make by doing so. What are they going to give back in jobs and housing, what opportunities will they create and how will it grow and improve the community?' asks Georgie. 'We want to set up apprenticeships and sponsor kids to study brewing and distilling, and for graduates to do their dissertation in the pilot plant so that they know that when they come back there is a job and a house. This is different, this is a place for the community.'

Port Ellen's bay is dominated by the Diageo-owned maltings, which, at the time of writing, had stopped providing malt for other distilleries and was only making it for its own estate. It takes skill to make smoky malt. Here there are large kilns, fuelled by peat cut from the moor, their smoke driven up through the barley by hot air. Peating levels are measured by the shovelful, the strength and angle of the wind, how much water to spray on to dampen down the fire and how long the fire needs to run. Such craftsmanship in this echoing, peat dust-filled place is often overlooked.

Above *The Port Ellen distillery reopened in 2024.*

Next door are the old **Port Ellen** distillery warehouses, but the distillery itself has gone. This was a pioneering place. Opened in 1824, its fame started in 1836 when 21-year-old John Ramsay took charge, who together with his business partner Walter Frederick Campbell (see opposite) started the steamer route between the island and Glasgow. Islay and Islay's whiskies were now more readily available. In 1869, W.P. Lowrie (see page 54) took over sales and export trade started. In 1925 it was folded into Distillers Company Limited (DCL)/Scottish Malt Distillers (SMD), only to close in 1930.

It's often forgotten that Port Ellen has had two long periods of silence. This first one lasted until 1967, while the second started in 1983 when, as the smallest player in SMD's Islay portfolio (with Caol Ila and Lagavulin), it shut down.

Like Brora, it attracted the new wave of malt aficionados. Port Ellen never had as high a reputation as its sister plants, or within Islay as whole. It was put into refill casks, to be used as a punch of unadulterated peatiness in blends. Leaving those old casks and allowing air to play its part steadily changed the make. It became more complex, strange, austere, dissonant, mixing sweetness with smoked fish and tarry tea. Stocks dwindled, prices rose. Then in 2017, extraordinarily, Diageo announced it was going to rebuild the distillery. In 2024 it reopened with a 'classic' side and an experimental one. Six wooden washbacks hold the wash for between 98 and 130 hours. The two large 'Phoenix' stills are run slow to get reflux and a new make with around 12ppm. The experimental side has two stills, one-fifth of the size. The main focus there will be on the nature of smoke.

CENTRE & WEST

One reason why Islay was more prosperous than other Hebridean islands was the result of a different approach to the Improvements by four generations of its owner, the Campbell family. Daniel Campbell the Elder (laird of Islay 1726–53) introduced flax, while his grandson Daniel the Younger (laird 1753–77) increased the sizes of farms, distributed free seed, built Bowmore town and developed linen manufacture and mining.

It is under his ownership that Bowmore's distillery may have started. Brother of Daniel the Younger, Walter Campbell (laird 1777–1816) had

Above *Bowmore's tun room.*

clamped down on illicit distilling, ushering in the building of licensed stills, while his grandson Walter Frederick Campbell built the villages of Port Ellen, Port Charlotte and Port Wemyss and created lotted lands. He was bankrupted in 1847 through using his own money to pay for people to emigrate.

Bowmore distillery is another sitting in the heart of a town, and in many ways defines it. When the kiln is running, you can't escape the fragrant reek of peat smoke in the air. It has retained its floor maltings, supplementing it with malt from Simpson's.

Bowmore is a shape-shifter whose smoke seems to come and go. It sits in the middle of the peaty world, being gentle and slightly more floral. Part of this could be down to the fact that, over the years, ferments have moved from long to short, distillation from slow to fast.

Another whisky that was relatively widely sold as a single malt before the late 1980s resurgence, it is generally agreed that its glory days came in the 1960s and 1970s when, aged (mostly) in refill casks, the smoke disappeared, leaving a hallucinatory tropical fruit character.

Suntory has owned the distillery since 1994, but it's only in recent times that it has taken full charge of production and the changes have been dramatic.

The aim is to return Bowmore to that legendary old style. The key, Suntory believes, lies in long kilning of the barley, greater wort clarity and a lengthening of the ferment to a minimum of 70 hours, and using two yeasts, one promoting fruitiness. It also comes from slower distillation and in the casks, some of which are stored in No.1 Vaults, one of whose walls is the seawall for the town. At high tide the warehouse is below water.

It's never easy for a distiller to admit that things change. If you have a 200-plus-year-old distillery, it's good to be able to say that it has always been made in the same way, but of course this cannot be true. Everything changes, constantly, but sometimes change means taking a step back in order to move forward again.

WEST OF LOCH INDAAL

Many of the ideas that have stimulated the Scotch industry in recent years coalesce at **Bruichladdich**. Closer links to farmers, organic barley, rye, biodynamics, landraces, sustainability, community. It has been a remarkable transformation for a distillery that was set up in 1881 as a supplier of fillings for blends. After a period in mothballs (1995–2001) it was bought by a consortium of Islay landowners, wine merchant Mark Reynier and Simon Coughlin and distiller Jim

PEAT RESTORATION

Peat bogs are one of the most efficient ways of locking in carbon. Cutting and burning peat is therefore the opposite of what needs to be done in the climate emergency. Ireland has already banned peat cutting and closed its peat-fuelled power stations. A ban on horticultural peat is coming into force in the UK in 2024.

The Scotch whisky industry uses only 4% of the peat cut in UK, but when the ban comes into force it will be one of the biggest offenders. Is it possible to balance the need for positive action on climate change and maintain one of Scotch's signature aromas?

There are ways in which to generate the same phenols using less peat, but the central issue is can peat bogs be restored? Thankfully, the answer to that is yes. By damming drains and ditches and removing trees and scrub, the bogs can be reflooded. Mosses and other plants can start to grow again and slowly the peat can 'grow' once more.

'Drainage of a peatland has been likened to a knife wound – the bog will only survive and begin to heal if the wound is plugged and the lifeblood of the bog retained,' writes Mike Billett in his book *Peat and Whisky*. Research has shown that peatlands can be restored in as little as ten to twenty years. Mike concludes, 'The whisky industry in Scotland is in a fortunate position because the country is already a global leader in peatland conservation, with research in restoration feeding into government policy with a strong buy-in from the landowning community. In short, the Scotch whisky industry can tune into the existing knowledge base.'

In 2021, Suntory Global Spirits started its Peatland Water Sanctuary initiative, investing £400,000 into the restoration and conservation of 1,300 hectares (3,212 acres) of peatlands by 2030, enough to produce the same amount of peat that Suntory Global Spirits harvests every year in making its Scotch whiskies on an ongoing basis. The aim is to have restored sufficient peatlands by 2040 to equate to twice the volume of peat that the company harvests to make its Scotch whiskies. It's a start.

BOWMORE

BOWMORE 15 YEAR OLD 43%

Melted milk chocolate, black cherry, sloe, pipe tobacco, wet leather and light smoke. Some gingerbread, plum and soft apricot. A warm hug by a wood fire.

RICH/DEEP

BOWMORE 18 YEAR OLD 43%

Light cedar, low smoke. Soreen malt loaf, then peat, and big weightiness alongside thick bramble jelly. Tropical fruits glimmer at the back. Layered and resinous.

RICH/DEEP

McEwan. In 2015 it was bought by Rémy Cointreau, which maintained the philosophy established by the 2001 team, including the growing of barley on Islay for the first time in a century, but increased the levels of investment. It's not the end of the story, though.

'Rémy was Chapter 2,' says head distiller Adam Hannett, 'Chapter 3 is where we are now and where we are going. We're a big beast now – there are 90 people on site – which means changes. We have to keep what we are now known for, but also get things into the mainstream and grow. It's an interesting phase of growth, but we do what we believe in.'

Plans for on-site floor malting are progressing, while crop trials are underway at the 12-hectare (30-acre) distillery croft. 'It plays a crucial role in research, grain development, soil health, regenerative agriculture and crop rotation', Adam explains. 'It allows us to test and learn on Islay and exchange knowledge and learnings with our farming partners.'

The distillery is making three types of spirit: unpeated Bruichladdich, medium-peated Port Charlotte and heavily peated Octomore. Then there are variations of all with different sources of barley by location or farm, cask types, finishing. It's a complex web. In total, 14 different spirit types are made.

Islay and its community sit at the heart of it all. 'We want to get as much of a sense of place into the whisky as possible. It doesn't make sense not to have the story. £1 million goes from here into the island a year in barley purchases. We have to have provenance and, yes, we lose in terms of yield, but we are doing that for the right reason. The island benefits. Now 50 per cent of the barley we need is grown in the island.'

'There is a danger that Islay could become a whisky theme park, like some Western movie set where the buildings are just façades. As distillers, we have to look after the whole island, so that means investing in it, which is why agriculture is leading us. Distilleries are part of the community – it's why they were here in the first place. It's much easier to make whisky on the mainland, so why here? Because of the people.'

Growth has increased pressure on the distilleries – not from the vagaries of the market, but the precarious nature of the island's infrastructure. Everyone I spoke complained of the roads falling to bits, the failures of the ferries and the chronic lack of housing for a growing workforce.

BRUICHLADDICH

★ BRUICHLADDICH 18 YEAR OLD 50%

Layers of sultry fruits, creamy toffee, pink peppercorn, honeysuckle and flambéed banana. Palate is all white flowers, red liquorice, cashew. Meadowsweet and vinous with water. Light cereal and honey. All grown up.

FRUITY/SOFT

★ PORT CHARLOTTE 18 YEAR OLD 54.3%

The Port Charlotte balance is there. Subtle smoke mixing orange zest, potpourri, tayberry, date, pomegranate molasses and pork crackling. The palate is long, elegant with star anise and a mineral touch.

SMOKY/PEATY

OCTOMORE 14.1 (5 YEAR OLD) 59.6%

Deep, earthy smoke: oils, coal tar, dubbin and spent fire. This is given lift and balance by intense mintiness. The palate is generous with some coconut cream, just-lit peat and dry grasses. Mineral on the end.

SMOKY/PEATY

Above *Bringing in the harvest at Kilchoman.*

Above *Bruichladdich is one of only a few distilleries to retain an old rake and plough mash tun.*

Above *Kilchoman's original stillhouse.*

A distillery helps to create and maintain a community. It's more than just a business, and the more remote you are, the more significant the distillery's role becomes. Whisky's evolution is taking Islay full circle. The reset goes deeper than bottles; it's about reconnecting.

When Anthony Wills founded **Kilchoman** in 2002, he was looking both to the past and the future. The distillery sits a mile (1.6km) inland on land that was once owned by the Beaton (MacBeathad) family, famed medieval doctors and conceivably the people who had brought the knowledge of distillation to Scotland. He was also tapping into the Islay wave that had seduced single malt lovers around the world. 'I always look at it as being in the right place at the right time,' he says, 'and Islay having this reputation has helped.'

A pioneer of the new wave, he was on speed dial for many new distillers. Generous with his time, it's been fascinating while writing this book to see how many distillers around the world either did a *stage* here or turned to him for advice. What does he tell them? 'If you think you need £10 million to build your business, you'll actually need £20 million! Financing Kilchoman was the biggest learning curve.'

One of legendary whisky consultant Dr Jim Swan's distilleries, Kilchoman has always taken a different slant to the Islay style. Jim's philosophy was making good-quality early-maturing whisky. Kilchoman is peated, but the smoke isn't heavy or medicinal, rather fragrant and lifted. Some of the make is malted at the distillery made from barley grown on the distillery-owned farm.

Demand has grown and now what was a small distillery has doubled in size and is looking to expand once more. Home-grown barley now accounts for 20% of the production. The maltings has expanded and there are plans to install Saladin malt bins. When the latest expansion is complete, production will be close to 1 million litres (264,000 US gallons) per annum.

'People relate to things that have provenance and tradition tied into them,' says Anthony, 'and place is part of that. One element of that for us is barley, not just growing it but finding the variety we think will work best on Islay and not just what the malting houses tell us to use.' Trials with yeast are ongoing. 'Things like that are part of whisky's natural evolution and we're in the fortunate position where we can be flexible and innovative. Smaller distilleries have a better understanding of what could make them different. We always have to be ahead of the game.'

ISLAY OF THE FUTURE

What is happening on Islay is a microcosm of the wider Scotch industry. It's looking at barley, production and the need to deepen or re-establish links with the community. Whisky has always been holistic. It's reliant on ties between land and science, science and craft. Sustainability is an easy term to bandy about, but it's the right one. For whisky to survive, it must sustain not just share prices but people and the earth.

Islay's biggest names are, indeed, peated, but it contains more variation than its reputation. As with anywhere in Scotland, its diversity is what's important. Islay is many things and that is what single malt is about. Islay 2.0 is underway.

KILCHOMAN

★ KILCHOMAN 100% ISLAY 50%

Gentle smoke. Quite estery: green pear, apple, vanilla-speckled banana, light mineral note. Smoke picks up on the palate, perfumed light golden raisin.

SMOKY/PEATY

KILCHOMAN SANAIG 46%

Distillery character: machair sand, flowers, scallop, sweet scented peat over apricot, berries and chocolate. More smoke on the palate, some clove and raisin.

SMOKY/PEATY

FIFE

Fife is, as its inhabitants remind visitors, 'almost an island'. There's more than a suggestion of self-determination in the statement, and a degree of self-sufficiency that's understandable when you consider it builds ships, had coalmines and has fishing ports and fertile farmland. In whisky terms, it has been home to a remarkable upsurge in distilleries. There are threads here that hold it together, making it worth its own section. Welcome to the Kingdom.

In 2003, Scotch whisky was still shaking off the after-effects of the 1980s collapse in sales and distillery numbers. The notion of someone building a new distillery was madness. If eyebrows had been raised about Arran, the idea of a farmer with no experience doing the same was incomprehensible. Yet that was what Francis Cuthbert, whose family had raised cattle and grown barley for generations at the 405-hectare (1,000-acre) **Daftmill** farm in Fife, did.

'Several things came together,' Francis says. 'At the time, barley was £70 a tonne, which was barely covering the cost of production. We were looking for ways to diversify and add value to what we were producing. Also, my barley was already being used to make whisky!'

If whisky started on farms, in time the farms sent barley to the now larger local distillery. By the 1960s, you grew barley and sent it to maltings where its origins would be lost. It made sense at a time when Scotch was about scale, but there was a disconnect between crop and whisky. Daftmill changed that.

In 2005, with a whisky course under his belt and a consultant in tow, Francis opened Daftmill. It remains a one-man operation. 'We survived by *not* opening to the public,' he says. 'Most new distilleries will have five staff, a visitor centre, sales and marketing. That means they have to try and recoup those costs, which means they'll have to release a whisky earlier.'

Above *When Francis Cuthbert started Daftmill on his farm, little did he know he'd be a pioneer of a new approach for Scotch whisky.*

FIFE

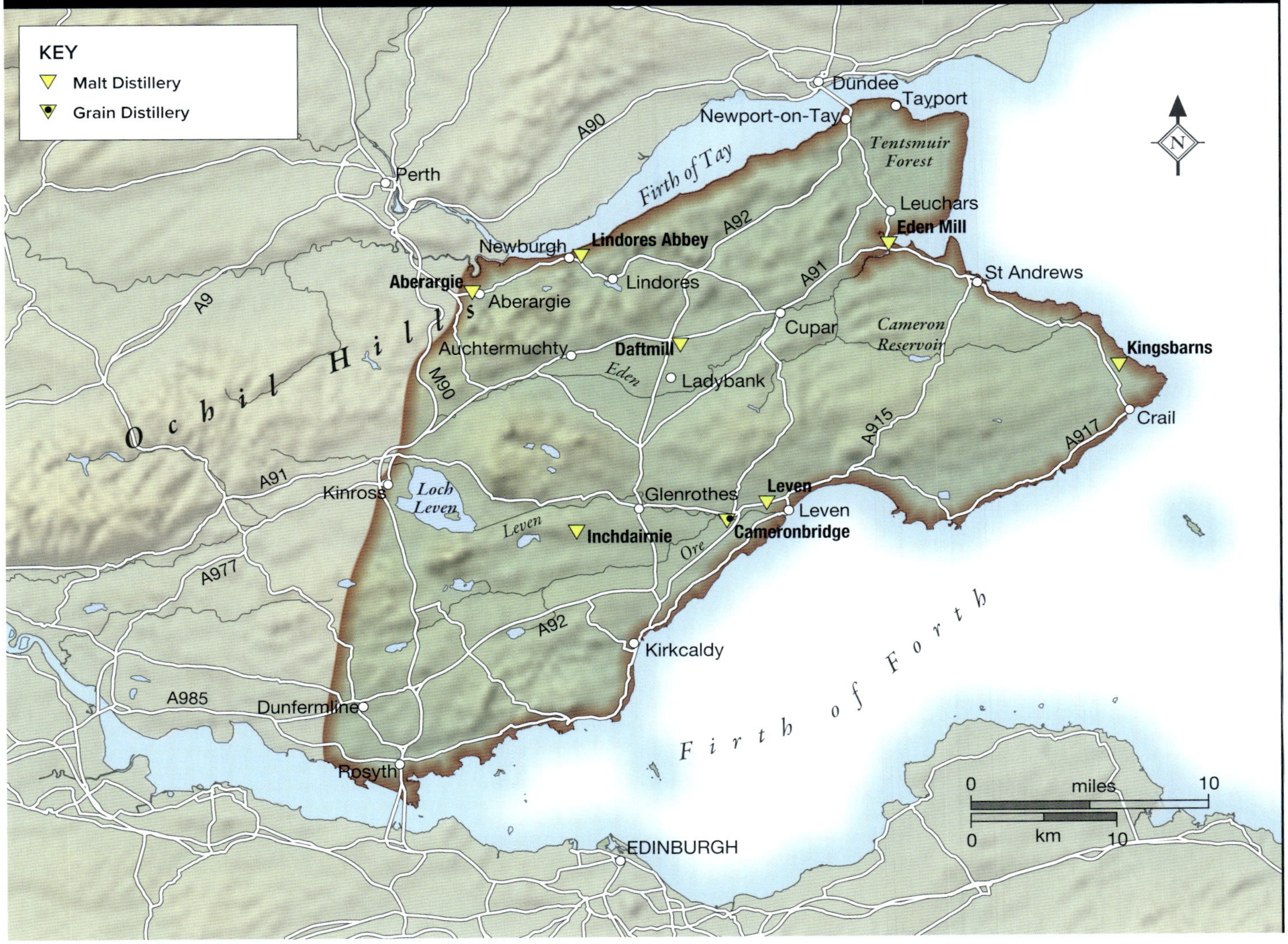

The style is 'Lowland in the line of St. Magdalene, Bladnoch and Rosebank'. Technically, it's about clear wort, long ferments and slow distillation (the roof is low, so it needs time to converse). Production is concentrated on the slower periods in the farming year, distilling between November and February and then a burst in summer before harvest in August. 'It's the old way. Everything is reused in some way. Manure on straw goes back on the field, locks in carbon and helps to grow the barley that makes whisky and gives draff to feed the cows. The straw is their winter bedding. We're green, but don't make any claims. We just do what we do and make the whisky that we want to make.'

It took him 13 years to release his first bottle. 'I just wanted to make sure it was right.' Now, deservedly, it's a cult.

DAFTMILL

DAFTMILL 2010 WINTER BATCH 46%

Delicious poached pear/apple purée freshness, then muscovado sugar, grapefruit, lemon curd, fresh muffins and nutmeg. The palate has ginger, jasmine and more sweet spices.

FRAGRANT/FLORAL

★ DAFTMILL 2009 (BOTTLED 2022) SINGLE CASK 60.2%

Rich and funky with (appropriately) a farmyard note. Pencil shavings, Jamaica ginger cake, canned prunes, candied walnut, chipotle. Lush, sherried and complex.

RICH/DEEP

GLENROTHES

With the building of **Inchdairnie** in 2015, industry veteran Ian Palmer has been finally able to answer many of the questions he'd asked throughout his 45-year career but was never able (or allowed) to answer.

This distillery, on the outskirts of Glenrothes, is taking the question 'What is Scotch?' and shaking it up. 'Just because that's how it has always been done doesn't mean it's the only, or right, way,' he says. There's no visitor centre, for example. 'We're in the Scotch whisky business, so we want to do that really well. Having a visitor centre means compromises.'

With that clean slate, he and distillery manager Scott Sneddon have looked at 'materials, method and maturation'. There are four different new makes for its own single malt, each created to reflect the character of the season, as well as StrathEnry single malt, for trading, made to a different recipe, and PrinLaws, the distillery's annual experimental spirit.

The first, in 2019, was produced from oats and has since been joined by wheat, sourmash and pot still.

The distillery's first release was RyeLaw, made from a mixed mash bill of 53% rye and 47% barley (both malted). All the grains are put through a hammer mill, mixed with hot water in a conversion vessel and then passed through a mash filter (see page 14).

Seven different yeasts are used in different combinations across the range. Spring and Summer use spring barley and wine yeasts, while Autumn and Winter use winter barley and ale yeasts. KinGlassie uses a different one to boost phenols, and StrathEnry's yeast is different again, as are those for RyeLaw and the PrinLaws makes.

Nothing is straightforward – maybe apart from a standard three-day fermentation, though as that takes place outside in huge temperature-controlled washbacks, it's hardly conventional either. The wash and spirit stills are large with massive condensers to increase the amount of copper availability. 'We create particular wort for each whisky, then use special yeasts to get a specific flavour,' Ian explains, 'so we don't want to make this sulphury. We're about complexity of flavour.'

INCHDAIRNIE

RYELAW FIFE SINGLE GRAIN 46.3%

Comforting mix of green peppercorn, fennel seed, candied ginger, lime marmalade and blackcurrant leaf. Lightly oily palate, buttery biscuits, a tingle of clove/allspice from the rye.

SPICY/HERBAL

There is also a 'Lomond Hills' spirit still made to Ian's design by Frilli (see box, page 81) used for RyeLaw and other expressions.

It sounds (and is) highly technical, but Inchdairnie is very much a local distillery. All of the grains used are grown locally (the whiskies' names are those of neighbouring farms), and the distillery is working with the Fife Growers Group on barley and agronomy trials, including one aiming to find ways of reducing the use of nitrates.

Provenance is important. 'If you're going to build a distillery in Papua New Guinea, it should speak of that place, not Scotland. The cereals that Inchdairnie uses are Fife grown, distilled and matured. Not Lowland. Fife,' says Ian. This most modern of distilleries is tied into whisky's oldest element, an identification with place.

'What's important with distilling today?' he asks. 'Location. We're free to do more things in Fife because there are fewer expectations...less dogma. Therefore there is room, therefore there is freedom.'

LEVEN

Fife might not have been known as a whisky-producing area in recent times, yet it produced more whisky than anywhere else in Scotland. That is down to **Cameronbridge**, by some distance the largest (volume) distillery in Europe, which not only makes Smirnoff, Gordon's and Tanqueray but 30 million litres (8 million US gallons) of grain whisky for Diageo and others.

It might be highly modernized – it has seen numerous expansions and upgrades in the past decade – but whisky has been made here since it was founded by John Haig in 1824. At that time, the Haigs and their relatives by marriage to the Steins dominated whisky production and were the earliest to adopt the innovations in distillation that led to grain whisky.

CRAFTY MALTSTERS

In the search for flavour and closer links to the land, 'alternative' barley varieties have becoming an increasingly intriguing area from both a commercial and flavour point of view. The problem has been that as the supply chain is set up to handle large volumes of the same variety, it is hard to process often very small batches of varieties.

Alison and Daniel Milne farm in Auchtermuchty in the northwest corner of Fife. Already growing heritage grains, in 2017 they started **Crafty Maltsters**. They are malting their own barley as well as some for customers, such as organic from Falkland Estate for Lindores Abbey and Salome for Raasay. They are now also working with the International Barley Hub (IBH) (see box, page 111) on commercial-scale field trials.

There is an ecology at work within whisky, but it is one that has become fragmented. Compelling though the growing for flavour element is, the cultivation of heritage crops or different grains has to benefit the farmer and the land, as well as the distiller, maltsters and the consumer.

'What we're trying to do is tell a story about the people and the place behind the barley. We feel there is real vibrancy to that story which is as valuable,' says Alison. It gives back provenance, which after all is what underpins the idea of single malt.

Above *Inchdairnie is one of Scotland's most creative distilleries.*

Robert Stein's column still was first trialled here in 1829, as was Aeneas Coffey's improvement to the design. A 'Coffey-esque' still remains in use to make Cameronbridge's grain spirit.

It's the home of Haig Club and its variants, and has also been one site where Diageo has produced its rye campaigns.

Close to Cameronbridge is Diageo's smallest and least-known distillery. **Leven** is the distiller's experimental plant where all its distillation trials take place.

EAST COAST

In 2009, Doug Clement, who worked as a caddie at the Kingsbarns Golf Links, looked at the ruins of an old 18th-century steading at the entrance to the golf course and had an idea. His clients loved whisky, but there was no distillery near any of Fife's famous courses. Doug envisaged that the ruins could, conceivably, be home to a distillery.

CAMERONBRIDGE

CAMERON BRIDGE 26 YEAR OLD 56.2%

Clean and lean. Spring flowers on the nose, but it's on the palate where it shows best – cinnamon buns, demerara sugar, honeycomb chocolate and wheat chaff. Long.

FRAGRANT/FLORAL

Doug spent four years slowly building finance, then the Wemyss family, who have extensive land holdings across Fife (they leased the land to John Haig to build Cameronbridge) and were already an independent bottler, bought into the idea. In 2015, with them now as full owners, the late Dr Jim Swan as a consultant and former brewer (and Heriot-Watt University graduate) Peter Holroyd as manager, the first spirit ran from the **Kingsbarns** stills. From day one local barley has been used.

Peter's brewing links are immediately apparent – there are two yeast strains used, distiller's for efficiency and a second one to promote berry fruits and boost esters. The style was fixed from those first runnings: light, gentle, floral/fruity but with sufficient guts to be able to carry lengthy yet relaxed maturation. This is seen in its second core release, Doocot, which blends ex-Bourbon casks with Swan's patented STR barriques (see box, page 120).

Clement still caddies across Fife and runs golf and whisky tours. He has a tattoo of the Kingsbarns bottle on his arm. His dream has come true.

Kingsbarns is only 7 miles (11.3km) from St Andrews where, in 1655, Robert Haig was accused of distilling on the Sabbath at his site. This would in time become the family's Seggie grain distillery, which closed in 1860 and was converted into a paper mill. In 2012, in sight of the old Seggie site, **Eden Mill** began production.

It started as a brewery, then came gin and whisky, the last of which was released in 2018. In 2023, now owned by private equity firm Inverleith LLP, production moved to a larger purpose-built site, also on the Eden Estuary.

Above *Lindores is on the site of an old monastery where spirit may once have been distilled.*

NORTHWEST OF CUPAR

Whisky's roots run long and deep in Fife. If Drew McKenzie Smith is correct, **Lindores Abbey** was where Friar John Cor, a Tironensian monk, was domiciled. It is Cor's name on the exchequer roll (expense account) of King James IV in 1494, where it was noted that eight bolls of malt were given 'wherewith to make aqua vitae'.

Was this whisky? Certainly not as we know it. Whatever spirit was made, it could have been for medicine, or gunpowder. If it was for social purposes, it would have been distilled along with herbs and spices into the drink that would become usquebaugh.

Distillation would have been practised before this time but, as is always the case, records only begin when the taxman gets involved. Still, it's a starting point.

Drew is now 'custodian of Lindores', looking after the Abbey's ruins. Knowing the potential link with distilling history, in 2013 he began a project to build a distillery. After archaeological investigations were completed on what is an important site, building started in 2016, and the distillery opened the year after.

Fife-grown barley is distilled in one wash and two spirit stills, producing a sweet and floral new make. Lindores has also been working with Auchtermuchty's **Crafty Maltsters** (see box, page 108) to use organic barley grown on Falkland Estate as well as bere barley. In a nice link with those early days, it also produces an aqua vitae using herbs and spices.

Although only 5 miles (8km) from Lindores Abbey, **Aberargie** is technically in Perthshire, but as its thinking is closely linked to what is happening in the Kingdom, it makes sense to place it here. Owned by the Morrison family (formerly of Morrison Bowmore), it started in 2017. Perhaps inspired by Daftmill, it's also following in the revival of farm distilling. Any discussion about Aberargie goes back to the barley

Above *Aberargie is another field-to-bottle distillery.*

INTERNATIONAL BARLEY HUB

It's only recently that flavour has been seen as part of the equation when barley is considered. As blended Scotch has been the dominant player for over a century, it had to be made at scale.

As a distillery's output was always dictated by the size of its malt barns, malting began to be outsourced to larger commercial maltings. As yield became increasingly important, so the number of barley varieties being used was restricted. Barley became a commodity, and flavour became homogenized.

With the price of barley not rising in real terms, the only way a farmer could break even was by upping yields, further restricting the choice of varieties that could be grown only with added inputs. The soil became degraded, and as yield became the driver, so the barley genetic pool shrank.

In 2022, as a result of a £35 million investment, the International Barley Hub (IBU) opened in Dundee. It's the global focal point for barley research, with partners including the James Hutton Institute and the Scotch Whisky Research Institute (SWRI). From a whisky perspective; its discoveries will have a profound impact on the future of not just Scotch but all single malt. 'The main issue facing barley in Scotch is managing the potential risk to current and future supply,' says SWRI's director Professor James Brosnan. 'In Scotland, we grow spring barley drawn from a narrow genetic pool, all of which is harvested at the same time. That's a lot of eggs in one basket, so one of our projects is trying to broaden the genetic diversity and spread the risk.

'That doesn't mean going down the GMO [genetically modified organism] path, but looking at genetic diagnostics. There's a lot of science that can help with improving malting or agronomy.'

The challenges are huge. Which variety can cope with climate change, how to reduce inputs, what role heritage varieties can have either as alternatives or potential crosses giving flavour and yield. How can this benefit the farmer and the planet, and can there be a way to widen rather than restrict the flavour potential of malt whisky?

The work being done will have a huge impact on 21st-century whisky, but also food production and the health of the planet.

and its provenance – everything comes from the Morrison's 121-hectare (300-acre) farm.

It uses a wide range of barley types – Concerto, Laureate and Tungsten – each of which is processed differently.

'Crop rotation changes a bit year by year,' explains Niel Hendriksz, its commercial director. 'Our mainstay is Golden Promise, but each year we will also take opportunities to grow a few fields of Concerto and Laureate. We'll try new crops from time to time. In 2023 we mashed for a few weeks with Tungsten. Each variety and each year is a unique reflection of the growing season, and our distilling season the following year. The real variable is in vintage and crop.' The malt that will go for commercial trading is often run as cloudy wort to get a specific new make character.

Aberargie's own single malt is Golden Promise-based, with clear wort and a ferment time that will be dependent on how that year's crop likes to perform. 'Generally speaking, we run three-day fermentations with a pair of five-days in any given week. However, if the washbacks haven't reached a good final gravity, we will hold onto them. Golden Promise from 2022, for example, was notoriously slow to ferment to completion.'

Short, dumpy stills with downward-pointing lyne arms and a low cut gives a heavier and more densely fruited style – the Golden Promise also adds in body and texture – than what might be considered typical for a Lowland whisky, but taste around Fife and it becomes obvious that there is no such thing.

Above *Heading towards the West Lomond hill in Fife (after which InchDairnie named its reflux still).*

THE SOUTH

The Lowlands were long dismissed as an afterthought in whisky. No one cared about grain whisky – and anyway, the distilleries weren't open to the public. Neither did it fit into the classic single malt picture of mountains, rivers and peat bogs. The lack of distilleries didn't help. Now it's one of the most dynamic areas of Scottish whisky production. Malt distilling has not only returned to its major cities but also to its farmland.

CENTRAL LOWLANDS

Lowland whisky can be highly technical, or represent a return to the past. It, like every other region, cannot be easily pinned down – and that's to its benefit.

EDINBURGH AND ITS PORT

Edinburgh was once a whisky city. By the end of the 18th century, as well as an estimated 400 illicit stills, there were 8 licensed malt whisky distilleries. The Stein distilling dynasty had Lochrin, while the equally powerful Haig family started Canonmills, the scene of a riot in 1784 when starving poor laid siege to the site believing it was hoarding grain and vegetables.

A clue to the styles of Scotch whisky at the time comes from economist Adam Smith's visit to Canonmills distillery, where he recounts the following mash bill being used: '⅓ malted barley to ⅔ raw barley or ⅓ malt barley, ⅓ raw barley, ⅓ wheat'. In other words, single pot still.

The Caledonian grain distillery (known as 'the Cally') ran between 1855 and 1988. Occasional bottlings do appear; Diageo's Special Release in 2021 was particularly noteworthy.

The Diageo links with the city run deep. It was in Edinburgh in 1877 that the grain producers who owned Cambus, Carsebridge Cameronbridge, Glenochil, Kirkliston and Port Dundas joined forces to become the Distillers Company Limited (DCL), which would ultimately become Diageo. Andrew Usher, claimed as being the first commercial blender, was also Edinburgh based.

Of the city's malt distilleries, none survived the downturn of the 1920s. Dean distillery shut in 1922 and West Sciennes three years later.

Leith was a more significant player. Until the start of the 18th century, this was Scotland's busiest port, second only to London in terms of the wine and sherry landed. It was home to Yardheads and Bonnington, among others, although it was as a blending centre that it gained its fame. Leith's warehouses were packed with whisky blended by firms such as Mackinlay, Crabbie, Macdonald & Muir and the revered William Sanderson (of Vat 69 fame). Now, Woven is reviving the art.

Above *All the nice sailors love an Usher's.*

Above *Holyrood whirls with a large porfolio of different malted barley types.*

Above *Holyrood is one of a growing number of urban distilleries.*

Edinburgh's silence as a malt whisky producer made the opening of the **Holyrood** distillery in 2019 a major event. Situated in the 180-year-old goods shed of the former St Leonard's railway station, it's hemmed in by houses, making it not the most logistically easy place to manage malt deliveries, cask transport, effluent...and noise. An ambitious project, even more so in terms of flavour. The distillery's early runnings were good in terms of quality, but there was an itch to scratch.

If you have a new site with a young team, and there is no baggage, while around the world distillers are doing exciting things, then surely this is an opportunity to go for it?

In the 18th century, Edinburgh was the home of the remarkable intellectual flowering that was to become known as the Scottish Renaissance. The distillery is located underneath the Salisbury Crags, where James Hutton looked into deep time and began the science of geology. It's a mile (1.6km) away from Edinburgh University where David Hume set out his philosophical theory of empiricism – the belief that all of our ideas come from our sensory experience of the world, meaning that knowledge can only be derived from experience. Given this, Holyrood's empirical testing of what whisky could be seems somewhat apposite.

HOLYROOD

HOLYROOD ARRIVAL 46.1%

Lots of leafy green barley, some straw, new trainers, Jacob's cream cracker. The palate is boiled sweets and a touch of smokiness, then sweet porridge.

MALTY/NUTTY

Edinburgh is also home to Heriot-Watt University, which runs degree and doctorate courses in brewing and distilling. Many of the world's smartest new distillers have come through its gates.

The brewing aspect of whisky production is often overlooked in Scotch, nowhere more so than in yeast. Until the start of the 20th century, distillers would have used brewer's yeast. This had an effect on flavour not just because of the breweries' own strains but also due to the distance between brewery and distillery. The further away, the lower the activity in the yeast. This also had an impact on flavour.

In a series of technical articles in the 1920s, *The Wine and Spirit Trade Record* claimed that the lighter Lowland style was the result of easy access to fresh yeast, but the more distant the distillery was from the source, the heavier the whisky style became.

Highland whiskies 'develop medium malt flavours', while Islay, whose distilleries could wait for two weeks for yeast supplies, made 'a heavy flavour...heavier than Campbeltown'.

Holyrood's ethos is: Test, Learn, Improve, Repeat, part of 'an iterative, holistic approach'. So far, beer and also sake yeasts have been used, as well as different mash bills, speciality malts and explorations of heritage grains, all of which are released in limited editions as new make.

All of this is not random, as it might appear. Behind the experiments, the distillery's recipe –its heartbeat – is being refined. Its first single malt appeared in 2023.

In 2020, Leith began to produce whisky once more in the form of a new **Bonnington** distillery – a pilot plant at Chain Pier had run for a year

THE SOUTH

prior to the distillery being commissioned. A limited edition, named Yardhead, was released.

With water from 147 metres (482 feet) below the still, Bonnington mashes 2 tonnes (2.2 US tons) of grist and ferments in half a dozen washbacks. Ferment times vary depending on the style being produced. Part of its ethos is to bring back the blending heritage of John Crabbie, whose name is now only known for Crabbie's Green Ginger Wine but who in the 19th century was a significant player.

In 2023, Leith got a second plant. **Port of Leith** distillery is located next to the Ocean Terminal where cruise ship passengers disembark. It's also where The Royal Yacht Brittania is berthed. A busy spot.

The brainchild of Leith-born Ian Stirling and Paddy Fletcher, it's the UK's first vertical distillery (the world's first is Sweden's Mackmyra – see page 178), a decision partly forced on the pair by the small site of the site. 'Why not build a contemporary industrial landmark in an industrial port?' as Fletcher puts it. The distillery opened in 2023.

They admit that they 'dabbled' with elements of whisky production in their back garden, which led to them asking the same question that many new distillers have also pondered: if whisky is brewed and yeast gives flavour, why has the entire Scotch industry used the same fast-acting, highly efficient but flavour-neutral strains? The answer is yield – more alcohol per tonne of barley.

If, however, you are a new single malt distillery where a character isn't fixed and you are happy to take that hit on yield to create new flavours, then yeast is a major lever to pull. Two dozen strains were trialled to find the combination that will give Port of Leith a complex and robust style that will stand up to ageing.

After the Cally's closure, Edinburgh's sole distillery was **North British.** It was established in 1887 by a consortium of blenders (Andrew Usher, William Sanderson, John Crabbie and James Watson) who wanted an alternative to the Distillers Company Limited monopoly. They chose the site for its transport links – next to the Union Canal, and the railway line.

The cooperative ethos ran until the 1990s, by which time it was controlled by Robertson & Baxter, International Distillers & Vintners (IDV), William Lawson, Macdonald Martin, Seagram and Teacher's. In 1993, Edrington and IDV took control, and when the latter was bought by United Distillers (UD) (which in turn became Diageo), North British passed into the hands of the firm it was set up to escape from.

It runs a 72-hour ferment and a mash bill with a minimum of 15% malted barley – up until the 1930s, rye was part of the mash bill. This produces a slightly fuller style of grain; there's a sulphury note in the new make that adds muscle to blends. With no official bottling, all releases have come from independent bottlers.

EAST LOTHIAN

Heading 20 miles (32km) to the east is **Glenkinchie**. We're in East Lothian, whose important role in the early development of Scotch is virtually forgotten. In 1723, the Society of Improvers in the Knowledge of Agriculture in Scotland was formed to promulgate new theories – often imported from the Netherlands – of a wider variety of crops, which should also be rotated, field drainage, establishing pasture on higher levels and introducing new machinery.

Farms grew in size and small towns emerged. In turn, this meant the larger farms had to produce more food, further stimulating the new practices and regional specialisms; East Lothian's was arable.

The first great innovator was Sir John Cockburn, whose village of Ormiston in 1726 was the first planned settlement in the country. Improvements were also about creating integrated settlements with new industries, and malting and distilling became part of the East Lothian approach. At its height, the county had 13 distilleries and multiple maltings.

Today, only Glenkinchie and the maltings at Pencaitland survive. The distillery was founded in 1837 by the Rate brothers, but it wasn't until the late Victorian blend-led boom that production restarted in a new and substantially larger plant.

Unsurprisingly, it was owned by investors with an eye on supplying fillings for the blending market – easy access to Leith was a real asset. In 1914, along with Rosebank, St. Magdalene, Grange and Clydesdale, it was one of the founders of Scottish Malt Distillers, which, in 1925, merged with DCL (now Diageo), becoming its malt distilling arm.

It's a solid, red brick Victorian site that, like Ardmore, seems slightly out of step architecturally with its surroundings. A quiet producer, its emergence into the malt spotlight came in 1988 when it became Diageo's single malt in the Classic Malts, winning out over Rosebank (see below).

Lowland is thought of as light and delicate, but at Glenkinchie you get a distinctly satanic whiff of sulphur. There are two massive pots giving reflux, but also worms that are run cold. Only after time does the sulphur dissipate, leaving meadow flowers and lemon, although the mid-palate has the thick texture you expect from a worm-tub site. In 2022, as part of Diageo's multimillion investment in whisky tourism, a new visitor experience opened.

Above *Falkirk now has two distilleries as well as its wheel.*

WEST OF EDINBURGH

Follow the Union Canal 30 miles (48km) east from Edinburgh and you reach Falkirk. In the past, this was the site of an important cattle market, and given the propensity for drovers to smuggle whisky along with beasts, a distribution centre for northern moonshine.

By the turn of the 21st century, the canal was disused, the 11 locks that linked it with the Forth & Clyde Canal in disrepair. In 2002, however, the locks were replaced by the world's only rotating boat lift, The Falkirk Wheel, which carries boats the 30 metres (100 feet) between the two waterways.

In the 19th century, easy access to the Union Canal was a factor in where to locate distilleries. North British and Caledonian in Edinburgh took advantage, as did **Rosebank** in Falkirk. The desperate state of the canal was also the reason why, in 1988, Glenkinchie was chosen over the more highly regarded Rosebank as one of the Classic Malts.

With one eye on whisky tourists, the decision came down to a distillery on farmland or one trapped between a housing scheme and a stagnant canal. Five years later, Rosebank closed. Its worm tubs were cut in half, copper thieves raided it and it sat mouldering. Then, in 2017, the site and all the remaining stock was sold by Diageo to Ian Macleod. In 2023, transformed, it reopened.

Rosebank was another of the cult 'ghost' distilleries, though never with the same following as Port Ellen or Brora. Among those who knew,

GLENKINCHIE

GLENKINCHIE 12 YEAR OLD 43%

Barley straw backed with light floral notes, oven-warm baguette. Sweet on the tongue with good length. Slight brassiness. Medium-dry, generous. An almost smoky back note.

FRAGRANT/FLORAL

ROSEBANK

ROSEBANK 31 YEAR OLD RELEASE 2 48.1%

A basket of super-ripe tropical and soft pulpy fruits. Hint of vanilla and then night-scented stocks and honeysuckle. Fresh and pollen-like. Clinging and juicy fresh acidity. Remarkable.

FRUITY/SOFT

Above *The appropriately named Clydeside is located in Glasgow's old docklands.*

though, nothing could touch its fragrance. Triple distilled, with a flat-topped wash still (of a similar design to Cragganmore's spirit still – see page 60), it spoke of light, almost tropical fruits and cut flowers, but with a distinctly palate-clinging quality, thanks to the use of those worm tubs. It's being made as close as possible to the original and the first casks were filled in autumn 2023. In the meantime, Ian Macleod is eking out older stock in limited releases.

Close by sits the **Falkirk** distillery, which started production in 2020 after a prolonged battle with Historic Scotland, which feared that the build might damage the remains of a Roman wall. It's set up along traditional lines with an open-topped rake and plough mash tun, a fermentation regime that tops 100 hours and slow distillation (the mash tun and stills both came from Caperdonich – see box, page 49).

In nearby Bathgate is **Starlaw**. This slightly mysterious grain distillery opened in 2010 and is owned by French firm La Martiniquaise-Bardinet, which also owns Glen Moray. It was constructed to supply grain (from wheat and corn) for its blends, which include the 2 million-plus case Label 5. It also produces grain neutral spirit. The distillery was designed by Ian Palmer, who went on to open Inchdairnie (see page 107), and can produce 25 million litres (6.6 million US gallons) per annum. The design of the columns allows it to make multiple styles.

GLASGOW

Our Central Lowlands tour ends in Glasgow. This too was a whisky city. The warehouses beside the River Clyde were filled with casks, brought by train from the north, or by puffer from Campbeltown and Islay. Major blending houses were domiciled in the city centre: Buchanan's, Teacher's, White Horse, Whyte & Mackay, Robertson & Baxter and more. Bottles were shipped out or headed by train south from St Enoch and Central stations – the former's cellars were used by MacDonald Greenlees as its bond.

My whisky life began at Black & White's bottling in Stepps in the north of the city, part of a shift of whisky to its periphery. Highland/Edrington moved R&B's blending lab and reinstalled it in the west, Diageo built a bottling plant in Shieldhall, Chivas and Allied's sites were in Paisley, then Dumbarton, and Dewar's opened in the East End. The site of the old Adelphi distillery is now the city's mosque. In 2011, the Port Dundas grain distillery on the side of the Forth & Clyde Canal closed. The centre was emptied, the docks silent. Whisky might have been blended and bottled here, but its history was strangely forgotten, pushed to the margins.

Then, in 2018, in an old pumping station by the Finnieston docks a new distillery started up. **Clydeside** is the distillery arm of independent bottler A. D. Rattray, owned by Tim Morrison (ex-Morrison Bowmore) and his son Andrew. Its distiller, Islay-born Alistair McDonald, spent 35 years with Morrison Bowmore, latterly at Auchentoshan (see opposite).

With an idea to make a gentle, sweet, fruity spirit, he has created a process that uses 72-hour ferments to get citric elements, a slow distillation rate and high spirit cut (it comes off at 71% ABV). Every year for two weeks heavily peated barley is run.

It's a prime spot, next to the Scottish Exhibition Centre (SEC) and other event spaces,

CLYDESIDE

CLYDESIDE STOBCROSS 46%

Fresh green fruits, *pastéis de nata*, cantaloupe melon, creamy oak, coriander seed, malt flour, then redcurrant. Hint of green grass and tobacco. Assured.

FRAGRANT/FLORAL

and even has a stop on the city bus tour outside. Despite this, Alistair says, 'We're not a visitor centre with a distillery, we're a distillery with a visitor centre. At every step it's been how can we make the best spirit.'

There is a certain irony hearing this, as the distillery gets 80,000 visitors a year, but the thinking becomes clearer inside. The tour is more about Glasgow's whisky heritage than a Disneyland approach to what is being made. Placing it in a continuum.

'People are here to learn more,' says Andrew Morrison. 'What's now forming nicely is the tie between the city's history and us as a new distillery. We could make this a successful venture just by selling to the people coming through the door, but the real success will be in building a whisky that's respected.'

This young team of six, out of a workforce of 35, 'makes it feel like we're playing a role in the area,' he adds. 'I go to distilleries making millions of litres with two people, but I think what we're doing is adding to Glasgow. People here are seeing them work and we're training them on the job.'

It's not been easy. Because of the location, only 50 casks can be matured on site. All the waste – any water, draff, pot ale and spent lees – has to be tankered off. The timing could also have been better. 'Covid closed us for two years, then the supply chain collapsed and now there's inflation,' says Andrew. 'But we're committed to our goals and are laying down stock.

'To increase production is a big statement because it's never been so expensive to make whisky. We could have five expressions and sales would triple, but in ten years' time we wouldn't have the stock needed to build the brand. You have to play the long game. The Glasgow way is not to shout about how good you are but just get your head down and do it quietly. We know where we want to get to, but it's been a hell of a journey!'

The closest malt distillery to Glasgow for many years was **Auchentoshan**, which sits close to the banks of the Clyde at Bowling, where the Forth & Clyde Canal meets the river. Though started in 1817, it wasn't until 1834 under John Hart and Alexander Filshie that it became established. It passed through many pairs of hands before greater stability arrived in the shape of Morrison Bowmore and now Suntory Global Spirits.

Until Rosebank's reappearance, it was the last redoubt of what was identified as the Lowland style – triple distillation. There are a number of theories why this appeared to be a regional speciality: the influx of Irish people, some of them distillers, in the 19th century; the rise in popularity of single pot still, meaning it was being made in Scotland; demand from blenders for lighter whiskies to add to their palette of colours.

Above *Auchentoshan remains the sole distillery in Scotland to only triple distil.*

What is overlooked is that triple and partial triple distilling was commonplace across Scotland up until the 1930s (Talisker, for example, was triple distilled). We'll never know the full story.

It's a chunky wee site, making 2 million litres (528,340 US gallons) per annum, with a 7.5-tonne (8.3-US-ton) mash tun and 70 hours of fermentation. Distillation starts in the wash still, which runs as normal. The spirit from the intermediate still is split into strong (heads) and weak (tails), with the strong going forward to the spirit still where they are mixed with the feints from its previous distillation. The tails are recycled with the next charge of the intermediate still. The spirit still is run with the cut made at 81% ABV.

In the past, short ferments added a nutty note to the mature spirit, but the Suntory team has been hard at work. Clearer wort, longer ferments and a mix of brewer's and distiller's yeasts are in place to up the fruity quotient. Its lightness means that Auchentoshan could easily be a passive contributor when put in cask. While the wide range of different woods used will still steer the spirit in new directions, it's the overall balance that's important.

AUCHENTOSHAN

AUCHENTOSHAN NAS AMERICAN OAK 40%
Light malt and dry leaf. Lime chocolate and pink coconut biscuit. Butterscotch sweets. Very light-bodied. Biscuity with some char on the finish.
MALTY/NUTTY

AUCHENTOSHAN THREE WOOD 43%
Walnut Whip, fig roll, then some grape jelly and dense raisin. Very sweet. Layers of chocolate and black cherry. Toffee fudge and raspberry syrup.
FRUITY/SOFT

GLASGOW

GLASGOW 1770 TRIPLE DISTILLED 46%
Light dusty florals, rose/rouge, then mango chews and fresh pineapple. Firm. Quite minty on the palate, then potpourri and almond cake.
FRAGRANT/FLORAL

GLASGOW 1770 THE ORIGINAL 46%
Full-bodied. Juicy Fruit chewing gum, vanilla pod, ice-cream wafer, slightly vinous. Firm.
FRUITY/SOFT

The **Glasgow** distillery has been quietly getting on with business since 2012. Located close to the city's airport, its 16-person crew makes three styles of new make in Christian Carl stills: an unpeated triple-distilled (81–75% ABV), unpeated double-distilled (72–63% ABV) and a double-distilled peated expression that comes on spirit at 72% ABV but has a low cut point of 57% ABV. Spin three types of spirit and a wide range of casks, and the potential range of flavours is increased exponentially.

The city has one other distillery, which has quietly been producing since the 1920s. **Strathclyde** was built in the Gorbals in 1928 and made neutral spirit for London gin as well as grain whisky. It was a bold move at a time when distilleries were closing, especially since it contained the most modern equipment, including a lab. It was a financial gamble that paid off.

For a short time it had a malt distillery, **Kinclaith**, which operated between 1957 and 1975, making a heavy, slightly smoky style. It closed when the grain and neutral side expanded.

It's now owned by Chivas Brothers and supplies its grain for Ballantine's, Chivas and Teacher's (and neutral for Beefeater), but will exchange stocks with other distillers. The spirit is wheat-based with 9% malt and distilled in two linked sieve tray columns. It can be found through independent bottlers – often at great age.

BORDERS & AYRSHIRE

You might think, reasonably enough, that the cereal-growing Borders region would be the perfect place to build distilleries, but whisky making ceased in the area in 1837. It wouldn't be until 2018 that spirit began to flow once more, thanks to the efforts of John Fordyce. His **Borders** distillery in the mill town of Hawick was, in his words, 'an opportunity to recruit and train a group of local "never before" distillers and integrate with the Borders' farming community'. John now works in conjunction with Berwick-based Simpson's Malt and 12 local farmers within a 30-mile (48km) radius of the distillery.

The location made perfect sense – access to grain but also, as he points out, 'Hawick is famous for manufacturing in industries that rely on water, and we are proud to be part of a 250-year-old tradition.' The distillery (a 19th-century factory) backs onto the River Teviot, which is used for cooling water.

It's a chunky producer. The mash tun holds 5 tonnes (5.5 US tons), the ferments go on for 80 hours and there are two pairs of stills. 'Our aim is to create an estery, fruity spirit for maturation largely in first-fill ex-Bourbon barrels,' says John. Capacity hits 1.6 million litres (423,000 US gallons) per annum. Rye is also processed and made up half of the blend for the distillery's first whisky release (see tasting note). As he says,

BORDERS

BORDERS MALT & RYE 40%

On the green, herbal side of things. Meadow hay, sugared almond, Dolly Mixtures, whitecurrant, envelope glue and straw. Good mid-palate weight. Spice from rye, softness from malt.

FRUITY/SOFT

Above *Whisky is being brought back to the Borders.*

'We see a new wave of Scotch whisky consumers who want to drink their Scotch their way and we aspire to become a brand that contributes innovation, fun and creativity to 21st-century Scotch.'

Hidden in the Borders town of Jedburgh is **Reivers**. Owned by Mossburn (see Torabhaig, page 90), this is the company's 'secret' distillery that, as of 2023, has yet to release anything. Flexibility is the key here, with pots that have 18 plates in the neck that can be engaged or disengaged depending on whether more or less reflux and copper is required for the spirit character.

'We have a three-year-old,' says Mossburn's CEO Neil Mathieson, 'we're just not selling it.' The core spirit is a rye whisky made from a 70:30 rye to malted barley mash bill. The key in its production, he says, was to find the right temperature in cooking to get the menthol but also spice.

There have been other styles produced, including a four-grain mash bill using wheat, barley and rye (all malted) and torrified oats. All the whiskies are mashed and distilled 'on grain' (no filtering after mashing).

There is also a finishing warehouse with two full-scale soleras run in the Jerez manner. The top layer of one is topped up with 6-year-old whisky, the other with 12-year-old. Neil is also trialling the use of 1,000-litre (264-US-gallon), sessile oak (*Quercus petraea*) blending vessels, as he believes that marrying whiskies in wood helps more with development and harmonization.

The distillery will move to another site in Galashiels where it will be joined by another plant with two pots and another pair of the hybrid stills. 'All these games we're playing makes us feel like we are a proper whisky company,' he jokes. 'We can't compete at the bottom end, so we just do the interesting stuff.'

DUMFRIES & GALLOWAY

The southwest counties of Dumfries and Galloway have also been strangely under-represented in terms of distilleries. In the late 19th century, Langholm's Glen Tarras made an intriguing 'birch whisky' – conceivably smoked with birch wood. There were also distilleries in

Above *Annandale – a labour of love and scientific rigour.*

Annandale and Bladnoch. The former, established by George Donald and his wife Margaret (who was also the still woman), remained in family hands for 50 years before being bought by John Walker & Sons, who closed it in 1918. It made peated whisky, further undermining this idea that all Lowland whiskies are light.

It sat silent, just a set of farm buildings with one of Charles Doig's original cupolas on the roof, until 2007 when Professor David Thomson and Teresa Church bought the site. They own MMR, a consulting firm that uses sensory science to build brands. Annandale was a labour of love. £17.5 million later, the first new make ran in November 2014.

The question for any new distiller is 'What do I want to make', then aim towards that, see what the distillery decides, tweak, adjust and find that line between science and art where craft lies.

ANNANDALE

ANNANDALE MAN O' WORDS 2015 BOURBON CASK 57.3%

Mellow. Brown butter, apple Danish and old banana. Jellied fruits on the tongue, roasted barley, *café crème*, hot sawdust, yuzu, overripe pear. An extra dimension.

FRUITY/SOFT

★ ANNANDALE MAN O' SWORD 2017 STR 60.6%

Thick. Ripe melon, tayberry and toasted cereal. Some marzipan. The woodsmoke is balanced by some lush red fruits and yellow plum. Clinging.

SMOKY/PEATY

The aim for Annandale was to make equal amounts of peated and unpeated spirit, at which point the sensory techniques developed at MMR truly came into play. Creating a sensory map of 60 widely available whiskies, Professor Thomson and his team looked for where the white spaces were. For peated whiskies, they saw a gap between Ardbeg and Kilchoman's smoke and estery lift. The same technique was then applied to the unpeated whiskies. Once their target was ascertained, they steered for it.

He called the late Dr Jim Swan, who created a liquid template and off they ran. The other specification was that when mature, the whisky should have complexity and it should be well balanced. Annandale is very much a child of Jim Swan's loins.

A 2.5-tonne (2.8-US-ton) mash gives 'fairly clear' wort, ferment is long and there's enough conversation in the still to help further esterification and remove any sulphur. The cuts are high. The smoke, for example, is apparent but not pungent, sitting on top of a buttery and fruity new make character.

Most of the early releases were aged largely in fresh Bourbon casks, but as releases progress, so more refill is coming into the mix, which is ideal for longer maturation. STR casks (see box, page 120) are also used. Unusually, all the bottlings are single casks. Recently, they've branched into blending under the eye of Keith Law, while 4 hectares (10 acres) of farmland have been bought for barley, with the aim of malting on site.

'Distilleries require a huge amount of capital outlay and large amounts of working capital before revenues start to flow back into the business,' David says. 'It takes a long time before the business will be cash neutral/cash positive. It's crucial, therefore, that the new distillery should be properly financed at the start.

STR CASKS

The late Dr Jim Swan was fascinated by wood chemistry and how aspects such as different toasts, temperatures and time could be used in cooperages to produce different flavours in the final spirit. Part of this research resulted in the STR cask.

This involves a cask that has previously held red wine having 3–5mm (⅛–¼ inch) shaved off its inner surface to remove tartrates and wine residue. Because the wine has penetrated past this point, the flavours remaining are the result of interaction between it and the oak.

The cask is then toasted to activate the flavour compounds within the oak and then re-charred. This further caramelizes the wood and wine, as well as opening the surface of the cask, allowing deeper penetration by the spirit.

Overt 'wine-iness' is reduced, but there's greater sweetness – akin to a red wine and vanilla reduction. By selecting different oak types and grape varieties, together with toasting and char levels, distillers can play many tunes on the principle.

'Also, it should have a clear and visionary marketing plan and strong commercial teams. Any new distillery must have a genuine and compelling unique selling point that sets it apart from the rest.' Annandale has all of that.

Bladnoch goes back a long way. This, the most southerly distillery in Scotland, was legally established outside the town of Wigtown in 1817. As the Lowlands began to move away from whisky production, it remained, isolated, stuck down in the far south when most whisky tourists headed north.

Maybe that is one reason for a rather chequered history with the distillery passing from one set of owners to another, all of whom seem to have had their own idea about what its style should be. Under Inver House (1970s) it made a light fragrant style; with Arthur Bell (now Diageo) it switched to a nutty/spicy make. It was mothballed in 1993 and then bought by Northern Irish property developer Raymond Armstrong a year later. It ran intermittently from then until 2014 – there was more money to be made in leasing out the extensive warehousing than there was investing in a plan and brand.

Its saviour came in the form of Australian David Prior, who bought it in 2015 and installed a cast of Scotch luminaries to get the distillery back on its feet. Among them was Nick Savage, former master distiller at Macallan. 'We're a brand-new, 200-year-old distillery!' he laughs. 'Everything from the mill on is new. We've effectively rebuilt the whole thing.' Production started in 2017, then came Covid. Not that a pandemic stopped them. 'We went onto 24/7 production and just bulldozed through,' he says. 'We never shut down and no one was in harm's way. Crazy times.'

Now production is heading towards 2 million litres (528,340 US gallons) per annum and the style has changed once again. 'We're making a floral, grassy, lightly fruity style,' says Nick. 'It's what the Lowlands is known for.' Peated is also run and the new makes are maturing in a mix of 40 different types of cask.

'We need to protect and nurture Bladnoch, and build the reputation of the Lowlands,' he adds. 'It's a region that is still untapped. There's so much scope for adventure here. I believe it's the next big thing in Scotch.'

Crafty, the third Dumfries and Galloway distillery, is located in Newton Stewart. Starting in 2017, its first release was a gin, but a whisky named Billy&Co after founder Graham Taylor's father has been made while, at the start of 2024, **Moffat** distillery began to produce whisky from its tiny wood-fired pot.

BLADNOCH

BLADNOCH VINAYA 46.7%

Pretty. Hawthorn blossom, light oak behind. Marjoram flowers. More depth with water. Yellow plum skin, fleshy Viognier-like fruits. Balanced.

FRAGRANT/FLORAL

BLADNOCH 11 YEAR OLD 46.7%

Shoe polish and cream cheese, a hayloft in a dairy. Palate is totally different: canned peach, and a slippery feel (with water), walnut oil and bitter orange on the finish.

FRUITY/SOFT

WEST COAST

Heading 30 miles (48km) south up the coast is the town of **Girvan**, whose distillery was built in 1963 after a dispute between Charles Grant Gordon (William Grant & Sons) and Distillers Company Limited (DCL) over the fact that the former wished to advertise Grant's Standfast blend on TV, thereby breaking the gentlemen's agreement that nothing so crassly commercial should happen. This led to DCL cutting off Grant's grain fillings contract. Charles simply built his own distillery.

Expanded in 1986 and again in 1992, it was the first in Scotland to use multi-pressure distilling, up until then only used for making neutral spirit production, and adapt it for grain whisky production. The project was led by John Ross, known as 'the distiller's distiller'.

'The stills we had were conventional, Coffey-esque two columns making a heavy-style grain,' John recalls. 'When you use direct steam injection, you can burn material onto the internal surfaces. If the still operates with a vacuum in the wash column, however, you won't get any burning, therefore there's no sulphur, no off notes, and you get clean, fruity esters – and greater consistency.

'I figured that if you got lighter and cleaner spirit by altering the pressure, you could do the reverse and make things that bit heavier.' Today Girvan has six columns, and its floral style is the key to Grant's blends and is widely used by other blenders.

Grants operated a malt distillery within the site. Ladyburn only ran between 1966 and 1975. Bottlings are ultra-rare, but the mature stock has a distinct linseed oil character.

In 2007, a new malt plant was constructed called **Ailsa Bay**. With Balvenie being ring-fenced as a single malt, malt of similar character was required for the firm's blends. Flexibility was built in so that Ailsa Bay's 16 stills can make nutty, fruity, meaty and varying levels of peated.

Some new distilleries tease you with updates every few months on how things are going, while others throw a big opening party. Occasionally, though, there's one that stays resolutely under the radar until it suddenly appears with mature whisky. Such is the story with **Lochlea**.

In 2006, the Ayrshire farm was bought by Neil McGeoch. In 2014, he switched it from beef to arable and then, realizing that his barley crop was going to make whisky, decided to bring production in-house, or on-farm. By 2018 he had a distillery. In 2021, with former Laphroaig manager John Campbell at the helm, the first Lochlea whisky was released.

Was moving from a large, cult malt to start-up a big leap? 'That was the point,' John says. 'It would have been easy to stay at Laphroaig. I felt I knew the answers. I needed to go somewhere with no reputation where I had to discover everything and reveal all the possible layers.'

LOCHLEA

LOCHLEA OUR BARLEY 46%

Fresh, light and bright pear with hay, lemon peel, light malt and mace. Some spiced-up oak (Cinnamon Oddfellows sweets), then vanilla ice cream. Juicy peaches. Balanced and assured.

FRUITY/SOFT

LOCHLEA SINGLE CASK 60.3%

Ripe apple, then charred wood. Some coconut, caramel pudding. Soft fruits (melon, white peach), redcurrant, lavender, wild flower, citrus and milk chocolate.

FRUITY/SOFT

All of Lochlea's production is grown and matured on the farm, though John is not yet sure if the location has an impact on flavour. 'We have an interesting microclimate here. We're at the top of valley, so we get this cooling wind from the coast. It will affect maturation, but we'll know more in a few years.'

Early releases follow a farming year: Spring is ex-Bourbon, Harvest, ex-port-forward, Fallow, ex-sherry, and Ploughing is aged in smoky ex-Islay casks. The series will last for four crops, then an aged statement range will begin.

Floor malting has started, albeit at small scale, though the ambition for the future is to have everything from barley to bottling on site, all powered by a bioplant.

The wort is clear from a 'chunky' grist, 'to stop turbidity', and ferments are between 64 and 110 hours to get the fruits that John wants along with some oils and richness on the back palate.

'Without that clear wort it would be too grainy,' he says. 'Also, I don't want it too dominated by grassiness. The short ferments give fruits and the longs just enough of that grass to give balance. We're building depth over the week through the ferments.'

There are wine casks arriving – he's particularly keen on the potential of old vine Southern French for full-term maturation to help create topical fruits by the time it hits its mid-teens. 'You're always planning flavours to come out at certain times as a whisky matures,' he explains. 'Nothing is left to chance! We're currently filling new make into refill sherry for a 25-year-old.'

The most ambitious plan is to isolate a natural yeast strain from the farm. 'It's going back 200 years to how it would have been done on farms,' he says. 'Now you're finding young distillers pushing that older agenda. We can learn from the past even if we're following a different process. Whisky is cyclical. Every 100 years it refreshes itself.'

Above & right *Lochlea is moving towards total self-sufficiency.*

IRELAND

Ireland stands as an originator in the world of whisky. It also stands as an example of the vicissitudes that can (and will) befall whisky. Politics, the battle between town and country, law and dissent, colonialism and identity. Its story has been seen as a classic three-act tale: rise, apotheosis and fall. But now comes the fourth: resurrection. Today, Ireland is once again one of the most exciting whisky-making countries in the world.

HISTORY & HERITAGE

Much ink and breath has been wasted on the debate as to whether Ireland or Scotland was the first to make whisky. The honest answer is that we don't know and, really, does it matter?

There is a reference in *The Red Book of Ossory* (1350) about 'aqua vitae' being consumed, but the likelihood is that this was a wine-based spirit. What is significant is that distillation was happening, but who were the distillers? The answer lies in Ireland's medieval medical heritage and its doctors who had studied in the major medical centres of Montpellier and Salerno, who then translated medical texts from Arabic, Latin and Greek into Irish.

In 1415, Tadhg Ó Cuinn's *An Irish Materia Medica* not only listed the curative properties and uses of local plants but contained an essay on the uses of aqua vitae. In time, wine would be replaced by grain as the base for these medicinal tinctures. The similarity with Scotland's early explorations of distillation is not coincidental. After all, the MacBeathad (Beaton) family (see page 128) were originally from Antrim.

It wasn't until the 16th century that spirits began to become social drinks. The first record historian Fionnán O'Connor has found for a cereal distillate is a wheat spirit made in Kilkenny in the 1540s. By 1556, consumption of wheat-based aqua vitae, 'a drink nothing profitable to be daily drunken and used [that] is now universally used throughout the realm, especially in the borders of the Irishry', was banned in the English-controlled part of Ireland called the Pale. It had moved from the product of doctors and their monastic counterparts into the wider world.

The echoes of its medical past lived on – a cereal-based spirit, redistilled with herbs and spices, that was known as *uisce beatha* (aka usquebaugh), which would become a noted Irish speciality for the next 300 years; praised by Elizabethan traveller (and possible spy) Fynes Moryson as a 'mingling of raysons, fennell seed and other things, mitigating the heat and making the taste pleasant'.

The imposition of martial law by Elizabeth I during the Nine Years War (1593–1603) explicitly mentions 'makers of aqua vitae' as enemies of the state. The rebellion defeated, in 1608 James I issued patents (monopolies) for distillation to four cronies, including Sir Thomas Phillips of Antrim (see page 128).

Right *The clouds have lifted over Ireland's whiskey story and a bright future beckons.*

Above *Poitín-makers cunningly disguised as nuns.*

Above *The Kilbeggan distillery was a major producer.*

By 1662, the patent system had been scrapped and distillation was coming under the control of the Excise. Licensing was optional, though a ban on production outside towns and cities came in 1732. By the 1750s, something resembling a commercial whisky industry had begun. Matthias McManus (Kilbeggan – see page 132) and Peter Roe (Dublin – see page 134) both established their distilleries in 1757. The ban on home distilling three years later created a split: legal distillers on one side; those who continued to make their own poitín on the other. The seeds of single pot still (see page 126) were sown here, as was the battle between legal (Parliamentary) whiskey and poitín.

The former tended to use mixed mash bills (malted and unmalted barley, oats, sometimes wheat and occasionally rye), with rural poitín more commonly made from a single cereal (malted barley), which would have been kilned over peat. Poitín's popularity grew for the same reason as illicit whisky in Scotland – the tax on still capacity forcing licensed distilleries to distil quickly.

RISE & FALL

By the 1820s, Ireland had only 40 legal stills, while Inishowen in Donegal boasted 800 illicit sites. What stills there were, however, were becoming increasingly large. The 1823 Excise Act further encouraged capital investment and production became centred in Dublin, Cork and other towns. Parliamentary whiskey won the day.

The single pot still recipes became further refined, and Irish whiskey entered its glory years. It outsold Scotch, and was recognized as the definitive style of whisky.

A popular myth is that Ireland's distillers rejected column stills. In fact, in 1848, Scotland had 13 columns, while Ireland had 15 – and not only its native son Aeneas Coffey's design but those from Stein, Saint Marc and Sir Anthony Perrier. Although there were only 10 Coffey stills in Ireland in 1900, they were producing 71% of the country's spirit.

Single pot still was under pressure. Grain production was rising, and Scottish blenders were becoming increasingly confident. Single pot still's popularity had also led, perhaps inevitably, to mass counterfeiting, with grain spirit being shipped to Dublin and either re-exported labelled as 'Irish' or blended with Irish spirit and re-exported.

In 1878, Dublin's four major distillers issued a manifesto called *Truths About Whisky* (see page 126), setting out their argument that only single pot still could be called genuine Irish whiskey, but wider forces were now at play.

Recession, World War I, Ireland's fight for independence (1919–21), Civil War (1922–3), partition, a trade war with Britain that cut off the Empire market, the refusal to play ball with bootleggers during Prohibition, the Great Depression, then fatal protectionist measures post-World War II, with an export cap on whiskey coupled with high domestic taxes. Less a perfect storm, more the ultimate hurricane.

Ireland's whiskey industry was devastated. By the 1980s, only two distilleries, Midleton and Bushmills, were in operation. Both were owned by Irish Distillers Limited (IDL).

A NEW BEGINNING

The first signs of an upturn came in 1989 when John Teeling started up Cooley (see page 132), but it wouldn't be until 2010 with the arrival of Dingle (see page 139) that the sense of an Irish revival began.

Major international players started to invest: William Grant & Sons bought Tullamore and built a new distillery, Beam (now Suntory Global Spirits) bought Cooley; Diageo, then Proximo, came in for Bushmills, while Brown-Forman has Slane and diSaronno has Royal Oak. Around them have blossomed a host of other independent ventures.

The reasons are varied. The majors saw the Jameson Phenomenon (see page 139) at work and recognized a business opportunity. The independents, however, while obviously aware of the commercial potential, have taken a subtly different tack.

For them, this is a creative opportunity, a chance to express who they are, reflect where they are from. The best of the new whiskies represent a re-engagement with Ireland, an exploration of its rich whiskey-making heritage and a marker for its future.

KEY
Malt Distillery
Grain Distillery
Malt & Grain Distillery
Bonders Distillery
N
ATLANTIC OCEAN
IRISH SEA
Celtic Sea
St. Georges Channel
North Channel
ULSTER
NORTHERN IRELAND
CONNAUGHT
REPUBLIC OF IRELAND
LEINSTER
MUNSTER
Baoilleach
Crolly
Bushmills
Coleraine
Derry
Lifford
Larne
Antrim
Ardara/
Sliabh Liag
Wild Atlantic
Omagh
Cookstown
Lough Neagh
Titanic
McConnell's
Belfast
Belfast Lough
Hinch
Echlinville
Donegal Bay
Sligo
Lough Gill
Lower Lough Erne
Enniskillen
Armagh
Shortcross
Downpatrick
Upper Lough Erne
Lough Allen
The Shed
Achill Island
Nephin
Clew Bay
Two Stacks
Newry
Killowen
Dundalk
Cooley
Great Northern
Lough Mask
Lough Ree
Drogheda
Boann
Slane
Boyne
Lough Corrib
Ahascragh
Kilbeggan
Pearse Lyons
Roe & Co
The Dublin Liberties Distillery
DUBLIN
Teeling
Galway
Micil Distillery
Galway Bay
Tullamore
Dun Laoghaire
Burren
Church Of Oak
Fercullen/Powerscourt
Glendalough
Wicklow
Lough Derg
J.J. Corry
Shannon
Shannon Estuary
Limerick
Kilkenny
Royal Oak
Ballykeefe
Tipperary
Slaney
Barrow
Suir
Dingle
Dingle Bay
Killarney
Blackwater
Waterford
Wexford
Rosslare
Midleton
Cork
Clonakilty
West Cork Distillers
Lough Foyle
Lough Swilly
Foyle
Bann
Liffey
Brosna
A6
M1
M2
M4
M7
N3
N5
N6
N8
N9
N18
N20
N21
N25
N72
0 miles 50
0 km 50

SINGLE POT STILL

While there might not be a single Irish style, there is a signature one. It's single pot still, that miraculous palate-sticking mix of fruits, grass, cream, oil and ebullient spiciness whose story parallels that of the country's whiskey making.

What is it? Classically, it's a whiskey distilled in a pot still made with a mash bill of malted and unmalted barley ('malt' and 'barley' here) with additions of 'other grains': oats, wheat or rye.

19TH-CENTURY ORIGINS

It was widely believed that mixed mash bills only appeared in the late 18th century as a tax dodge by distillers after the imposition of the Malt Tax in 1785. If malt is required for converting starch to sugar, then as long as there is a sufficient percentage of it in the mix, it can work its enzymic magic on unmalted grains and the tax burden would be substantially reduced.

It's a great theory, and undoubtedly the tax issue played a part in the widespread shift to single pot still whiskey, but there are multiple accounts of mixed mash bill recipes pre-1785. These probably started for reasons of expediency – distillers using what cereals were available.

When the modern Irish whiskey industry started in the 19th century, it would be the single pot still whiskeys made by the major urban distillers that would make it the gold standard of its time.

As distillers refined and fixed their own interpretations, so it became a recognizable style. The fact that Scottish distillers began to make 'Irish style' later in the 19th century was the greatest form of flattery.

In their 1878 manifesto *Truths About Whisky* – note spelling – Dublin's four main distillers argued that real whisky was 'distilled either from malt, or a mixture of malt and unmalted corn, from barley or oats, or malt, or from a mixture of them…in a pot-still'. It should also be aged in ex-Spanish sherry casks.

Grain whisky, on the other hand, was, 'plastic to the hand of the compounder, and capable of being converted into sham Whisky…or into good varnish'.

20TH-CENTURY DECLINE & FALL

At the start of the 20th century, single pot still was at the heart of the 'What is Whisky' debate – and subsequent Royal Commission – where it was debated whether grain spirit from column stills could be classed as whisky.

If grain, in the parlance of the time, was 'silent', then single pot still was (and is) noisy, conversational, garrulous.

As each Irish distillery closed in the decimation of the 20th century, so the number of single pot still whiskeys fell inexorably. By the 1970s, this, the definitive Irish style, had virtually disappeared, with only Green Spot and Redbreast being bottled, and even they had only a precarious grip.

It still existed, however. Single pot still was still made for Irish Distiller Limited's (IDL's) blends, such as Jameson and Powers. If you knew what to look for, you could find it in the luscious texture of Crested Ten, in Powers' weighty, syrupy fruits, but its dynamism had been blunted.

21ST-CENTURY REVIVAL

In 2005, IDL took a punt on a new whisky market that had fallen in love with single malt Scotch, hefty Bourbons and rye, and launched Redbreast 15, Powers' John's Lane and, in recognition of the man who had been single pots still's greatest guardian, Barry Crockett Legacy. The timing was perfect. Single pot still was back.

But was it? The IDL recipe – and therefore the only one used – was made up of malt and barley, though history showed that oats were used up until the 1960s. In fact, if you knew where to look, you could appreciate how complex and varied the recipes were. Single pot still wasn't one thing but many. Every distillery had its own take.

Above *John Jameson & Son became one of Dublin's largest producers.*

Above *The famous three swallows of Power's.*

Above *Midleton's pot stills, where the style was preserved.*

In 2021, as part of his PhD thesis, historian Fionnán O'Connor assembled a group of distillers and writers (me included) to blind taste a selection of ten new make spirits made at Boann distillery (see page 132) from historic single pot still mash bills dating back to the 18th century – malt, barley, oats, wheat and rye in various proportions. There was fluidity to the recipes rather than an adherence to a single option. The results were mind-expanding.

Oats gave creaminess and sweetness, rye added spice and a meaty herbal element, wheat a buttery pastry-like quality, but the effects weren't as simple as more oats equals more sweetness. The internal balance mattered as much as the ratios. The qualities of each interacted in ferment and distillation, creating new textures, aromas and flavours. Oats acting as a glue, holding things together, and as an enhancer for barley's spice or malt's cereal and fruity florals. Distillation is a dynamic process, about transference of energy. Here was evidence of that.

Ireland's new wave of distillers are digging deep into the possibilities of this singularly Irish style, finding the truths about whiskey once again. They didn't have to follow the mash bills that O'Connor discovered. They could take this concept and apply it in whatever way they felt like. Vary the mash bill, maybe peat some elements, malt others, double distil. It was an open field. The only stumbling block was that the 2014 Technical File, which set the rules, only permitted a minimum of 30% malt, the same of barley and a maximum of 5% 'other grains'.

In 2021, a submission was sent to Ireland's Department of Agriculture to allow up to 30% of 'other grains', the most common recipe used at the end of 19th century and start of 20th century. At the time of writing, the decision has yet to be given.*

The creative thinking behind single pot still made Irish whiskey. Without similar freedom, its potential will be muted. It's also needed to demonstrate that Irish whiskey isn't just smooth blends, or light shooters; that it's more than just single malt or grain. Single pot still is the foundation. Its sticky, oily, grassy, spicy loquaciousness is integral to Irish whiskey's future.

* *In anticipation of this being passed into law, I've used the term single pot still for all examples using a mixed mash bill of malted and unmalted barley and other grains.*

THE NORTH

You might observe that this 'region' doesn't follow a political boundary but a geographical one. Donegal, while not in Northern Ireland, is in the north of Ireland. The story here, mirrored across the rest of Ireland, is that of a super-bloom after years of parched land. No (legal) whiskey was made in Donegal for 200 years before Crolly and Sliabh Laig (Ardara) started in 2020, yet this is the part of the country that can claim one of its oldest distilleries, Bushmills. The north's whiskey-making heritage runs deep.

COUNTY ANTRIM

As ever, the best way to chart its early days is through tax records, specifically the 'patents' (monopolies on production) given in 1608 to supporters of the British Crown. Here, the beneficiary was Sir Thomas Phillips, deputy for the occupation of Ulster, who was given the monopoly 'to distil within the county of Colrane otherwise known as O'Cahane's country or within the territory known as the Rowte in County Antrim'. This allowed him to make '*acquavitae, usquebaugh* and *aqua composita*': whiskey, whiskey redistilled with flavourings and whiskey with flavourings macerated in it.

Although Bushmills was within Sir Thomas's demesne, there's no evidence of a distillery at that time. Whiskey would, however, have been made in this part. After all, 'O'Cahane's country' was also home to those early doctor-distillers the MacBeathads (Beatons).

By the late 18th century, the north was a hotbed of illicit distillation, with Donegal's Inishowen peninsula its nexus point. At times, the charges levied against Donegal's moonshiners were greater than the rest of Ireland. The Excise officer sent to try and control these flagrant breaches of regulations was Aeneas Coffey, who would have approved of the founding, in 1784, of Bushmills' legal distillery.

The 19th century saw operations starting up in Belfast, Derry and along the Antrim coast. Whiskeys came in a number of styles. Single malt from Bushmills and Coleraine; double-distilled single pot still at Comber. In Belfast, Avoniel, Connswater and Royal Irish all had made single pot still, malt and blends. Derry had Abbey Street (column) and Waterside (single pot). In 1902, all of the two cities' plants (bar Royal Irish) merged to form United Distillers Limited (UDL). The combination of political unrest and a downturn in the market forced

Above *Bushmills has kept the whiskey flag flying in Ireland's north.*

Above *The shiny new extension to the Bushmills distillery.*

the new conglomerate to sell to its major shareholder Distillers Company Limited (DCL) in 1922. The Scottish conglomerate closed them all down. Despite still being profitable, Royal Irish, spooked, closed in 1933. Comber followed in 1953, while Coleraine and Bushmills merged, with the former turning to grain production in 1964 before closing in 1978.

Then there was one.

Bushmills is a survivor. The question is why did this distillery, close to the Antrim coast and the Giant's Causeway, survive when all of the north's other sites fell silent? Good fortune played its part, but so did a canny knack of always being either slightly ahead of the curve, or maybe sufficiently off to one side that the travails of the rest of the industry didn't affect it.

It made double-distilled malt in the era of single pot still. Because it never made grain whiskey, it remained out of the DCL's death grip. It subsequently switched to triple distillation and then blending, giving it another point of difference. It exported to the USA. It kept a little peat in its malt until unsmoked blends became more popular. It had a triple-distilled malt when single malt took off. Also, by being in Northern Ireland, it escaped the political impact of independence that put paid to so many of the Republic's stills. It shimmied through every trial, changing its shape but always, somehow, remaining Bushmills.

In 2005, Proximo Spirits (maker of Jose Cuervo) bought it. Since then, £60 million has been invested – most spectacularly in a new distillery that has doubled capacity.

Its style has remained fixed for a long time now, a triple-distilled single malt, as well as its blends, Black Bush and fresh, gingery Original, which add a Midleton-distilled grain to the mix (see page 140).

There are now 20 pot stills running a highly complex distillation. The wash stills operate as per normal, but the spirit from the intermediate stills is split into three. The heads and weak feints go into a low wines receiver, the middle cut into a strong feints receiver that makes up the charge for the spirit stills.

The malt's youthful notes of cut grass and light fruits have an instant apple flavour, but give Bushmills time and an incredible tropical fruit

BUSHMILLS

BUSHMILLS BLACK BUSH 40%

Fresh barley, dry grass, light mineral, then canned mandarin and bread. Very soft palate, vanilla, chocolate milk, Turkish Delight, marzipan. Easy.

FRUITY/SOFT

BUSHMILLS 16 YEAR OLD 40%

Rounded. Walnut. Touches of sherry, plum jam on toast, meadow grass. Soft fruits on the palate. Dried apricot. Easy-going and quite sweet.

RICH/DEEP

Above *The Shortcross stills allow for multiple experiments.*

Above *Killowen, where single pot still is being reimagined.*

character emerges. It is discernable in the 25- and 30-year-old, and in some old single malts being bottled by third parties.

'The fundamentals had to be revived, supported and reintroduced to a world that had forgotten Irish whiskey's difference, appeal and excellence,' says global brand director Donagh Mc Henry. 'We're all participants and key protagonists in this new thinking. It gives us the licence and latitude to offer a larger and ever more diverse choice.'

COUNTY DOWN

With distilling returning to Belfast with the Titanic distillery, the north is shaking itself back into a 21st-century version of its early 19th-century shape. Variety is key.

That's the watchword of David Boyd-Armstrong who, with his wife Fiona, owns the 202-hectare (500-acre) Rademon Estate in County Down where **Shortcross** whiskey is produced. If you don't experiment at this stage, then when are you going to do it?

At Shortcross, there's single pot still (unusually double distilled), as well as rye and malt (both malted), and there's heavily peated double-distilled malt. In another challenge to convention, the ferments are between 160 and 180 hours in order to give a 'velvety' texture. Let it sit to allow those esters to develop. Boyd-Armstrong cuts by aroma: 'We come in on pineapple and cut before soggy bread.'

He's not sure if the full extent of the new Ireland is yet visible. 'There's lots of malt whiskey because that's what's available on the open market, but pot still will take time. Only then will the diverse picture become clear. Real innovation comes from distillers producing their own liquid. It's therefore about trying to also communicate a sense of place, by learning what we can make, how we make it and where we are from.'

Brendan Carty is, if anything, pushing the idea of new Irish whiskey even further. At his **Killowen** distillery, on the shores of Carlingford Lough, he specializes in single pot still (though not to the 2014 spec – see page 127). There are ancient mash bills, high rye and bulcan (50:50 peated oats to peated malt). His ferments can run for a fortnight, including a two-day wild infection period, the stills are direct-fired and the spirit condenses in a worm. Flavour is, unsurprisingly, rich and complex – Light Irish this ain't. 'We just make what we want to create a taste that we want to drink,' as he puts it.

It's easy to call him a maverick, but this is a serious distiller. He is summoning an ancient heritage – 'wanting to taste whiskey that could no longer be bought'. There is a sense of duty to this. 'We owe this to our culinary heritage,' he says. 'We have a responsibility to put it right.' It's not being wild for the sake of it; it's preserving, it's balancing.

'Our distilling heritage is still alive. It allows us to tap into the recent past and create thick, oily and spicy whiskeys once again. When Irish whiskey was King, it was made in this way. If its to be King again, we need the same standard.' What's been lost has been found.

COUNTY DONEGAL

If Bushmills is looking at 200 plus years, then **Ardara** (Sliabh Liag Distillers) is looking at a lot less. An important distillery, though, as it was one of three

SHORTCROSS

★ SHORTCROSS RYE & MALT 46%

Quite citric with a little of malted rye's nutty character. Some maple syrup, baked peach, bergamot and green pear. Rich and sweet on the palate with a little nutmeg on the end.

FRUITY/SOFT

KILLOWEN

★ KILLOWEN BARÁNTÚIL CASK #KD001 60.7%

A peated single pot still aged in PX. Lustrous fruits with strong smoke mixing turf and light tar, then clove, treacle and barbecued meats. Oily and complex.

SMOKY/PEATY

bringing legal whiskey making back to Donegal. A crowdfunded, 500,000 litres (132,000 US gallons) per annum plant with a 2-tonne (2.2-US-ton) mash, it makes mostly single malt, but a quarter of the output is single pot. In keeping with Donegal tradition, smoke wreathes through it all – either medium- or heavily peated. The pot still mash bill is 50:30:20 peated malt, barley and peated malted oats. To ease processing, there's a mash converter and cooker.

The peated malted oats might seem a radical jump, but as owner James Doherty explains, 'Ireland is an island of contrasts and challenges, mischief and melancholy. A category that makes a post-1960s style that is smooth, sweet and easy is inherently unrepresentative of what Irish whiskey was pre-Independence.' This is not the time, he argues, to limit possibilities and close down the mind. Quite the opposite. The next step for him is building identity. At the core of Ardara's approach isn't just mash bills but place and how it reflects a forgotten heritage. He's faced many setbacks from local and national politics, disinterest in smaller operations, pushback from retailers. It's not an easy road, but he sees regionality, depth and diversity opening up the category. 'Ireland,' he points out, 'is the oldest and the newest whiskey-making nation. It has to be more than just one, samey style.'

DUNVILLE'S

★ DUNVILLE'S 10 YEAR OLD SINGLE MALT PX CASK 46%

Positive, sweet sherried character, hawthorn jelly with a back note of cooked apple, nuts and an estery top note. Marmalade and dates with water.

RICH/DEEP

OTHER NORTHERN DISTILLERIES

Donegal's **Crolly Distillery** began in 2020, while **Baoilleach** in Carrickart uses direct-fired small stills with short columns to make a variety of peated poitíns and peated single pot still whiskeys. Belfast and environs is now home to three new sites: **Hinch** in Killaney, and **Titanic** and **McConnell's** in the city itself (the former on an old dock, the latter in the old Crumlin Road Gaol).

COUNTY SLIGO

The **Lough Gill** distillery on the outskirts of Sligo can be seen as the perfect metaphor for Ireland, and Irish whiskey. Situated in the grounds of a decayed Palladian mansion, housed in an abandoned textile factory but boasting state of the art equipment and overseen by former Bushmills master blender Helen Mulholland – that's landed gentry, Celtic Tiger and whiskey revival on one site. Owned by Brown-Forman, it started production in 2020. Originally intended to produce the Paddy blend, it has switched to single malt. A highly complex triple-distillation system is being used, with five separate fractions being taken from the spirit still.

ECHLINVILLE

County Down is also home to **Echlinville** on the 600-hectare (1,480-acre) estate on the Ards peninsula, started in 2013 when Shane Braniff and wife Lynn bought the house. All of its grain is grown on the estate and distilled in a mix of pots and column. It has two functions. The couple revived the Belfast brand Dunville's, which is currently made with sourced whiskies, while the farm-grown Echlinville range – grain, malt and single pot still – is maturing.

Above *Ardara – the return of whiskey-making in Donegal.*

THE MIDLANDS

There is scale, tradition and new thinking in this part of Ireland, a perfect reflection of a whiskey-making industry beginning to flex its muscles once again. Scale is important. While smaller distillers often feel that the larger players are dictating image, style and law, the lesson of President Éamon de Valera's error in assuming Ireland could consume all of its production cannot be repeated. There has to be balance.

John Teeling's **Great Northern Distillery** (GND), which took up residency in the vast former Harp brewery in Dundalk, is producing grain malt and pot still for the still thirsty bonder/bottler/sourcer market. Teeling moved there after the sale of his Cooley Distillery in 2011 to Jim Beam (now Suntory Global Spirits). The **Cooley** story itself is not just about the repurposing of an industrial alcohol plant in 1989 to make whiskey and break the Irish Distillers Limited (IDL) monopoly (see page 126), but how it offered an alternative. There was no single pot still, nor any triple distilling. Cooley made double-distilled single malt and grain. It even peated one of its brands (Connemara). It bought up old brand names, supplied bulk. Today, the Cooley (and GND) stock owned by Teeling has also helped build the new category, even if, as many small distillers complain, it's skewing the Irish offering away from the traditional and towards a style (single malt/blends) that played a lesser role in the heritage. Balance again.

The distillery still makes its gentle sweet style helped by cooling tubes in upward-angled lyne arms. Even Connemara, at its heart, has a delicacy that acts as a counterpoint to turf-like phenols. The grain is fat, and corn-based. Once called Greenore, it's now been renamed (along with the rest of the unpeated range) as Kilbeggan. It's a single grain from a single distillery, but it's not named after the one in which it is made.

Let's untangle this Celtic knot. **Kilbeggan** is also a distillery (originally owned by Cooley and now by Suntory Global Spirits). Established in 1757, it closed in 1953 before reopening thanks to a local initiative in the 1980s as a whisky museum. Cooley bought it for warehousing and the Locke's brand name, but in 2007 distilling restarted. Inside is a mash-up of water-wheel and steam-driven engineering. The pots are small – one tiny still is the oldest and maybe smallest in Ireland.
In 2010, mashing and fermenting restarted. The first

Above *Kilbeggan is now producing once again.*

100% Kilbeggan was a surprise, though – a double-distilled single pot still with a 30% percentage inclusion of rye. Because of the Technical File (see page 127), this could only be called 'Irish whiskey'. It was followed by a single pot still with 2.5% oats. More will appear. Balance again.

Some 30 miles (48km) south of the Cooley peninsula is Drogheda. Close to the massive Neolithic site at Newgrange, it's also where, in 1556, a law was passed, restricting the production of uísce beatha to 'peers, gentlemen and freemen'. Over 160 years after Drogheda's last-remaining distillery, Woolsey, closed, the Cooney family opened **Boann**. What they've created is a centre for the single pot still's renaissance. The new makes for the historical mash bill project (see page 127) were all distilled here. If they looked

COOLEY

CONNEMARA 40%

Light and grassy with mossy smoke. Quite crisp. A cold wind over the moor. Rooty, barley sweetness, light smoke, white pepper.

SMOKY/PEATY

CONNEMARA 12 YEAR OLD 40%

Linseed oil building. The higher strength boosts the smoke. Oily start, coal tar over a sweet base.

SMOKY/PEATY

KILBEGGAN

KILBEGGAN 40%

Has the easy-going Cooley gentleness. Lots of American oak, sweet grain and crisp malt. Summer fruits, lemon posset and vanilla, light dry pastry on the end.

FRUITY/SOFT

KILBEGGAN SINGLE POT STILL 43%

Green fruits in the light pot style. Fresh bread, then gingery spice, queen of puddings, black pepper, green tea and spice.

SPICY/HERBAL

Above *Tullamore's visitors' centre.*

OTHER MIDLANDS DISTILLERIES

The **Slane** distillery was started by Alex Conyngham, his family owning the castle on whose grounds it sits. It was bought by Brown-Forman in 2015. Triple-distilled malt and single pot still are made, with virgin, ex-Jack Daniel's and oloroso sherry casks being used. The whiskeys are blends of different percentages of all three cask types.

The **Church of Oak** distillery in Ballykelly Mills, County Kildare started production in 2023 for one reason, legacy. 'I believe that phrase connects us all,' says founder Garrett Pitcher. 'The journey for me began when I found out I has having my daughter. I wanted to make her proud. I wanted to build something creative, remarkable and something larger than me. That became Church of Oak.'

From the outset, he and distiller Greg McEntee have embraced a multi-grain approach, making single malt, single pot still, sourmash and a malted rye/malted barley mash bill. Distillations are either double or partial triple. All bar the single malt are made grain-in. All, are currently maturing, something that could be said to apply to the wider new Irish scene. 'It will be a few more years yet before you see the full Irish whiskey picture emerge,' says Garrett. 'Many of us don't have mature stocks and we are yet to realize the potential of Irish pot still. As the category matures, I think you will see a lot more diversity including bonders and bottlers. Ninety-nine percent of Irish whiskey is blends, so I for one would welcome some more single malts! There are also a few stereotypes to be broken.'

DiSaronno built its **Royal Oak** distillery in Carlow in 2014 and expanded it between 2021 and 2023. It now produces its Busker range of single pot still, single malt and two blends (grain/malt and grain/malt/single pot) under the one roof.

into the past, then Boann's distillation is futuristic. A mash convertor helps to ease processing of unmalted grains by allowing the specific temperature that each grain requires for full modification. The ferments are long, and the stills are the first in any whisky distillery to use a copper-coated nano-crystal coating in each of the three stills' lyne arms.

Developed by the Italian company Green Engineering in association with the University of Siena, it increases the amount of available copper by six times, thereby removing heavy, sulphur-laden elements from the new make. Reflux controls can be engaged to make the spirit lighter or disengaged to produce a heavier result. Reviving a tradition doesn't necessarily mean only using the most ancient approach. Not for the first time, here is Irish whiskey facing both ways.

Perhaps appropriately, **Tullamore**, which sits in the geographical heart, draws from three different whiskey traditions. Pot still is made, as is single malt and, since 2017, grain. This is because the Tullamore D.E.W. brand was the first Irish brand to be a blend of all three styles, a clever move by then owners the Williams family to try and tap into the lighter palates of the American market. The wheat-based grain gives grassiness, the malt fruit and the pot still spice. After the original distillery closed, the whiskeys were made in Midleton. William Grant & Sons bought the brand in 2010 and within four years had a new malt distillery operational.

The **Powerscourt** estate in Wicklow, just 20 miles (32km) from the centre of Dublin, installed a distillery in 2016 and began operations two years later. Its **Ferculllen** brand was originally exclusively using ex-Cooley stock (its first distiller was Noel Sweeney who had made it there). Now its own makes are beginning to appear and in common with an Irish theme the net is being spread wide. There's double- and triple-distilled malt and pot still being made from Irish barley (that from the estate is used for the single pot still).

BOANN

★ BOANN WINTER SOLSTICE SINGLE POT STILL 59.9%

Rich and concentrated. Raisins and roast pumpkin. Oily, meaty with coffee, chocolate and raisin. Lightly charred. Some Moscatel perfume. Powerful.

OILY/SPICY

FERCULLEN

FERCULLEN SINGLE MALT 46%

This is the inaugural release of the distillery's make. Yellow plum, apple skin, fresh quince and crème anglaise. Ripe melon and vanilla. Lightly sour fruits on the end give balance.

FRUITY/SOFT

TULLAMORE

TULLAMORE D.E.W. 40%

Light. Aromas of barley, light char, a little fig at the back. Light and mixable.

FRAGRANT/FLORAL

TULLAMORE D.E.W. 12 YEAR OLD 40%

Huge levels of citrus: orange peel, bergamot, pink grapefruit juice, then some green apple and creamy oak/milk chocolate. Pot still oils on the end.

FRUITY/SOFT

DUBLIN

Distilling arrived in Dublin in the 18th century. Its commercial distilleries were established in a 'liberty' outside the old city walls, ungoverned by the City Corporation. The Liberties contained tanneries, silversmiths and weavers, and in time the drinks trade. As well as breweries (most notably Guinness) and distilleries, there were numerous ancillary trades: mills, mashing houses, cooperages and warehouses.

In 1757, Peter Roe started distilling on a site in Thomas Street that would be inherited by his son George. James Power followed in 1791, who began making whiskey in the back of his public house on the same street.

By then, the Scots had arrived. Robert Haig opened his Dodderbank distillery in 1795, and John Stein Jr founded Bow Street and Marrowbone Lane distilleries in 1780, passing on the running of the former to Haig's son-in-law John Jameson. The latter took over the distillery in 1805 and started John Jameson & Son (JJ&S).

Dublin's 19th-century dominance and style emerged through a number of factors. In 1785, John Beresford, then head of the Excise, proposed that rebates be given to distillers if they installed extra-large stills, effectively restricting trade to those with ready capital. This, and a subsequent law stating that 'no Irish still should have a diameter greater than its depth', didn't just change the look but the character of the spirit. Big, fat-bellied, short-necked stills gave reflux and oils.

Urban whiskeys weren't reliant on peat fires to dry their malt. Coal heated the closed kilns to dry the malt. Result: unsmoked malt.

The final piece in the Dublin style came with the adoption of mixed mash bills, whose use was sealed with the passing of the Malt Tax in 1785, coupled with the growth of these larger plants with higher running costs. Any saving would be worthwhile.

THE BIG FOUR

In 1822, John Jameson's son William married Stein's daughter Isabella and took over the running of Marrowbone Lane, trading as William Jameson & Co. The Big Four were in place.

Dublin post-1823 saw a contraction of the number of distilleries, but a significant increase in capacity. James Power's son John took over the John's Lane distillery (the firm moved there in 1822) and installed larger stills.

Above *The enormous Jameson distillery back in the 19th century.*

Above *Built at scale – Power's pots at John's Lane.*

Above *Roe & Co is part of the Dublin revival.*

By 1833, the distillery was making 300,000 gallons (1.36 million litres/360,000 US gallons) and owned the Wexford estate providing its barley. By the end of the century, George Roe's Thomas Street distillery covered 17 acres (6.9 hectares), boasted 8 pots and could make 2 million gallons (9.1 million litres/2.4 million US gallons) a year (twice Jameson's Bow Street capacity), and was the largest in Europe.

In the 1870s, they were joined by the consortium that built the Dublin Whisky Distillery Company (DWD) and Scotland's Distillers Company Limited's (DCL's) Phoenix Park, which rose from the ashes, appropriately enough, of the old Chapelizod plant. The consortium of Scottish grain distillers had begun to make single pot still in a 'If you can't beat 'em, join 'em' move. It wouldn't be long before that changed to, 'If you can't beat 'em, beat 'em up'.

Irish whiskey, and particularly Dublin single pot still, was at its height. It was selling three times as much as Scotch, and outselling it in Scotland.

BONDERS

Whiskey bonding is an old Irish tradition. Distillers would sell whiskey direct to customers – grocers, wine merchants, publicans – around the country. Some like the Mitchell family of Dublin and Gilbeys would then create their own blends, such as the former's Spot range and the latter's Redbreast.

Now bonding is re-emerging, though the designation is slightly confused. It's different to the sourcing that many of the new distilleries have been doing as they wait for their whiskey to mature. Brand owners also source whiskey and bottle it under their own brand.

There are some firms returning to the original concept of bonding, such as Louise McGuane from West Clare who revived the J.J. Corry brand* (Corry was a bonder and bicycle inventor from nearby Kilrush). For her, bonding is controlling everything post-distillation – buying in stock, maturing it and then blending it to her recipes. W.D. O'Connell Whiskey and Two Stacks also take this approach.

* *Full disclosure – I have worked on liquid analysis with J.J. Corry.*

The city's distillers were supplying bonders across Ireland and British merchants, the warren of tunnels under the streets stuffed with casks.

Whisky is a notoriously cyclical beast, however none of them had experienced a downturn. The first breach of Dublin single pot still's dominance came with the column stills designed by Aeneas Coffey (who briefly ran a distillery in Dublin) and the arrival of 'silent spirit' (aka grain whisky).

By the 1880s, the second and third generations of Scotch blenders were beginning to change recipes to suit a range of different palates and serves, embracing branding and advertising. They had learned from Dublin the importance of consistency of character and quality, and having the name of the distiller (or blender) as a bond of trust.

The multiple pressures resulted in the still remarkable *Truths About Whisky* (see page 126), which gives an insight into the fears and lofty arrogance of Dublin's Big Four, as well as how their whiskeys were made. It was a noble rearguard action.

In the face of recession, falling prices and rising stocks, in 1891, Dublin Whisky Distillery Company, William Jameson and George Roe formed the Dublin Distillery Company (DDC), but the last two closed in 1922 while DWD's Jones Road shut in 1926, with occasional distillations until 1942.

The two remaining Dublin distillers (JJ&S and Powers) survived until 1966 when they merged with Cork Distillers Company to form Irish Distillers Limited (IDL). Five years later, Jameson's Bow Street closed, with John's Lane following five years after that. Dublin was no longer a whiskey city.

DUBLIN'S WHISKEY STYLES

What was this whiskey that so excited the 19th-century consumer? Unpeated, triple or in some cases partial triple distilled, mixed mash bills and aged in ex-sherry casks. Each distiller had a different mash bill (often

Above *Teeling has led the new wave of Dublin distilleries.*

more than one) that varied the amount of malt and barley. All had oats, some included rye. Dublin styles rather than style.

They would have been rich in oils and dried fruits, with touches of leather and a green edge. Spices would have fizzed. If blended Scotch built itself by being amenable to lengthening with mixers, Dublin's whiskeys were the ones you sipped.

One clue lies in IDL's John's Lane 12-year-old, which, though made at Midleton, follows the old Powers' style – ridiculous tropical fruits, oiled cricket bats, pepper, tanned hide, cherry and cigar humidor, which was a revelation when it appeared in 2011. Redbreast's sherried depth and spiced finish hints at it, but the one-off Jameson 15-year-old single pot still suggested that the old Dublin style was heavier again. Oils, autumnal fruits, wet earth, a headfirst dive into mixed spices. However, these are only clues. We look at mash bills, we think of casks, imagine what happened under Dublin's streets, but in the whiskey world 1+1 never makes 2. Secrets remain just that.

There's a happy ending though. In 2014, distilling began again in The Liberties with the arrival of Teeling's €10 million distillery. In 2017, Pearse Lyons moved into what was St James's Church in James's Street, while the same year, Diageo used some of the money from selling Bushmills to open a new distillery opposite Guinness's St James's Gate brewery. It's called George Roe. Two years later came Liberties. The question is, are they making 'Dublin whiskey' or whiskey in Dublin?

THE LIBERTIES

When John Teeling sold Cooley to Jim Beam in 2011, the big question was what was he and his family going to do next? It was unimaginable that the team who had so stirred up the Irish industry would move away from the business. The answer wasn't long in coming. John bought Dundalk's Harp brewery and turned it into the Great Northern Distillery, and began supplying the thirsty requirements of the bonders and bottlers. His sons Jack and Stephen had other ideas, heading to Dublin to build a new distillery and create a new brand. Where else would they buy a site but The Liberties. And from there the spirit first ran in 2015.

Initially, they drew on the family stock made at Cooley, which also included whiskeys from

TEELING

★ TEELING SMALL BATCH 46%

Vanilla and banana to start with, dried mango at the back. Spice-driven palate. Verbena, lemon balm, light oak and toffee. Some melon and nutmeg on the finish.

FRUITY/SOFT

TEELING BLACKPITTS PEATED SINGLE MALT 46%

Immediate smoke. Honey-glazed meat, then oil of clove and a little tarriness. Runny dark chocolate, then soft tropical fruits come through.

SMOKY/PEATY

ROE & CO

ROE & CO SOLERA SINGLE MALT 46%

A complex solera made up of nine cask types results in a silky feel with stewed apple, canned peach, maple syrup, flaked coconut, black cardamom, dry roast spice, pepper and chocolate. Will gain even greater resonance as the solera matures.

FRUITY/SOFT

A.N. Other northern distillery (no names, but guess), but the aim with **Teeling** was always to look at the future, forge a new style, be individual.

'The way we approach creating whiskeys is looking at every aspect of the production process,' says brand ambassador Rob Caldwell, all guided by distiller Alex Chasko.

What is Dublin whiskey? In Teeling's case, it can be pot still, it can use different kilnings of malt, it can stir in rye and corn. Alex was a brewer by trade, so no surprise that there's plenty of work with ferments (an average of four days), yeast (white wine and distiller's) and kilning temperatures.

They're in Dublin, so inevitably and rightly there's single pot still as well as single malt. They worked at Cooley, so there's peat (Blackpitts), triple distillation, naturally, but also double, giving light and heavy peat, for example.

The result? Even within those variations that push and nudge flavour this way and that, there's a core of estery and tropical fruits. The use of 150 different cask types builds on the past with wine and beer casks, but with other spirits like tequila and Calvados and woods as well – oak of course, but also chestnut, acacia, cherry and the Brazilian wood amburana. 'It paints a very vibrant future of flavour for Irish whiskey,' says Caldwell, 'and opens up the conversation regarding whiskey with a wine drinker through the shared flavour notes. Then they may be more inclined to take a step further down the whiskey rabbit hole.' It's that Janus-faced nature of new Irish whiskey at work once more. In 2017, Bacardi bought a 40 per cent stake, which was upped to 79 per cent in 2024.

A scientific approach lay behind the strategy of returning son of the city, the late **Pearse Lyons**. He came from five generations of coopers on his mother's side, but studied a master's in brewing – he was part of the design team that developed new Midleton in 1975. On moving to the USA, he founded yeast specialist Alltech, as well as The Alcohol School in Lexington, Kentucky, finally making the jump into production by opening the Lexington Brewing Company and then in 2013 the Town Branch distillery. The same year he started up trial operations in Carlow, having already bought a distillery site housed in an old church in The Liberties. Restoration took four years, opening in 2017. Sadly, he died the year after. Single malt was first followed by pot still.

In 2021, the distillery began to draw from spring barley and oats and winter barley, wheat and rye grown at the 48-hectare (120-acre) Lyons family farm, 16 miles (26km) from Dublin. Unusually for the country, the fermenters are open-topped, while the one pot and one hybrid stills sit on what was the church's altar.

The revival of Dublin's whiskey-making life continued with Diageo's refurbishing of the old Guinness brewery's power station, which it renamed **George Roe**. In early 2024, the first of the distillery's own triple-distilled make was released. The new site has three pots – the intermediate with a double boil ball and a head from an old Tanqueray gin still. Long (79–80-hour) and longer (120–130-hour) ferments are used, activated with two yeasts. The new makes are blended at different ratios at birth, allowing distiller/blender Lora Hemy a wide range of flavours to work with. Maturation is predominantly in ex-Bourbon but a wide range of oaks, woods and varieties is also used. Roe & Co was launched using sourced whiskey, but its first own release gives an indication of Lora's forward-thinking. It is aged in a classically set-up solera system, with a hugely complex make-up: refill and first-fill American oak, ex-Bourbon, new Alligator-char barrels, chestnut barrels with two custom toasting profiles and four different types of sherry casks.

The most recent arrival on the Dublin scene is **The Liberties**. With a 2-tonne (2.2-US-ton) mash and 700,000-litre (185,000-US-gallon) capacity, it runs triple distillation and like most started with a blend made from sourced whiskey.

Above *The pots at Dublin Liberties, ready for action.*

THE SOUTH & WEST

Whisky making at its heart – both philosophically and practically – is about being able to be different enough to cut through in what is already a cluttered category. This is even more important in a country that is relearning its heritage while at the same time moving that tradition onwards. Is Ireland an old country making new whiskey, or a new country making old whiskey?

COUNTY WATERFORD

There are many approaches, but few have taken it as far as **Waterford**. The distillery, in the city's former Guinness brewery, works with 35 farmers, but keeps their harvests separate out of the belief that each of them produces something unique to that place.

The whiskeys are released as single farm origin, or blended into *cuvées*. Wine thinking runs through the whole project: barley as grape, nods to eau-de-vie's purity of intent. Other Irish distillers look to mash bills and heritage, and while Waterford appears to look in a different direction, it is tapping into the oldest way of whiskey making – one community, one field.

Organic and biodynamic are being examined; there are forays into the character of Irish peat and the potential of forgotten varieties, such as 1950s variety Hunter.

There's a mash convertor and mash filter at play, ferments of a minimum of 120 hours, with a high setting temperature of 30–32°C (86–90°F) – twice the norm to encourage esterification. This is a barley distillery rather than a whiskey one.

Historians are helping to lead this new wave. One of them, Peter Mulryan, then made the leap to distiller, opening the **Blackwater** distillery in Ballyduff. He makes an excellent gin, and single malt, but it's pot still that most exercises him. Peter's been vocal about the need for smaller distillers to have an equal voice, pushing for the changes to the 2014 Technical File that permitted only 5% other grains as a sop to the existing mash bills of pot still and bore no resemblance to what had gone before. Blackwater uses Irish grains, 100-hour-plus ferments and is double distilled.

WATERFORD

WATERFORD HERITAGE HUNTER 1.1 50%

Punchy, but clear slightly earthy barley notes. More of the hay/stalk on the palate with light stewed fruits and chilli heat on the end.

FRAGRANT/FLORAL

Above *Grain storage at Dalton's Mill, Waterford Distillery.*

Above *A mash filter helps process 'difficult' grains.*

Above *Dingle's draff goes off to feed local cows.*

Above *The washback tanks in the tun room at Dingle.*

COUNTY KERRY

In the (deserved) excitement about the new Irish whiskey scene, it's easy to forget that the pioneer, **Dingle**, has been making it since 2012. Here, conceivably, is the start of the trend to long ferments and slow distillation seen elsewhere. Local barley is used, while the single pot still is a malt/barley mix (with rye and oats in 2020). The distillation is run very slow, including, unusually, for the feints – even they need copper contact.

It gives a light, fruity/citric new make malt, and a fuller savoury single pot still. Both intriguingly have a minty touch. Dingle slowly reduces the whiskey to bottling strength to avoid soapiness while upping mouthfeel and preserving volatiles.

'It's important for Irish whiskey to be diverse,' says distiller Graham Coull, 'but also retain our traditional roots. There needs to be a balance struck between experimentation and traditional methods. It's better to be a slow burner than a flash in the pan.'

COUNTY CORK

One way of looking at the story of the great Irish whiskey collapse is that its last act was the formation of Irish Distillers Limited (IDL) in 1966. Another way, and the one that has proved to be true, is that the saving of Irish whiskey and its long, slow revival came as the result of the founding of that company.

It's easy to look at scale and story. For almost 30 years, IDL was the only whiskey producer in the whole of Ireland (it owned Bushmills from 1976 to 2005), it owns Ireland's largest brand, Jameson, and is a giant in terms of production. Yet that is to miss the real story.

IDL may be massive, but it was also the firm whose old triumvirate of master whiskey makers – blender Barry Walsh, distiller Barry Crockett and head of maturation Brendan Monks – preserved Irish whiskey's tradition, specifically single pot still, as well as giving the world a new-look Irish blend (Jameson was reformulated in 1970). They pioneered wood management before Scotch and in Pernod Ricard were backed by an owner that believed in long-term, consistent messaging from its takeover in 1988.

The recent Irish whiskey boom has been driven by the Jameson boom, and the figures are extraordinary. 'In 1988 we sold 588,000 cases,' says master distiller Kevin O'Gorman. 'A decade after, we were at just over 1 million. Now [2023] we're selling 10 million.'

BLACKWATER

★ BLACKWATER 'SHENANIGANS'
(FIVE-GRAIN SINGLE POT STILL) 46%

Juicy. Some linseed oil, peaches and cream, mango, white pudding/proving sourdough, then coffee and thickening oils. Ginger and grass on the end.

OILY/SPICY

DINGLE

★ DINGLE SINGLE MALT 46.3%

Pungent, powerful and robust. Earthy with dry spice, vanillin, smoky wood and Bourbon biscuit. *Vin jaune*, leather and myrrh. Chewy and mouth-filling.

RICH/DEEP

DINGLE BATCH NO. 5 TRIPLE DISTILLED 46.5%

The use of PX and Madeira adds a glossiness. Balance of blue fruits and liquorice. Petrichor. Succulent, energetic, with sweet ripe fruits.

RICH/DEEP

Above *Midleton's pot still distillery.*

Above *The hand-made approach at Micil harks back to poitín-making.*

The forecasts promise further growth, so much so that the already massive **Midleton** distillery is about to double capacity once again.

'We've been running close to capacity for a number of years,' says Kevin, 'and as the aim is to sell 16 million cases of Jameson by 2030, that means increasing output.' The new 22-hectare (55-acre) site will be operational by 2025, and the whole complex will be carbon neutral the year after.

Midleton is a fascinatingly complex distillery, and made in that way to give diversity. Triple-distilled single pot still comes in three main styles: light, 'mod' (medium-bodied) and heavy. The mash bill is the same, but cut points in the intermediate and spirit stills vary, producing more oils and fruits, or lighter green apple esteriness. Other variants are made. Pot still distillate can also be redistilled in column stills, or vice versa, while grain whiskey is also made. The number of streams is huge.

This was deliberate and part of the preservation act not just for Jameson but also Powers and, from 2011, the relaunched single pot still range. 'That was another big step,' says Kevin, 'and was about identifying and embracing our heritage with Redbreast, but also Powers' John's Lane, Midleton and the Spot range.'

As well as steering the needs of a consistent global blend, forward thinking is part of the mix for Kevin and his ever-growing young team. Midleton's 'pilot distillery' (bigger than most full-sized ones) is home to, among other things, the Method and Madness series, which asks the 'What if we do this?' question.

Out has come rye/malt, oat/malt, rye and single malt as well as yeast trials ('an amazing area', says Kevin), and ongoing work with wood (Ireland's more liberal rules on using any type of wood rather than only oak helps). They have trialled chestnut, mulberry, hickory, maple, acacia, cedar and mizunara oak, and then there's the Dair Ghaelach Irish oak series. A firm that might be seen as a monolithic enterprise is actually remarkably nimble.

'I agree that this is a reset for Irish whiskey,' Kevin concludes. 'Today's whiskey is about provenance and sustainability. Everything has to change. We are going to the forests, sawmills and cooperage to get the best wood. With Origin Green we're working directly with farmers.

'It's exciting, and it's about the future, but also the past. Part of this is going back to the 19th century and seeing how whiskey was made then. Having Carol Quinn here with the archive is an incredible

MIDLETON

JAMESON 40%

Green apple and pear with some light cream, coconut, then cedar, green plum, blackcurrant leaf. Starts sweetly, then tightens into fennel, spice and green apple.

SPICY/HERBAL

POWERS GOLD LABEL 40%

A touch of shoe polish, super-ripe apricots and peach, ripe apple, mace. Palate is soft and slinky with some grapefruit peel and light ginger on the end.

FRUITY/SOFT

★ REDBREAST 12 YEAR OLD 40%

Dubbin and leather oil, green fig jam, black tea and raisin, then a lift of green pear, sandalwood, coriander seed and chocolate. Oily, then fresh with ground ginger.

OILY/SPICY

★ METHOD AND MADNESS RYE & MALT 46%

High rye content. Some nutmeg, dry lemongrass, white pepper, sweet malt and nutmeg. Then green pepper, light rye tempered by crisp malt, a hint of chilli heat.

OILY/SPICY

benefit. I also love seeing all these different distilleries all doing different things. We have to support each other and build Irish whiskey.'

Another pioneer of the new Irish scene is **West Cork Distillers**, which was started in 2003 by John O'Connell (whose family has a long tradition of poitín making) and cousins Ger McCarthy and Denis McCarthy. Since 2015 it has been producing a wide range of spirits from its new distillery in Skibbereen. It's a not inconsiderable site, capable of producing 4 million litres (1 million US gallons) per annum from 8 pots and 2 columns for whiskey, and a further pair for gin.

The size makes sense. West Cork began at a time when many start-ups were in need of sourced whiskey but only able to obtain it from either mature ex-Cooley (and some Bushmills) stock owned by John Teeling or new make from Great Northern. Being able to draw single malt and grain from another source allowed distillers a greater range of flavours for new brands or the intermediate whiskeys that were to be sold while their own stock was maturing.

West Cork's own brands include blends, single malt and single pot still, as well as different finishes such as casks that have been infused with peat and bog oak smoke.

COUNTY GALWAY

One of the notable developments is how whiskey is returning to 'lost' areas. Now Galway has its first legal distillery for over a century with **Mícíl**, run by the Ó Griallais family whose distilling expertise goes back six generations. They have been making poitín for 170 years. Now head distiller Jimín Ó Griallais makes single malt and single pot still and non-GI (Geographical Indication) compliant 'pure pot still' styles of Irish whiskey.

In keeping with tradition, peat is used for most, while Micíl taps into the old trade with Spain and uses wine, fortified wine and spirit casks, some in the traditional 50-litre (13.2-US gallon) size. 'We're showing that Irish whiskey is more than soft, sweet and smooth,' he says. 'We have great flexibility to use a variety of cereals, distillation and ageing methods, and allow consumers to access regional and historical differences in Irish whiskey.'

This 'drawing from the well of heritage and tradition' is the manifestation of the old versus new scenario. Having a 170-year perspective helps. For him, Micíl is not just a distillery but a revelation of what was made in Galway and Connemara. Tradition can be overly glib. This is deeper – and true. It adds to Ireland's diversity.

I asked all the distillers and writers if this was a renaissance, a revival or a reset? Most thought one of the last two. 'We're currently in the revival period,' says writer Matt Healy. 'We've not seen the renaissance. Single pot still will be the flavour driver. It's important that we have diversity in order to carve a niche in a global industry reared on single malt Scotch. We have a steep mountain to climb in global popularity and we won't get there by just trying to be Scotch Lite®.'

There is an infectious optimism across the category and a profound belief in rediscovering Irish whiskey's roots. The questions are financial but also existential. Who are we? Where are we from? What does that tell us? The answers will come and they will be fascinating.

MÍCIL

MÍCIL INVERIN SMALL BATCH 46%

Nicely earthy and powdery mix of cereal, light cask. Vegetal, dry black tea leaves. Palate-clinging: light banana, then crackling peppery spice.

FRUITY/SOFT

OTHER SOUTH & WEST DISTILLERIES

Tipperary Distillery was founded by Scotch whisky veteran Stuart Nickerson and his daughter Jennifer in 2016. It uses barley from a single farm, Ballindoney. In 2020, the full plant, including four stills, was installed. Until its own whisky is available, it has used sourced whiskey and more recently it has had its barley distilled to its spec elsewhere.

Clonakilty (see left) in West Cork is one of the many food towns in Ireland's west coast. In 2019, the Scully family added to the offering by opening a €10 million distillery. Three Barison stills produce 250,000 litres (66,000 US gallons) per annum. It uses locally grown grain from fields next to the Atlantic. Will any salinity come through? We'll see.

Noel Ó Lochlainn, protector of The Burren's rebellious heritage, has been making whiskey under the watchful eye of legendary industry consultant Jack O'Shea at the appropriately named **Burren Distillers**. He's producing it in the 'sean gael' (old Irish) style using local barley, floor malted at the distillery, mashed and then fermented in Irish oak vats.

Our Irish tour ends on Achill Island in County Mayo at Ireland's only island distillery, where **Drioglann Oileán Acla** (Achill Island) installed two pots in 2019. At the time of writing, only sourced whiskey had been used.

ENGLAND & WALES

Quite why the concept of English whisky was so strange for many people is hard to untangle, but for years there was an element of cognitive dissonance when the words 'whisky' and 'England' were placed together. It was easier, somehow, for the majority to accept that whisky could be made in Japan or Australia than it was just down the road from Scotland.

In recent times, the two pioneers of whisky in England and Wales, the English Whisky Company (see page 149) and Penderyn (see page 150), have been joined by an ever-growing array of other distillers, each of which is taking a slightly different tack when it comes to whisky making.

With an estimated 60,000 casks of whisky maturing, distillers in both countries have recognized the need for a set of technical standards.

While this might sound like the most exciting topic in whisky, regulations exist to protect drinkers – not having a framework or definition only ends up with confusion, infighting and a lack of trust. A whisky world that is moving so quickly exposes itself to opportunists who can easily take advantage of loose or nonexistent regulations.

The English Whisky Guild (EWG) was founded in 2022 and at the time of writing its set of guidelines for a UK Geographical Indicator (UKGI) is with the Department for Environment, Food & Rural Affairs (Defra).

If passed, the GI would define English whisky as one where the entire process, including maturation, takes place in the country. There was a debate as to whether all the ingredients must also be English. While the use of local grains is admirable in terms of achieving greater sustainability and helping to reinforce roots, some producers argued that permitting imports of grains would be sensible in case harvests were low, or failed. Wales has already gone down that route, and a UKGI has been approved (see page 150).

Regulations such as this should not restrict creativity, but act as a further bond of trust between producer and consumer.

Right *From lakes to cities, England's distilleries are scattered throughout the country.*

KEY
Malt Distillery
SCOTLAND
North Sea
Irish Sea
Isle of Man
ENGLAND
WALES
English Channel
Newcastle upon Tyne
Carlisle
The Lakes Distillery, Cockermouth
Spirit of Yorkshire, Filey
Cooper King Distillery
York
Lancaster
Leeds
Manchester
Liverpool
Sheffield
Penderyn Llandudno
Aber Falls, Llanfairfechan
Bangor
Wrexham
White Peak Distillery, Belper
Angelsey Môn, Gwynedd
Norwich
Birmingham
The English Whisky Company, Roudham
Adnams Copper House Distillery, Southwold
In the Welsh Wind, Cardigan
Dà Mhìle, Llandysul
Cotswolds, Shipston-on-Stour
Coles, Carmarthen
Penderyn Brecon Beacons, Aberdare
Gloucester
The British Honey Company, Worminghall
Fielden, Oxford
Penderyn Hafod Copperworks, Swansea
East London Liquor Company
Bimber
LONDON
CARDIFF
Bristol
Canterbury
Winchester
Brighton
Portsmouth
Exeter
Hicks & Healey/ Healeys, Penhallow
Plymouth
Truro
Severn
Trent
Thames
A1
A1(M)
A64
M6
M62
M1
M54
M40
M4
M25
M20
M5
A30
N
0 miles 50
0 km 50

ENGLAND

While England might not have the same depth of whisky-making heritage as Ireland and Scotland, it was made in the country. The first mention of worts being distilled comes in Chaucer's *Canterbury Tales* as part of a long (satirical) list of ingredients used by alchemists. By the time whisky's first chronicler Alfred Barnard toured the distilleries of the United Kingdom in the 1880s, England had four large grain distilleries.

Within a couple of decades, however, the closing of London's Lea Valley in 1905 saw English whisky cease to exist for a century. But since then, 45 distilleries have started up and any confusion in people's minds has cleared. English whisky is here. And it's good.

THE NORTH

The Lakes Distillery started distilling at the end of 2014. Harold (Hal) Currie was a legend in the Scotch trade (MD of Chivas Brothers, founder of Arran – see page 86). When his son Paul was contemplating where to build another new site, the Lake District made sense: there was land, water, high visitor numbers and, at the Bassenthwaite site, farm buildings suitable for conversion. Also, like many tourist destinations in Britain (Scotland's west coast, Cornwall), the Lakes region has high unemployment. A distillery, complete with restaurant, brings much-needed year-round jobs.

Its first whisky maker, Dhavall Gandhi, instigated a complex flavour-creation process by approaching oak as a range of flavours and textures. In 2023, Sarah Burgess (former lead whisky maker at The Macallan – see page 54) took over.

She has now expanded the flavour matrix further. Ferments remain long, but the existing mix of yeasts has increased. The two stills have copper shell and tube condensers, but she is also using the stainless-steel condenser on the wash still to add a meaty new make to the fruit-driven 'old' style.

THE LAKES

THE LAKES THE WHISKYMAKER'S EDITIONS DECADENCE 49%

Plum pudding, light suet, flamed orange peel, almond butter and gooseberry fool. Soft tannins, heavily scented flowers, redcurrant and ginseng tea, praline. Complex and ordered.

RICH/DEEP

★ **THE LAKES SIGNATURE** (BENCH SAMPLE) 47%

Firmer wood with more malt. Stewing plums, black cherry and apple pie filling. The Lakes' lushness and depth: rowanberry jelly, rosehip syrup, light leather and balancing drying malt.

RICH/DEEP

Above *Washbacks and stills at The Lakes Distillery.*

Above *David Thompson in the fields that will give Spirit of Yorkshire.*

In addition to these changes, one month a year has also been set aside for the distillers to play further with all the variables.

The complex maturation regime remains. The Lakes is a believer in élevage: active involvement during maturation where maturing whisky is transferred into different types or ages of casks, expanding the flavour palette. The bulk of the Lakes' new make is matured in casks seasoned with seven different types of sherry, each of which has a role to play: manzanilla for aromatics, fino for creating space, amontillado for texture, palo cortado to link and add complexity, while cream corrals the flavours together. Unusually, oloroso and PX are here used discreetly to add a layer of dark fruits. The final blends are understanding of dynamics and nuance.

It was in 2011 that Tom Mellor, a farmer, and David Thompson, who worked for agricultural supplier Cargill, were standing at the bar at the former's Wold Top Brewery in Yorkshire. The conversation had turned to how growing barley was becoming less profitable – Tom was already using some of his crop for the brewery. 'Then we said together, "Why stop there?"' recalls David. 'Why can't we also make whisky?' Everything was already going up to Scotland to make whisky, so why not make it here instead?

The pair's **Spirit of Yorkshire** aims to frame that barley in the final product. 'There was an element of taking control,' says David, 'and getting the message across that barley isn't a commodity or a source of sugar. If you plant, grow and nurture it, it's more than that. We grow it using cover crops and direct drilling to protect the structure of the soil and retain carbon.'

To get the distillation side in order, they turned to renowned whisky consultant Dr Jim Swan. As well as fixing the parameters for flavour (mashing regime, long ferments, yeast selection – all done at the brewery), he suggested adding a small four-plate rectifying column to the Forsyth pots at the distillery 2 miles (3.2km) away. This would add a lighter (and more rapidly maturing) new make to the slightly heavier, oilier one from the two pots. As well as widening flavour opportunities, it allowed an already mature young whisky to be brought to market quickly – helping with cash flow. The first distillation started in 2016, with Filey Bay First Release appearing three years later.

With 'only' 100 hectares (250 acres) of the 263 hectares (650 acres) available at Tom's Hunmanby Grange Farm currently given over to barley, significant expansion is possible. Rye is now part of the cereal mix and a 51% rye/49% barley whisky is ageing, while the heavy and light new makes are now blended together.

'English whisky isn't one thing,' says David. 'We are all making our whiskies in a different way, but we're all pulling in the same direction.'

Inspiration in whisky can come in many ways and from many places. You might think that with a relatively well-established whisky-making country close by that Chris Jaume and Abbie Neilson would have turned

SPIRIT OF YORKSHIRE

FILEY BAY IPA FINISH BATCH #3 46%

Sweet oak, white flowers, pear drop, then the citrus burst of hops (balanced well), freshly cut grass and light pear. Akin to a cloudy IPA with more fresh barley sweetness.

FRAGRANT/FLORAL

FILEY BAY PEATED FINISH BATCH #3 46%

Has the distillery's lightness of touch. Green malt/barley chaff, some American oak and a bready back note. Distant campfire on the back alongside hot gorse.

SWEET/VANILLA

north for inspiration when they decided to start their **Cooper King Distillery** 12 miles (19km) north of York, but it came from the other side of the world.

When Abbie completed her PhD in Biomedical Science, the couple decided that they needed a complete break from academia (Chris had qualified as an architect) and hit the backpacking trail. When in Tasmania, they visited Peter Bignell's Belgrove (see page 320) and were bitten by the distilling bug. 'From that first visit I knew I wanted a piece of this,' says Abbie. 'The thing was, in the UK to progress in distilling you first needed a degree and the last thing either of us wanted was to go back to Uni.' Instead they trained at Redlands Distillery in Tasmania with Dean Jackson before returning home to Yorkshire and starting to raise money.

A crowdfunding campaign helped to persuade the bank that a distillery was a sound investment and the new make began to flow in 2016. The distillery was net zero from the start.

There are long ferments, while distillation is in Knapp Lewer stills. 'We loved that apple, fruit-forward style you get from these stills,' says Chris. Maris Otter is the chosen barley and maturation is in 100-litre (26.4-US-gallon) casks. 'It's all governed by flavour, not yield,' he adds, with a Tassy approach to problem solving also in the mix. 'We still say if we come up against something, "Can we be Bignell about this"', says Abbie.

The first release came in autumn 2023 and £5 from each bottle goes to support rural apprenticeships in the Yorkshire Dales. 'We're still backpackers,' says Abbie. 'We're still as excited about tasting new whisky as we were then.' Except now they're making it.

THE MIDLANDS

Max and Claire Vaughan moved back to Derbyshire's Peak District in 2011. Keen to create something that reflected the location around the Derwent Valley, distilling seemed to be a logical step and in line with the area's intriguing mix of an industrial heritage within a rural setting. When they saw the old Johnson & Nephew wire works in Ambergate, they knew they had found the home for their **White Peak Distillery**. The river water that had once cooled the wires could now be used to turn vapour back to liquid. Different alchemy. Their first Wire Works whisky arrived in 2022.

There's a sense of freedom in the mindset of England's whisky makers. Here is a fusion of disparate strands of Derbyshire history, old ways of making whisky and innovations. The barley from Warminster Maltings is unpeated; that from Crisp Malt is peated. They're usually blended together (80:20) to give a light peated result, but the possibilities of many other base starting points are there.

Yeast comes from the local Thornbridge Brewery, helping to promote estery fruits. Max has

COOPER KING

COOPER KING FIRST EDITION FRUIT + SPICE 48.1%

Well named. Spiced citrus, cassia bark, light clove, with a sweet mash central core alongside pear skin, mirabelle plums in syrup and soft toffee. A great start.

FRUITY/SOFT

Above *Checking the wash at Cooper King.*

tweaked the rakes on the mash tun to get the right wort clarity, the washbacks are insulated because of the six-day ferments, while the spirit still has an elongated neck and upward-facing lyne arm, but cuts are adjusted depending on what style is being made.

There are ex-Bourbon casks, but also STR, some rum (their own) and, for Necessary Evil, PX casks that have held Thornbridge's 13% stout. In all, the White Peak character is there, gently smoked, refined with a precision that nods to the Nordics, but with slightly bolder fruits. A leading light.

Many distilleries are examining heritage grains but none in quite the same way as **Oxford Artisan** and archaeobotanist John Letts, who combed gene banks to reintroduce landrace populations of wheat, rye, barley and corn adapted to low-input methods. Some were found locally in thatches of old cottages. Einkorn, emmer, spelt and oats are also grown, either individually or as field blends.

The crop is widely distributed geographically not only to spread risk but, says master distiller Francisco 'Chico'

WHITE PEAK DISTILLERY

★ **WIRE WORKS ALTER EGO** 51.5%

Deep, rich fruits and some roasted, toffee-like malt. Spice cupboard (ras el hanout), then dried mango and apricot, milk chocolate-covered hazelnut and smoke. Precise.

RICH/DEEP

★ **WIRE WORKS CADURO** 46.8%

More lifted and estery. Green banana, pear drop, sweet tangerine, light vanilla, honey and sweet mash. Lightly oily feel. Touch of smoke and caramelized malt. Balanced.

FRAGRANT/FLORAL

Above *The new Cotswolds distillery opening its doors.*

Rosa, 'to develop hardy, resilient and locally adapted crops that we need in a rapidly changing climate'. It's a major project with trial plots and new agricultural methods or adapted ancient ones, all grown with minimal tillage and no inputs.

For some whiskies, the mash bill starts in the field. 'For the field blends, John takes 30–100 varieties of grain and grows them together, similar to what would have been seen in medieval times, but with science backing it up. We're allowing nature to translate the sense of a specific place in a bottle.' In a radical move, in 2024 the distillery was renamed Fielden. It closed the old site and reopened a new one, complete with a mash filter, and Kothe hybrid stills. The new Fielden whisky is a 90% maslin blend of rye and wheat grown together with 10% malted barley.

Further variants are planned: malted rye, malted barley and unmalted rye and wheat as well as those speciality grains listed earlier and mixed populations like dredge (barley and wheat) and maslin (rye and wheat). Is it a new distillery or an adapted old one? Is your great-grandfather's axe that has had five shafts and three heads still the same axe?

Cotswolds, which was founded in 2014, doubled capacity in 2022 with a new distillery four times the size of the original, making it the largest in England. Set in the middle of a barley-growing area, Cotswolds' supply comes from 15 minutes away and is floor malted at nearby Warminster Maltings.

It's another where the hand of the late Dr Jim Swan is seen, but the opening of the new distillery has brought a subtle shift. 'Our new make has always been fruity and floral,' says founder Dan Szor. 'In the new distillery we're making a slightly more full-bodied, grain-forward whisky thanks to the better conversion from our new mash tun.'

The ferments are unchanged at four days, as is the use of Anchor and Fermentis yeast, while the cut remains high. Cotswolds, like so many of Jim's whiskies, was crafted to hit maturity early. Now with extra weight, Dan feels, it will develop for longer. With a spread of casks – ex-Bourbon, sherry, ex-Islay and 'pressure-seasoned exotic wine casks from Alter Oak', there's also now refill coming into the mix.

Dan quit his work in the City of London to start this new career 'as an excuse to be able to stay here full time and not go

OXFORD ARTISAN

OXFORD RYE 2018 HARVEST 50%

A highly scented and fruit-filled rye. Hibiscus, sloe, light lavender and sage flower, then dark chocolate and mint cream. Hot tiles. Moves to medicinal side, then blackberry yogurt.

RICH/DEEP

OXFORD RED RED RYE BATCH #5 46.4%

Cherry (cough syrup) fruits and light black pepper. Sweet and slightly jammy, but the palate is structured. Black grape (juice, skin, stalk), chocolate and more rye spice.

FRUITY/SOFT

COTSWOLDS

★ COTSWOLDS SIGNATURE SINGLE MALT 46%

Sweet. Mulled fruits, butter cookies, orange blossom honey, kumquat and firm (new) wood with nutmeg, cinnamon, Turkish Delight, lavender/rosemary and toasted barley/praline.

FRUITY/SOFT

COTSWOLDS FOUNDER'S CHOICE SINGLE MALT 59.1%

Off-dry. Cooked fruit and some cooperage toastiness. Nut bowl, roast chestnut, acorn and hessian, lavender, nettle, light banana and sweet pear. A serious side.

FRUITY/SOFT

home to London at the end of the weekend! It's like going back in time, to a nicer and more beautiful place.'

There is something prelapsarian within whisky, a notion that this is something that's natural and timeless, a sense that, for all the equipment and talk of process, it is at heart simple, innocent.

The pioneer is often in a lonely place. When the first bottlings from the **English Whisky Company** in Norfolk appeared, it was a tough sell in a Scotch-centric world. Even now, with so many new arrivals, what they have achieved is easily overlooked.

It remains owned by the Nelstrop family, who have been farming since the 1300s, though it wasn't until 2007 that James Nelstrop decided to start making whisky from some of his crop. His son Andrew is now chairman. They continue to question what English whisky could be. The core release is aged in ex-Bourbon, but there's also a smoked variant, triple distilled and a 'single grain' (actually a complex mash bill), as well as an English take on Bourbon called Popcorn.

LONDON

London's last whisky distillery Lea Valley was large, something that couldn't be said about **Bimber** (the name is Polish for 'moonshine') which, since its founding in 2016, has been wedged in a maze of car-repair shops opposite an industrial laundry.

This is a hands-on, sleeves-rolled-up kind of place. The mill doesn't grind the malt but cracks the husk in order to get a clearer wort. The American oak washbacks (made at the distillery's cooperage) are open-topped and charred on the inside to open up the surface, giving more crevices for lactobacillus to lurk.

All of Bimber's barley comes from the Fordham & Allen farm in Hampshire and is floor-malted at Warminster. Half a tonne (0.6 US ton) a time is mashed with filtered London water – 'mineral-rich and good for fermentation' – before those washbacks get to work with a seven-day ferment with two yeasts, plus time for the lactobacillus to get to work. Distillation is in a pair of 1,000-litre (264-US-gallon), direct-fired Hoga stills that have been modified to add reflux.

The new make is bright and fruity with a pleasing fatness to the centre – perhaps from that direct fire. Ageing is primarily in ex-Bourbon, but there is an examination of sherries, beer, port etc. The fruits run through all of them with a racy acidity that balances the weight. Another variant on this thing called 'English whisky'.

In 2015, Alex Wolpert left the bar world to found the **East London Liquor Company** (ELLC) in a former glue factory in Bow Wharf. As the name suggests, whisky is only part of the portfolio – its versatile Holstein stills (a pot for wash and hybrid for the spirit run) allow a wide range of spirits to be made.

The barley, rye and wheat for the whiskies are malted by Crisp, and ferments are allowed to extend to 140 hours followed by resting 'to encourage funky, tropical notes'. There's also a library of yeasts to draw from, and a wide range of wood types. As well as oak from USA, France and Hungary you will find acacia, chestnut and mulberry, and also previous fills of wine, whisky and brandies. The key is creating as many options for blending as possible – after all, this distillery was set up by bartenders.

The single malt uses three yeasts including Belgian saison and Abbaye, and is aged in rejuvenated casks, new American oak and ex-Bourbon, while the rye (55% malted rye, 45% malt) also uses those Belgian yeasts to drive up ester levels. It's then aged in cask seasoned with ELLC's own brandy, as well as the same mix as the single malt. That freedom of thinking at work again.

Above *The East London Liquor Company are relentless experimenters.*

BIMBER

BIMBER EX-BOURBON OAK CASK BATCH NO. 4 51.2%

Light canned apricot (with juice), custard tart and raspberry. A thick mid-palate feel with some poached pear, and a light, balancing, bitter touch on the end.

FRUITY/SOFT

BIMBER PEATED CASK BATCH NO. 1 54.1%

Green, vegetal, bay leaf/bay rum/bog myrtle. A sweet start to the palate with fat fruits, pineapple and a roasted malt element. Juicy Fruit chewing gum.

SPICY/HERBAL

EAST LONDON LIQUOR CO.

EAST LONDON LIQUOR CO. SINGLE MALT 47%

Immediate and upfront. Green apple, pineapple, banana and ginger and honey. Sultana loaf on the palate adding balance, while citric spice enters at the end.

FRAGRANT/FLORAL

EAST LONDON LIQUOR CO. SINGLE RYE WHISKY (WHISKY SHOW 2023) 62.1%

On the nuttier side of rye to start, then into dried red fruits and chocolate maltiness. Touch of roasting coffee bean. Bittersweet.

SPICY/HERBAL

WALES

The first legal Welsh whisky came from Frongoch village, near Bala town, Gwynedd, whose distillery opened in 1890. Its pots could produce 700,000 litres (185,000 US gallons) a year, but like so many it folded in the downturn of the early 20th century. It wouldn't be until 1975 when Dafydd Gittins and Mal Morgan began blending Scotch in a basement in Brecon (much to the annoyance of the Scotch Whisky Association) that whisky was returned to Wales.

Dafydd and Mal planned to distil and procured a still, but when their plans fell through, the enterprise was then acquired by the owners of what was to become Wales's first new single malt producer, **Penderyn**. With a site in the foothills of the Brecon Beacons, they hired Dr Jim Swan and got to work.

Designed by Dr David Faraday, their still comprises a pot with rectifying plates in the neck, leading to another column with more. The new make, drawn from the seventh plate on the second column, comes off at 92% ABV. Jim's advice was to age this light new make in ex-Bourbon, then finish in Madeira to add a sweet-fruited depth. The first whisky came out on St David's Day, 2004.

A second set of Faradays was installed in 2013 and a pair of standard pots a year later. The pots allowed new flavour streams to be introduced. The wash distillation either starts in the Faraday, or one of the pots, with the second run in the second pot. The result is a new make at 74% ABV. Penderyn Hiraeth is a blend of the two still types (75% from pots, 25% Faraday).

The Brecon site is in a national park and couldn't be expanded, so in 2021, a second distillery was built in Llandudno. This also has two Faradays, but only runs peated malt. Even that wasn't enough, and in 2023 Penderyn's third site, Copperworks in Swansea, opened with a pair of pot stills.

Welsh whisky was granted a UK Geographical Indication (UKGI) in 2023, which defines it as being 100% produced in Wales in a batch process with a 'lightness' of character.

PENDERYN

PENDERYN MADEIRA FINISH 46%

Scented. Light Madeira bittersweet notes, buttery cheesecake base with blueberries on top and a balancing singed back note. Apricot jam, hyacinths and bluebells.

FRUITY/SOFT

Above *Penderyn's third distillery is housed in a former Swansea copperworks.*

Above *Under the stills at Penderyn.*

OTHER WELSH DISTILLERIES

Cardigan's **In the Welsh Wind** was started in 2018 by Alex and Ellen Jungmayr. They use iStills, in which everything from mashing to distillation happens in the same vessel. There are three different sizes in the distillery, with the whisky coming from a 1,500-litre (396.3-US-gallon) and a 5,000-litre (1,321-US-gallon) one. Malting is also on site, using a mincing machine as a mill.

Dafydd Thomas began **Anglesey Môn** in 2017 using a wash from his family's barley and eight tiny stills with stainless-steel bodies, copper domes and lyne arms. Each produces 4 litres (1 US gallon) per run.

Lladarog's ancient White Hart Inn is home to **Coles**, which is distilled in a 2,500-litre (660.4-US-gallon) still that publican/distiller Cain Cole designed himself. It's another hybrid with a pot linked to a pair of columns.

In 2012, Penderyn was joined by **Dà Mhìle**, though founder, farmer John Savage-Onstwedder, already had experience in whisky. An organic producer, it was his barley that was used for Springbank (see page 83) in 1992, the first organic single malt Scotch whisky of the modern era. He also supplied to Loch Lomond (see page 81) for its organic single malt and the world's first organic grain whisky. It was perhaps inevitable that he would also go solo. 'Sustainability has always been at the heart of the company,' he says. 'Organic farming practices will be very important for soil health and to maintain the environment and wildlife.'

Today, two styles are distilled in a hybrid Kothe still with six plates. Single malt is matured only in fortified wine casks, and a 'single grain' (from a barley, wheat and rye mash bill) is aged in ex-Bourbon casks. However, he has further plans to work with heritage Welsh grains and create new lines – as well as building the company's own maltings. Still pioneering.

These two distilleries were joined in 2017 by **Aber Falls**, part of Halewood Distillers, whose pair of pots from McMillan have all sorts of bells and whistles attached. The spirit still has a cooling ring similar to that at Fettercairn (see page 72) and there are also stainless-steel and copper condensers. This allows light, medium and heavy to be made, although it's not clear how the last will fit in with the 'lightness' of the UKGI.

Above *Dà Mhìle is a pioneer of organics.*

EUROPE

How can you encapsulate an entire continent's whisky making? While there may be loose connections within each country or region in terms of grains used, the impact of climate and distillation technique that can impact on flavour, it's the distillery and the distiller and their interpretation of making something of their place that's the most important and endlessly fascinating thing about whisky. European whisky? There are whiskies made in Europe for sure. Are they all similar? No. Good.

Whisky is about finding your roots. In the case of many European distillers, they are to be found not only in using what grows around them but in the distillation of other spirits and how that muscle memory influences their whisky making. We will see it in France with the different approaches being taken by the Cognaçais, Breton and Alsacien distillers, but this thinking extends over the whole of the continent.

Distillers in the belt that stretches across Germany, Austria, Switzerland and northern Italy have had centuries of experience distilling fruit spirits. Now whisky is building on that expertise – the astonishing number of distilleries in Germany (estimated to top 300) that make whisky at some stage in the year is witness to that.

There's a shared approach of taking the technique of accessing the essence of each fruit and applying it to grain. Because of its climate, Europe's distillers can draw on a wider range of cereals, not just barley and wheat but corn, spelt, einkorn, emmer, oats and, of course, rye. Its wine and beer industries both offer areas for exploration – casks for maturation but also techniques for fermentation and malting. A strong organic movement is another notable trend particularly in France and Germany.

Rye itself has been making waves in the Netherlands and Belgium where genever distillers have moved into whisky making. This is less of a new departure and more of a reclamation of an ancient heritage. Many of America's early distillers were émigrés from the Low Countries who brought rye with them. The whiskies they made in their new home replicated those that they made in their old one.

Until recently European rye whisky has existed as a constituent element in genever – itself the missing link between whisky and gin. Rye has also had a central role in the new whiskies emerging from the Nordics. This time, the inspiration is coming not from spirits but from food culture. Folk roots, new routes.

Right *European distillers are heading off on new tracks as they embrace whisky.*

FRANCE

It's no surprise that France makes whisky. The country is, after all, continually vying with the USA as the biggest importer of Scotch in the world. Over 20,000 people visit Whisky Live in Paris every year – the largest whisky show globally. What is surprising, though, is that it's only in the past decade that France has started producing whisky in any volume, but it's now making up for lost time.

At the time of writing, there are over 100 distilleries in the country making whisky. Some are dedicated solely to the spirit, while others have diversified their existing distilling ventures (fruit spirits, Calvados, Cognac) to include it in their portfolios.

When speaking with France's new whisky distillers, the parallels with the thinking of the country's new generation of winemakers is clear – a move to organic, exploring old varieties and older ways of making; a rediscovery of land and expertise.

BRITTANY

With a different language and cultural identity closer to Scotland and Ireland, it's no wonder that single malt whisky has taken root in Brittany. The pioneer was Gilles Leizour, who began distilling whisky at his **Warenghem** distillery in Lannion in 1983, making it not only the first whisky distillery in Brittany but in the whole of France.

His decision is one mirrored by many of France's new distillers – the need to diversify from a declining core business, in his case liqueurs, mead and apple brandy, that the distillery had produced since 1900. Gilles's son-in-law David Roussier is now in charge.

Using the existing stills, their blended whisky, WB, was launched in 1987. A purpose-built distillery with Scottish-style pots made in the Charente was built in 1994 and **Armorik** single malt appeared four years later.

The most significant change took place in 2010 with the hiring of esteemed whisky consultant Dr Jim Swan, marking a shift from a supermarket brand to premium. 'We needed to make better whisky to be able to start exporting,' David says candidly. 'Clear wort, different yeast and a tighter cut resulted in extra fruits. The richness came from the shape of the stills, slow distillation and longer fermentation.'

Armorik is now made from unpeated organic French malt (peated is run once a year for Yeun Elez), an increasing amount sourced locally,

Above *There are many twists and turns in the story of French whisky.*

Above *Brittany's Armorik – the starting point.*

with wheat and rye for the column-distilled grain. Ageing is mainly in ex-Bourbon casks, with some oloroso and STR, and the remainder in Breton oak.

It begs the question, are they making Scotch in Brittany? 'We were inspired by Scotland, but our "Bretonness" can be found in the local barley, our own cooperage making casks from Breton oak – and the weather,' says David. 'We try to ensure that Armorik is a great single malt, and then twist it to be a bit more Breton. I think who we are, and how we think as Bretons, probably influences more than location: being authentic, down to earth, probably a bit more pragmatic than the rest of France!'

It was in 2016 that Stéphane Kerdodé began his **La Mine d'Or** distillery in Ploërmel close to the Brocéliande Forest, famed in Arthurian legend (the distillery's whisky, Galăad, is Breton for Galahad).

He uses unpeated French malted barley, splitting the wort between stainless-steel (five to six days of fermentation) and French oak washbacks (four days). There are also two designs of still running wash and spirit: a direct-fired lamp glass pot with long, thin downward lyne arm and worm that, after 14 hours for the spirit, gives a fruity/oily new make.

The second still is a Stupfler hybrid pot and linked column that runs in a single pass into a worm. The top of the column can be cooled to create reflux. The result is a lighter fruity/sweet new make.

Ageing is in damp cellars to encourage slow maturation and takes place in new French oak casks that have been heated in three ways to give different levels of toast. The duration of the toasting also varies, producing further flavours. There are also ex-Cognac

ARMORIK

ARMORIK CLASSIC 46%

Sugared almond, sweet spice, yellow wine gums, agave. Palate shows prickly pear, green banana, acorn, bookbinder glue, then violet flower. Broad-ranging.

FRUITY/SOFT

★ SINGLE CASK ARMORIK DEIZ 2023 53.8%

Sherried, thick fruit paste, medlar becoming meaty with sweet dark fruits. Coffee butter icing, then raisin, mixed pepper, overripe banana and light orange peel. Savoury.

RICH/DEEP

LA MINE D'OR

GALĂAD À LA TABLE DES CHEFS 47.1%

Needle sharp. Dry and malty, inner tube, then lemon shortbread, dried flowers, custard. Palate has light spice, blood orange and demerara sugar. Good feel.

FRAGRANT/FLORAL

GALĂAD ORIGINE MILLÉSIME 18 44.5%

Soft, perfumed. Orange blossom, green grass, rhubarb and custard, cooked fruits, nougat and cinnamon. Rounded.

FRUITY/SOFT

barrels, as well as acacia, chestnut and cherry. Reduction to bottling strength takes place slowly in the cask.

Jean Donnay was another of Brittany's whisky pioneers who, in 1997, founded his Glann ar Mor distillery in Pleubian on the well-named Côte Sauvage. It's been renamed **Celtic Whisky Distillerie,** with Glann ar Mor as one of the brands.

Jean was a traditionalist whose vision has been maintained by distiller Aël Guégan. The barley remains Maris Otter, the washbacks are wooden and the pots are small, plain and heated with flame, and run slow with condensing in worms. Maturation is predominantly in ex-Bourbon casks.

That seashore location gives an undoubted salinity to the unpeated Glann ar Mor and peated Kornog Roc'h Hir. In 2020, the distillery

FRANCE

FLEMISH BORDER/HAUTS-DE-FRANCE/PARIS/NORMANDY

The **T.O.S.** distillery in Aix-Noulette first fired up its stills in 2017 through the wall (T.O.S. stands for 'the other side') of the Brasserie Saint Germain. For the firm's Artesia whisky, distiller Katy Gravina uses a Holstein hybrid to distil a wash made from local malt that's been given an eight-day ferment. With the purchase of the old Wambrechies distillery near Lille, more experiments with grain are promised.

Étienne d'Hautefeuille started distilling whisky on his family farm in Picardy in 2017. Using his own barley, the spirit is distilled in a Stupfler hybrid.

Distillerie de Paris, from brothers Nicolas and Sébastien Julhès, began making whisky in 2015. Using the versatility of a Holstein hybrid, a variety of malts is employed as are types of oaks (and other woods), with extreme elevage producing layered and often herbal results.

In Normandy, noted Calvados producers **Château de Breuil** and **Maison Busnel** have both embarked on a whisky path.

Above *The menhirs of Eddu's distillery.*

was purchased by Maison Villevert, which has added a 'Celtic blend' Gwalarn (comprising Kornog, Scotch and German whiskies) to the range.

The late Guy Le Lay dug even deeper into Breton traditions when, in 1986, he began to make whisky at **Distillerie des Menhirs** in Plomelin. The family had been distilling *lambig* (apple brandy) and making pommeau since the 1920s, but with sales on the slide he had the brainwave of making a Breton whisky with buckwheat – the grain with the closest ties to Breton cuisine. His three sons Kévin, Erwan and Loig now run the distillery.

Buckwheat isn't easy to distil. The yields are catastrophically low, as is the strength of the wash. It can turn to concrete in the mash tun, and burn in the wood-fired still (itself a fusion between Scottish pot and Charentais). However, Guy persevered for 15 years, and in 1999, Eddu appeared. It's a remarkable whisky, one that challenges your expectations of what this spirit can be. With 15 farmers now supplying requirements, and 21-year-old Graal on release, all the hard work has paid off.

DISTILLERIE DES MENHIRS

★ EDDU SILVER 43%

Beech leaf buds, wakame, savoury...buckwheat! Tarragon, green mango, antiseptic. Complex. Dry with coconut, light acids, apricot and almost rye-like oiliness. A must-try.

SPICY/HERBAL

★ EDDU GRAAL 21 YEAR OLD 43%

Mature. Orchid, baked apricot, light *rancio*. Lush and elegant with honeyed tropical fruits, then spices fried in brown butter. Mango, pineapple and buckwheat's drive.

RICH/DEEP

ALSACE

Rivalling Brittany as a centre of whisky production, the distillers of Alsace, such as the Bertrand family of Uberach, have drawn from centuries of tradition making fruit spirits. Whisky became part of the **Distillerie Bertrand** portfolio at the start of the 2000s, with the Uberach single malt brand appearing in 2003.

Alsace, like Brittany, has an IGP (Protected Geographical Indication). To qualify, the whisky can only be made from non-GMO malted barley, water and yeast, and aged in oak casks of less than 700 litres (185 US gallons) for a minimum of three years in unheated cellars. Other woods can be used after three years.

Both Uberach and the organic St Wendelin single malt are distilled in a Holstein still – for Uberach two passes with plates engaged, for St Wendelin on the second pass they are disengaged. Uberach is aged in ex-Banyuls casks, while St Wendelin is in new French oak and is slowly reduced to 46% ABV.

With the St Wedelin Souffle collection, the endlessly experimental mind of master distiller Jean Metzger explores the possibilities of using local wine casks, Le Souffle Calcaire given secondary maturation in Pinot Gris Grand Cru Vendanges Tardives. Another of his creations is Biersky, a blend of an aged beer eau-de-vie with a malt spirit, which ties together

T.O.S.

★ ARTESIA FÛT NOIR CHAR #3 46%

Graceful. Apple purée, night-scented flowers, osmanthus, lily of the valley and a little hint of sap and light char. Delicate.

FRAGRANT/FLORAL

Above *Miclo has embraced whisky in its many forms.*

the fruit spirit's aim of capturing the essence of a single ingredient with the more expansive world of malt. All embed the distillery in its surroundings and distilling culture.

The family-owned **Miclo** distillery is located in Lapoutroie, in Alsace's Welche country, which also gives its name to its whisky. It's another that has diversified from eau-de-vie, beginning to distil whisky in 2011 in collaboration with the local Meteor brewery.

Unpeated, smoked and peated single malts all arrived in 2016. Distilled in Holstein column stills, Welche is aged primarily in ex-wine casks (Sauternes, red and white Burgundy). 'Our cellar master worked in wine before and always tries to think about the combination of whisky and wine,' says Bertrand Lutt-Miclo. Using the flexibility of the IGP, Miclo's limited edition series has included an expression finished for 18 months in barrels that had previously contained cherries macerated in brandy.

WELCHE'S SINGLE MALT (EX-SAUTERNES) 43%

Starts with barley, then thickens. Honeyed, apricot fruits. Botrytis adds a layer of complexity. Light mineral, chamomile tea, whitecurrant jelly. Soft and clinging.

FRUITY/SOFT

★ WELCHE'S SINGLE MALT FUMÉ 43%

Gentle smoke, slow wood fire. Bold but balanced. Cedar, pine, sour cherry. Luscious and soft. Integrated gentle red fruits.

SMOKY/PEATY

The **Lehmann** family started distilling eau-de-vie in 1850 and their distillery is now run by the sixth generation. Its venture into whisky, Elsass, is double distilled in Charentais stills before being aged for seven years in Bordeaux white wine casks. The Gold variant has an extra year in Sauternes, while Premium is just in Sauternes for eight years.

CENTRAL/EAST

Craft brewer Christophe Fargier decided to start making whisky in the town of Tarare (Monts de Beaujolais) in 2015 having founded Lyon's **Ninkasi** brewery in 1997. The name comes from the Sumerian goddess of brewing. Now with Alban Perret in charge of two Charentais and a pair of vacuum stills, Ninkasi's main line is a blend of peated and unpeated malted barleys (a 50:50 blend for the peated expression). Alban uses new casks followed by a secondary maturation in a mix of ex-wine casks, which he then blends together in different proportions. Slow reduction takes place in the cask.

NINKASI

NINKASI CHARDONNAY 46%

Fine-boned. Low wood, toasted malt, bread crust and some caramel. Becomes vinous, leesy. Lemon cream, cooked fruits. Light grass and cultured butter.

FRAGRANT/FLORAL

Famed for its quality knives, the town of Laguiole is also now home to the enigmatically named **Twelve** (it was founded in 2014 by 12 friends). The distillery is housed in the 300-year-old Presbytère des Anges with maturation taking place in its former convent.

At 1,000 metres (3,280 feet), the Aubrac plateau is a hard place to grow barley, so some is sourced from Tarn and Aveyron and distilled in a single pass in a Stupfler still. The whiskies take their names from geology and semi-precious stones, and utilize a wide range of casks. Basalte, for example, is ex-Burgundian Pinot Noir, Pineau de Charentes, Monbazillac, rhum and new wood. Casks are also split between humid and dry cellars – the latter giving quicker maturation.

The Grallet family has been farming in the Meurthe-et-Moselle village of **Rozelieures** since 1877. They began making fruit spirits, most specifically with mirabelle plums, since 1890. Whisky is a far more recent arrival, with production starting in 2007 using barley from its 121 hectares (300 acres) of cereal crops. It was the braincild of Hubert Grallet's daughter Sabine and her whisky-loving husband Christophe Dupic who, usefully, is also an engineer.

A brewhouse was built, followed by the building of Malterie des Hautes-Vosges capable of processing (and peating) 1,000 tonnes (1,102 US tons) a year. As peat extraction is banned in France, the fuel comes from either Scotland or Poland. As a result, other smokes are being trialled.

Barley is key to the thinking. 'As grain farmers, we're convinced that our terroir has a determining impact on the whiskies,' says marketing manager Jonas Vallat. 'We have several types of soils on the farm: volcanic, silty, clay, clay-limestone, and have found that the whiskies express themselves differently depending on each type.' A series of plot-based whiskies is now part of the range.

Distillation is in steam-heated Charentais stills and the whiskies have the silky texture typical of the design. A wide range of casks is used: new oak from local forests, ex-Bourbon, sherry, port, rum, Cognac, Pineau des Charentes and wine.

TWELVE

TWELVE AVENTURINE 48%

Silage, wet grass and barley. Sweet porridge, flinty/mineral, then hedgerow berries, amber malt. Relaxed with light spice, lemon pepper, poached pear and red fruits. Gentle.

FRUITY/SOFT

★ TWELVE BLUE ONYX 57%

Restrained. Light woodsmoke. A walk in the country in autumn, distant burning logs and, oddly, pineapple mint, sage, pine/juniper, then toasted cereal. Elegant.

SMOKY/PEATY

ROZELIEURES

ROZELIEURES SINGLE MALT EX-BOURBON 46%

Golden syrup and a slightly resinous back note. Citrus, peach, caramelized fruits, butter. Palate has a herbal vermouth-like element and light coconut.

FRUITY/SOFT

ROZELIEURES SINGLE MALT EX-VOSNES ROMANÉE 46%

Intense, some cereal, stewing fruits, citrus peel, marmalade, marzipan, baking spice. Slightly phenolic. More vinous. Sour cherry for balance.

RICH/DEEP

Above *Rozelieures is a farm-to-glass producer.*

Above *Cellar work at Domaine des Hautes Glaces.*

CORSICAN WHISKY

Domaine Mavela launched its, and the island's, first whisky in 2005, a joint venture between the estate and the Pietra brewery – hence the name P&M. It has taken time to find the right grain for the climate and two whiskies are now made, one from a Corsica-friendly strain of corn, as well as barley. Ageing is in Muscat casks, giving the whisky intense herbal and botanical notes that take it into the island's hinterland.

The mention of terroir is never far away when speaking with a French distiller, but few have it playing such a defining role as **Domaine des Hautes Glaces**, which sits at 900 metres (2,953 feet) in Trièves. Founded by Fred Revol in 2009, it's now part of Rémy Cointreau's whisky portfolio, joining Bruichladdich (see page 103) and Westland (see page 241) – two other distilleries that share Fred's belief in whisky being expressive of place. For him, the raw material is central – these are not whiskies where oak will ever dominate.

Barley, rye, spelt and hull-less oats are grown, all old varieties from the late 19th and early 20th centuries. Trialled in a nursery in collaboration with INRAE (National Research Institute for Agriculture, Food and the Environment) and the CRB (Biological Research Centre), they are then grown by 19 local organic farmers on a no-till, rotation basis.

When required, smoking is done over chestnut, while fermentation is induced by wild yeasts trapped from the fields, which Fred says not only helps to express terroir but allows the distillers to ferment for longer and at lower temperatures. Like so much in the new French whisky, the links to new French wines are clear.

The stills are wood-fired and Charentais-style. Triple distillation is used for the 100% rye Vulson. Every aspect of production has been examined to work towards sustainability but also a maximizing of where the spirits come from.

Located close to the town of Saint-Jean-en-Royans in the Drôme region, **Distillerie du Vercors** is housed in a 16th-century fortified farm. Its stills were lit by Eric Cordelle and his wife Anne-Hélène in 2016. Ferments of organic barley (the Drôme is France's largest organic farming area) last four days and are distilled in a pair of stills.

The wash still (called 'Nautilus') is made from stainless steel and works under vacuum, distilling at 50°C (122°F). 'My idea was inspired from chemical and perfumery processes,' explains Eric, 'either to save energy and/or to preserve fragile molecules/fragrances. To manage the boiling temperature, you need to manage the distilling pressure and stainless steel is required to avoid collapse at low pressure.' After trials he built Nautilus – and then heard of how a similar design is used for shochu. A Charentais spirit still ('Beautiful') adds its gentle texture to the distillery's Séquoia brand, named after the tree that shades the farm.

DOMAINE DES HAUTES GLACES

★ DOMAINE DES HAUTES GLACES EPISTÉMÈ R18P23 47%

Greengage, crisp white paper. Cool with a mix of pepper, pear and earth. Woodsmoke. Green vegetal palate: celery, artichoke, bamboo shoot and Sichuan pepper.

FRUITY/SOFT

DOMAINE DES HAUTES GLACES VULSON 43%

Fresh fennel and dill along with rain on hot stones. Adobe. Becomes sweet, green capsicum, tomatillo. Complex, fresh. Great acidity.

SPICY/HERBAL

DISTILLERIE DU VERCORS

★ DISTILLERIE DU VERCORS SÉQUOIA SIGNATURE 43%

Fresh summer fruits. Clean, precise, chervil, parsley, lemon. Some cereal dust. Clean and restrained with lemon, fruit kernel and green strawberry. Poise.

FRAGRANT/FLORAL

★ DISTILLERIE DU VERCORS SÉQUOIA 10.23 44%

Slightly fuller. A light herbal element. Barley, cream, more floral. Viognier-like fleshiness. Whistle-clean. White peach, ginger, lipstick and strawberry chews.

FRUITY/SOFT

Above *Castan's remarkable still set-up.*

SOUTHWEST

Benoît Garcia founded his **BOWS** (Bold Occitan Wild Spirits) distillery in 2016. Originally located in Montauban, in 2021 he moved to a new site on a farm in 'a dry and windy climate in Laure Minervois, to get closer to my roots'.

All the barley (Fatima and Planet) is either organic or grown with minimal intervention, and he is further extending his thinking of what his (and French) whisky could be by planting sorghum, partly because 'it doesn't require much water, doesn't impoverish the soil and allows better rotation'.

While peated expressions were originally made, the sustainability issues over importing supplies have led him to start exploring other smokes. Using different roasts of barley and a mix of dried and wild yeasts, his ferments run from seven to ten days.

He is also using koji (see page 289) to reduce the amount of malted barley per mash, again for 'eco-responsible' reasons. Distillation is in two different stills: an all-copper hybrid, and a 25-litre (6.6-US-gallon) with stainless-steel outer and copper inner that runs under vacuum. Ageing takes place in wine, ex-Cognac and toasted new oak, and uses a slow reduction practice. It all adds up to deeply flavoured whiskies coming from deep thinking.

BOWS

★ BENLEOIC ORIGINAL 45%

The aroma has touches of oilcloth/new waxed jacket, but the palate switches to effusive fruits, almond extract, oily cereal and a peppery finish.

FRUITY/SOFT

CASTAN

★ VILANOVA TERROCITA 43%

Rich malt, toasted nut, enveloping caramelized weight. Dark fruits with autumnal depth. Envelope glue, crystalline minerality, soft flummery and light pear. Good balance.

RICH/DEEP

Another Occitan distillery, **Castan**, is in Villeneuve-Sur-Vère, 10 miles (16km) to the north of Albi. It was founded in 1941 by Gilbert Castan and his son Jacques, who would travel through the Tarn region with a mobile still at harvest time to make each farm's fruit spirit and marc.

With the practice outlawed, his grandson Sébastien and wife Céline re-examined the business and in 2010 started to produce whisky, the first in Tarn, named Vilanova. It's a self-sufficient operation with all the (organic) barley grown on the family farm.

A total of 50 hectares (124 acres) have been laid aside for old barley varieties, including some from Scotland. Ferments take five days, and distillation is in an adapted mobile still from the 1920s with a boiler, two linked cylindrical pots and a small rectification column. A single pass is all that's needed for a new make at 80% ABV. Another example of adaptation, evolution and new/old thinking that typifies the French approach.

COGNAC

Cognac's distillers are increasingly looking to add whisky to their production. With distillation of Cognac only permitted between harvest and the last day of the following March, it means that the region's stills lie silent for five months. With distilling expertise, cooperages and blenders on hand, whisky makes sense. In fact, you wonder why it's taken them so long.

Alfred Giraud started with his father as a cooper in 1910 before becoming cellar master at Rémy Martin until 1954, passing the role on to his son André, whose son Jean-Pierre and grandson Philippe founded **Alfred Giraud** as a whisky house in 2012, though it's whisky à la Cognac. The barley comes from Malterie des Hautes-Vosges, which the firm co-owns with Rozelieures. It's distilled at Distillerie de Saint-Palais, a joint venture between the Giraud and Nau families.

The use of wood-fired Charentais stills gives a sleek, elegant new make, but it's in maturation where the heritage of one spirit can be seen to have been adapted by another to produce something that's not just different but of its place.

Blending is under the control of Gaétan Mariolle, who in turn was trained by the firm's cellar master Georges Clot. Maturation is in a mix of old Cognac casks and new casks coopered from trees grown in the firm's forest.

A mix of open- and tight-grained oak is used – the former for oxidation and more tannin, the latter allowing gentler maturation. Some American oak and wine and

ALFRED GIRAUD

ALFRED GIRAUD HARMONIE 46.1%

Apricot and blossom. A touch of barley. Refined, almost fragile, but there's plump vanilla, golden syrup and fruits. Layered, silky texture. Complex.

FRUITY/SOFT

★ ALFRED GIRAUD VOYAGE 48%

Waxy/beeswax, medlar jelly, quince, pollen. Luscious and rich yet elegant. Fruit syrups, decadent, jasmine absolute oil. Long.

FRUITY/SOFT

French spirit casks are also used. Reduction is slow and done by nose rather than to a formula. It allows a wide palette of whiskies for Gaétan to blend with – ages, strengths, cask types. Horizon, for example, uses 14 different cask types. For some blends such as Harmonie, distillates from other producers are used. Maturing, blending and elévage, rather than ageing and finishing. The key is elegance, complexity and length.

Situated in the town of Jarnac is **Tessendier & Fils**. A long-established Cognac house, it first added whisky to its portfolio in the 1990s when sales of Cognac collapsed. What started with brothers Jérôme and Lilian blending Scotch became a French whisky distillery in 2019. Three years later their Arlett brand (named after their mother) was launched. Using malt from Les Maltiers d'Issoudun, it's distilled in gas-fired Charentais stills. Ageing is predominantly in ex-Bourbon, but mizunara explorations have started. Maturation takes place in three cellars, one dry (for fast maturation) and two damp (for slower).

Angeac's **Maison Boinaud** was another producer that started making whisky as a means of diversification. The largest family-owned wine estate in the Cognac region, it owns Cognac De Luze as well as supplying eau-de-vie to other houses. It was Rémi Boinaud who decided to broaden the remit, and as well as making its Hériose whisky, he works as a consultant and contract distiller.

Malting, mashing and fermentation is done by third parties to Boinaud's specifications (aroma, strength, ester levels and volume) and can be as short as 36 hours or extended to 72 hours.

Cognac distillers either distil on- or off-lees. Boinaud takes the former approach, which gives a more powerful eau-de-vie, and replicates this in its whisky by using an unfiltered wash. Distillation and maturation is on site, with the former taking place in Charentais stills, giving a new make that has touches of pears and a hint of malt.

Ageing is primarily in first-fill ex-Bourbon casks, with other casks made from French or American oak. No new oak is used in order to avoid the distillate being dominated by wood.

The Palisson brand comes from family-owned **Les Brûleries Modernes** in Brie-Sous-Archiac, whose Vinet-Delpech distillery was founded at the beginning of the 20th century, and was started up by Jean-Baptiste Delannoy in 2019.

A considerable producer of a range of spirits, it has 13 stills and global reach. 'We wanted to try our hand at distilling French whisky using the traditional method,' he says. Palisson is distilled in Charentais pots with Batch 01 aged in ex-Cognac casks, 02 in rum and 03 in ex-Bourbon. 'Whisky gives us more latitude than Cognac,' he adds.

Above *Tessendier & Fils is one of the producers using Cognac stills.*

'It was an opportunity to expand our list and sales in France and export. Consumers are looking for high-quality products whose traceability is obvious. French products enjoy a good reputation and French whisky has all the cards in hand to make a place for itself.'

Former ballet dancer Allison Parc was the first to take a Cognac-distilled whisky to market with her brand **Brenne,** which appeared in 2012. Using heritage barley and double distilled in Charentais stills, it's a fruity distillate that's matured in ex-Cognac and new Limousin oak. Fresh as a young expression, its complexities are revealed when it is allowed to rest in casks for a decade. Once part of the Samson & Surrey portfolio, in 2022 it was bought by Heaven Hill.

TESSENDIER & FILS

ARLETT SINGLE MALT ORIGINAL 45%
Dry and malty with liquorous sweetness behind. Proving dough, tayberry, oak, barberry and Victoria plum. Firm with a sweet finish.
RICH/DEEP

ARLETT SINGLE MALT TOURBÉ 43%
Smoke has a herbal/green bonfire quality. Lightly nutty, mineral notes. Young and feisty.
SMOKY/PEATY

MAISON BOINAUD

HÉRIOSE LE PETIT TOURBÉ 46%
Light white fruits. Delicate and estery. Pear, vine flower, fresh linen, pineapple cream, vanilla and a little drying catch of graphite on the palate.
FRAGRANT/FLORAL

HÉRIOSE LE CLASSIQUE 46%
Ripe. Russet apple, patisserie, lemon flower. Broad. Wholemeal bread with a light touch of compressed fruits and coriander seed.
FRUITY/SOFT

Above *The fermentation eggs at Maison Lineti.*

What might happen if a distillery adapted winemaking techniques (rather than those from brewing or other spirits) into its philosophy? The answer can be found in St-Émilion at **Maison Lineti** where the trio of Dr Magali Picard, her husband Alex Cosculluela and Xavier Payan are doing just that. The wort (from organic French barley) is fermented for a minimum of one week in concrete eggs that both regulate the temperature and create a convection current, allowing the lees (dead yeast cells) to give complexity. Some ferments have lactic added; others sit on lees for an extra fortnight.

Distillation, in two direct-fired Charentais stills, is complex. Heads from the wash and spirit distillations are recycled in the next run of the wash still. The tails from the second distillation are distilled slowly in one still, then recycled in the next spirit distillation. In the other still, they are distilled quickly and added to the wash distillation to contribute structure.

Magali's experience as an international expert in wood chemistry shows in a maturation regime encompassing 20 different modalities across oak type, toasting, char levels, size and location in the warehouse. Each year, the team uses a different variation in production, and each vintage will then be released in five increments to show the evolution of flavour. The whole approach is a significant new development in this new world of whisky.

'We've been brewing beer since Gallic times, and we've mastered the various forms of distillation,' says Aymeric **Roborel de Climens**. Another oenologist turned whisky maker, he uses sourced spirit from Alsace that he then matures in a variety of ex-wine casks in his *chais* in St-Émilion. The new make spends 30 months in a mix of 20% new French oak and 80% second-fill French oak barrels to give a base character. To finish comes a year in *grand cru* wine barrels – Sauvignon (ex-Graves), Merlot (St-Émilion) and Semillon (ex-Sauternes) – in a humid underground cellar.

There is also a wine connection at oenologist/brewer/distiller Philippe Laclie's **Bercloux**, home to his own brand as well as Bellevoye, owned by his partners Alexandre Sirech and Jean Moueix (of Château Pétrus fame).

There is no single French style, but many – 21st-century whisky is, after all, a diverse place. It's hard to avoid the thought, however, that in France there is a shared character of length and elegance. Is it the Charentais still? Is it yeast, the use of French oak...or might it be its mindset and the willingness to look at location and the country's wines and older spirits for guidance? Maybe it's all of these. The potential is huge.

LOIRE VALLEY DISTILLERY

Vincent Lelièvre and Yann Leroux's **La Piautre** distillery is in the Loire village of La Manitrié. Beginning in 2004 as a brewery, they branched into whisky (along with distiller Gilbert Boudier) in 2013.

They have their own malting facility with smokehouse where they can dry the barley over (imported) peat, and local beech and birch. They are also cultivating their own yeast. Distillation for their malt and rye whiskies is in four wood-fired Charentais stills, while maturation starts in new oak followed by a second period in either ex-Cognac, local white wine casks or beer casks from the brewery.

LA PIAUTRE

LA PIAUTRE TOURBÉ 46%

Woodsmoke, snuffed candle, old coin. Freshly baked sourdough and polished wood. Palate has pigskin leather, redwood needles and chestnut. Well balanced.

SMOKY/PEATY

BRENNE

BRENNE 10 YEAR OLD 48%

Light oak. Apples and walnut and some oxidized elements plus dried pineapple. Floral. More guts and mature distillate-driven rather than oak. Elegant.

FRUITY/SOFT

ROBOREL DE CLIMENS

A. ROBOREL DE CLIMENS FINITION SAUVIGNON 40%

Gunflint to start, then jasmine, passion fruit, scented. Light cereal. Delicate. Meadow flowers and herbal elements. A little gentian dryness.

FRUITY/SOFT

★ A. ROBOREL DE CLIMENS GRENACHE NOIR ÉPHÉMÈRE (EX-BANYULS) 50.2%

Vinous. A layer of perfumed fruits. Sultry dark berries, a mix of sweet and tart black fruits. Honeyed. Light grip. Wild herbs on the end.

FRUITY/SOFT

GERMANY

KEY
Malt Distillery
0 miles 100
0 km 100
N
SWEDEN
DENMARK
North Frisian Islands
East Frisian Islands
POLAND
NETHERLANDS
BELGIUM
LUXEMBOURG
FRANCE
CZECHIA
AUSTRIA
SWITZERLAND
LIECHTENSTEIN
GERMANY
Brauhaus am Lohberg, Wismar
Hamburg
E26
Müritz
Elbe
Preussische, Mark Landin
Grumsiner, Angermünde
Bremen
Lübbehusen
Ems
Weser
Aller
Oder
Mittelland Canal
Hannover
BERLIN
Glina
Potsdam
Stork Club, Schlepzig
Old Sandhill
E51
Bielefeld
Rhine
Hercynian, Walkenried
E45
Gelsenkirchen
Essen
Dortmund
Duisburg
Hillock/Heinrich Habbel
Märkische
Wuppertal
Düsseldorf
Leipzig
Saale
Cologne
Dresden
Zeitzer
E40
Birkenhof, Nistertal
E41
Ore Mountains
E44
Eifel, Koblenz
Höhler, Aarbergen
Frankfurt
Simon's Feinbrennerei
Henrich
Wiesbaden
Moselle
Main
Stonewood/ Schraml, Erbendorf
Würzburg
Ziegler
St. Kilian
Blau Maus
E45
Bohemian Forest
Elch
Mannheim
Neckar
Ayrer's/Hausbrauerei Altstadthof, Nuremberg
Drexler, Arrach
E50
Bayerwald Bäwurz/ Brennerei Liebl
E35
Rieger & Hofmeister
E56
Danube
Stuttgart
Schwarzwaldbrennerei Seeger
Tecker, Owen
Bellerhof
Wild Kinzigbrennerei
Finch
Augsburg
E52
Fitzke, Herbolzheim
Feller
Munich
Black Forest
Steinhauser, Kressbronn
Slyrs
Lake Constance

GERMANY

Given the robustness of its single malt Scotch-loving community, it seems logical that Germany has over 150 distilleries where whisky is either exclusive or a significant element. However, rather than following a Scottish template, most of its distillers have adapted techniques from producing fruit spirits and brewing. The country's strong organic following has also had a marked influence on how its leading distillers are approaching whisky making.

One example is the Stetter family. Fruit spirits had been made in their Lantenhammer distillery since 1928, but after Florian Stetter's trip to Scotland in 1995, he was inspired to try his hand at whisky. Four years later, he distilled the first **Slyrs** whisky. Production, using Bavarian grain, is now on a dedicated site in Schliersee. The climate, with its hot summers and extremely cold winters, increases the intensity of what is already a full-bodied whisky. To demonstrate the impact of climate, the beech-smoked Mountain Edition is matured near the summit of Mount Stümpfling.

A more classical approach has been taken by **St. Kilian** in Rüdenau. A collaboration between whisky lover Andreas Thümmler, David Hynes (ex-MD of Cooley in Ireland – see page 132) and master brewer/distiller Mario Rudolf, its whisky is fermented in wooden washbacks for around 70 hours and distilled in two pots from Forsyths. Mario's brewing background emerges in the range of speciality malts used: Pilsner, organic Heimat Malt, two-row Barke, Munich, Viennese, crystal and caramelized.

Above *Revealing the treasures at St. Kilian.*

ST. KILIAN

ST. KILIAN CLASSIC MILD & FRUITY 46%

Light, discreet. Sweetened apple, melon, quince. A perky palate, some peach stone, crisp malt, dry tobacco. Soft and easy-going with a peppery finish.

FRUITY/SOFT

ST. KILIAN SIGNATURE EDITION TWELVE 50.8%

Orange juice, redcurrant jelly and vanilla. Parsley and angelica with a mix of light molasses and cow gum, then cedar. Fernet-like bitters on the finish.

SPICY/HERBAL

ST. KILIAN PEATED RICH & SMOKY 46%

Smoking bonfire. Pork crackling with banana and white fruits behind. The palate has a slight vegetable quality. Ashes and cinders with gentle sweetness.

SMOKY/PEATY

10 YEARS ST. KILIAN ULTRA HEAVILY PEATED 57.7%

Resinous nose with rich oak. Palate has a slightly gamey quality, then coal tar, animal hides, incense, medicinal and sooty, but somehow balanced.

SMOKY/PEATY

Above *Stoking the fire at Simon.*

Above *Keeping things clean at Stork Club.*

First appearing in 2019, the range is based around a core of Classic (unpeated) and Peated expressions made with barley sourced in Germany and Scotland, 70% matured in ex-Bourbon and 30% in ex-sherry casks (a mix of PX and oloroso), though more than 370 different cask types and sizes are used overall, including virgin wood. Beechwood smoke is also used.

The **Simon** family has been making spirits in Alzenau, Bavaria since 1707. Today, Severin Simon runs operations with sommelier wife Susanne, making a range of wines, fruit spirits and a whisky that's embedded in its locale. 'I was looking for a way to make exciting distillates with purely local resources,' he says. 'We grow our own grain, have our own water, heat the stills with wood from our own forest, use barrels from Spessart oaks.' Any new casks are seasoned with their own wines.

The aim is to build in aroma from the start by using a range of speciality malts. 'There's no boring Pilsner malt!' he laughs. 'We always have a high proportion of roasted malts, caramelized malts or beech-smoked malts in our mash bill.' Then there's the range of grains used: barley, emmer, rye, oats and surprisingly Bloody Butcher corn (see Grain Initiatives box, page 229).

Distillation is in a 300-litre (79.3-US-gallon) pot still with a small three-plate column and is done, slowly, in a single pass. 'We're really concerned with aroma conservation; we don't want to put a thermal burden on our local raw materials and thus our terroir.' Place is everything.

In common with the majority of Germany's distilleries, **Birkenhof** is both family-owned and long established – in their case since 1848. It was the eighth generation of the Klöckner family, Lukas

SIMON

★ SIMON'S BAVARIAN SINGLE RYE WHISKY IS ART 58.3%

Captures rye's scented elements: watercress, parsley stalk, dill, light florals. Palate is pumpernickel, roast pumpkin seed, geranium and chocolate. A warm blanket.

SPICY/HERBAL

BIRKENHOF

★ FADING HILL (BIRKENHOF) SINGLE MALT 5 YEAR OLD 46%

Nutty, with spearmint, yellow fruit, gooseberry, hay/wet grass, dessert apple and char. Focused and creamy, balancing nuttiness. Perfumed blue fruit on the finish.

FRUITY/SOFT

FADING HILL (BIRKENHOF) SINGLE RYE 7 YEAR OLD 45%

A sweet rye. Clove oil, rye flour, dried flowers, silage, turmeric/curry leaf/caraway, then rosewater. Palate is delicate, slightly powdery. Good.

SPICY/HERBAL

and Jonas, who made the move to whisky. The result is the Fading Hill range, comprising two single malts (one unpeated, one peated), a 'classic' single malt and rye aged in sherry casks.

Another who added whisky to an existing speciality is **Heinrich Habbel**, though its entry into whisky is the reverse of the norm. The distillery in Sprockhövel was originally a grain distillery that added fruit spirits to the roster in the 1970s, before in 1977 beginning to make whisky. Now Michael Habbel and daughter Michaela distil their **Hillock** single malt in a purpose-built distillery.

The Brandenburg region south of Berlin is the biggest producer of rye in Europe. Some of its grain even ends up in American distilleries. It was inevitable that when a local whisky was dreamed up by the team behind **Stork Club** it would be a rye. Its owners Steffen Lohr, Sebastian Brack and Bastian Heuser had gone to the existing distillery to buy a cask, but ended up buying the whole operation. That was in 2015.

The distillery sits in the Spreewald Biosphere Reserve. Only locally grown rye is produced, but in various iterations – unmalted, malted, roasted – and from a range of varieties. The grain is pulverized in a hammer mill, mashed and given a five-day ferment before distilling in a pot, then a small five-plate column. Subsequent ageing is mainly in American or (German) Napoleon oak. Germany's only dedicated rye distillery, it is adding to the ever-growing flavour world of this fascinating grain.

OTHER GERMAN DISTILLERIES

In 1894, Christoph Karl **Schraml** produced a 'brown' (oak-aged) grain spirit, potentially Germany's first whisky. It didn't catch on. In 1954, Alois Schraml tried once more with Steinwald Whisky. Again it foundered, though the family firm persevered with what it then called Farmer's Spirit. Now the time is right. The Schraml's distillery has been expanded and the range repackaged as Stonewood 1818. To emphasize how long they have been persevering, a 10-year-old aged in old Limousin oak brandy casks has appeared, as has Gregor Schraml's innovation, a malted wheat whisky, Stonewood Woaz.

A sixth-generation distiller, Martin **Steinhauser** makes his Brigantia whisky in a renovated 19th-century distillery on the shores of Lake Constance. He uses a 19th-century Pistorius design of column still, which produces a 70% ABV new make.

Elch from Thuisbrunn in Franconia is made by Georg Kugler who likes smoke – specifically peat from the north of the country, which he feels gives a salinity to the whisky. A week-long ferment with brewer's yeast, then distillation in a linked pot/column with dephlegmator. Ageing is predominantly in ex-Bourbon, sherry and acacia. All of the casks sit in a converted barn, and with a 50°C (122°F) temperature swing through the year, the maturation cycle is somewhat dynamic.

Nuremberg is home to the ancient Hausbrauerei Altstadthof, which was reopened by Reinhard Engel and his son Max in 1984 to brew local specialities from organic barley. The duo branched into whisky distilling in 2005 with the building of the **Ayrer's** distillery. The whisky, also organic, uses the same red malt for the brewery's RotBier. The new make starts in new AO blended whisky casks before being 'ripened' in a mix of sherry, Bourbon and port 30- and 50-litre (8- and 13.2-US-gallon) casks.

HILLOCK PARK DISTILLERY

HILLOCK 12 YEAR OLD 44%

Low cask influence. Green strawberry, touch of mint. Seems fragile. Moves into Taiwan oolong and sakura. Palate is light, sweet, buttery and very well balanced.

FRAGRANT/FLORAL

STORK

★ STORK CLUB RYE 50%

Medium weight. Light oils, caraway, five-spice, dill oil. Red fruits develop. The palate is thick and velvety. Pink peppercorn, dark chocolate, lime. Char with water. Very good.

SPICY/HERBAL

Above *The Habbel family's Hillock Park Distillery.*

NETHERLANDS & BELGIUM

Distilling's roots in this part of Europe are frequently to be found in genever, the native spirit of Netherlands and Belgium as well as Northern France. Genever is made from a double-distilled 'malt wine' base (a malted rye, corn and barley mash bill), a portion of which is then redistilled with botanicals before being blended back in, and subsequently often aged. It's closer to whisky than gin.

NETHERLANDS

The **Millstone** distillery in Baarle-Nassau was started in 1975 by Fred van Zuidam when he left his position as head distiller at genever producer deKuyper. Now under son Patrick, it's one of the most highly regarded of the new whisky distilleries across the globe. At a time when the world believed rye was a North American speciality, Millstone 100 Rye reclaimed the spirit's European origins while also revealing new facets of flavour.

The firm has long outgrown the original distillery and now grows most of its barley and rye requirements, mashing around 5 tonnes (5.5 US tons) a day. Ferments (grain-in) are a minimum of seven days, and distillation is in bain-marie stills.

'We're at capacity,' Patrick says, 'and there are plans to expand further.' It all seems a glorious success story, but he is quick to disabuse you of the notion that it was easy. 'You have to accept that it will be hard. You have to accept the hardship, and accept there are no shortcuts.'

The keys to his success? 'Invest in the future; it's more important than the now. Build relationships with your importers. Don't copy other people's ideas but find your story, then stick to it. Take more risks, invest more money earlier on...oh, and buy bigger equipment at the start!'

For him it comes down to quality and the consumer. 'My father told me "Always bottle products that people will come back for", and I'd add, make whiskies that are value for money for the consumer. There is no future for €150 3-year-old whisky!' Wise words.

The **Cley** distillery owned by Paul den Dulk and Maria de Sena Neves also has big plans. What started as a small operation in the couple's back garden will, in 2025, have expanded dramatically with new large pots (from McMillan of Prestonpans, Scotland) replacing the tiny Christian Carl still in which he ran spirit through three times to get his new make. The new distillery in Overschiese Kleiweg will continue to use only organic grains, with malted barley providing 80% of production, the rest being rye. Though the style will remain light and soft, there could be a change, given the switch from small still to large and triple to double distillation, but his unusual mashing regime will say the same – beginning at a high temperature (74°C/165°F), then reducing the heat, which helps, he claims, to increase sweetness. Ageing will still be in ex-Montilla and ex-Bourbon casks to start, then into different toasts and chars of new American oak.

The clearing and draining of peat bogs in the 1800s created new fertile soil for farms such as that supplying the **Kalkwijk** distillery in Vroomshoop. An established grain supplier, it began to make genever in 2009, with Eastmoor single malt arriving in 2015, another example of a farmer taking control of their own crops. Distillation is in a Holstein pot and linked column. The range has been expanded to include the Higgledy Piggledy 'grain' whiskies, a malted/unmalted barley, a rye and a corn.

Above *Belgium's Filliers has added whisky to its renowned genever range.*

BELGIAN OWL

BELGIAN OWL SINGLE MALT 36–41 MONTHS 46%

Sweet, gentle, fresh. Summer picnic, white flowers, patisserie, apple purée, pineapple, pear, crème anglaise. Succulent. Balanced with peppercorn, nutmeg and carrot tops.

FRAGRANT/FLORAL

★ BELGIAN OWL SINGLE MALT 48–59 MONTHS 46%

More depth from the cask. Sage, Sichuan peppercorn mouth-numbing heat/citrus. Oolong and more spice. Sugared almond, hot cross bun, wholemeal bread, mace and orchard fruits.

FRUITY/SOFT

BELGIUM

In light of the country's aforementioned distilling roots, it's no surprise that the esteemed Belgian genever distiller **Filliers** now also makes whisky – you could argue that it always has. Benoit Filliers is the sixth generation of his family to distil here in Bachte-Maria-Leerne. 'I believe whisky is the future of the company,' he says, which might seem startling given its 140-year heritage. 'Our motto is "honour tradition and embrace change". Tradition is genever, but it is declining every year. If we are to grow, it means widening the range of spirits we make.'

Quietly, in 2008, two pot stills from Forsyths were installed and distilling of malt and rye began. 'It meant we could launch in 2019 with a 10-year-old single malt and then an 8-year-old malted rye aged in virgin oak. A mash filter allows rye to be handled more easily.'

Now the investment is focusing on ageing capacity with a fourth warehouse built in 2023. This is long-term thinking.

A more Scottish-style distillery is found in the farmlands around Fexhe-le-Haut-Clocher, home to Etienne Bouillon's **Belgian Owl**. In 2013, his original distillery with its antique pots was replaced by a new larger enterprise, complete with the stills from Rothes's Caperdonich (see page 49).

The dramatic increase in capacity means he's now working with more farmers growing barley across 1,619 hectares (4,000 acres), with 20 (50) of those being organic. 'All this has been done in the respect of our terroir – Crétacé Secondaire,' he says. 'My vision is to see that, in 150 years, this specific terroir will still be there with the farmers to work on it.'

For him everything comes back to the barley and where it's grown. 'The terroir gives a very fruity, flowery new make. You can only make an outstanding product if you have an outstanding ingredient!'

MILLSTONE

★ MILLSTONE 100 RYE 50%

Hot cross buns, allspice, Jamaican five-spice powder, clove oil, tobacco leaf, Peychaud's bitters, maple syrup and bright orange zest. Encompasses everything about rye (and more).

SPICY/HERBAL

MILLSTONE SINGLE MALT OLOROSO CASK 46%

Massed fruits. Seems cask-driven, but the distillate is rich and thick. Nutty with suet pudding/spotted dick. Needs water to balance. Plummy, charred raisins, a hint of pepper.

RICH/DEEP

FILLIERS

★ FILLIERS 8 YEAR OLD SINGLE RYE 46.5%

Herbal rye to start: bouquet garni, parsley, dill pickle juice, rosemary, then caraway, then hints of tangerine. Good grip on the palate with dark sweeter fruits emerging.

SPICY/HERBAL

★ FILLIERS 10 YEAR OLD SINGLE MALT 43%

Rich caramelized fruits. On the silky side of sherry. Sweet walnut, figgy pudding, then beeswax polish, tack room and treacle toffee. Soft, sweet and long. Worth the wait.

RICH/DEEP

AUSTRIA, SWITZERLAND & ITALY

Austria has benefited from a strong contemporary whisky-making community and one that is now becoming increasingly organized. Italy might have turned up somewhat later to the whisky-making party but is another country on the continent where the flexibility of its distillers' minds is bearing new fruit – or in this case grain. Whisky made in old grappa stills? Why ever not?

AUSTRIA

With 50 whisky producers and an association with a strict set of rules – use of fermented grain only, ageing in wood for a minimum of three years, no GMO, sugar or artificial aromas – Austrian whisky has grown up. Quite a change from 1995 when farmer-distillers Johann and Monika **Haider** decided to start producing the country's first whisky.

'My parents had successfully run the farm,' says managing director Jasmin Haider-Stadler, 'but with Austria's accession to the EU, they had to look for additional income. So they started distilling whisky, making the best from the raw materials of the region – rye, barley, peat, water and oak.' Now, the distillery includes a Whisky World of Experience attracting tens of thousands of visitors a year.

Roggen is Austrian for rye, and with the farm being in the town of Roggenreith, there's little wonder what Haider specializes in, with 70% of their production coming from rye in different formats: 'raw', malted, roasted and peated. The rest is from malted barley. All are distilled after four passes through a Christian Carl still and matured initially in sessile oak casks from the Manhartsberg forest. The wide grain adds structure quickly to younger age statements, while refill and re-charred casks allow a more relaxed maturation – the oldest release so far is 18 years. 'Whisky doesn't have to be old,' says Jasmin, 'it just needs to be the right age.'

Above *Haider: where Austrian whisky started.*

If rye was the no-brainer for the Haiders, then Josef **Farthofer** has given himself a wider remit. With his wife Doris, he farms 45 hectares (111 acres) organically in Öhling. Self-sufficiency and sustainability are fundamental. The distillery is heated with miscanthus biofuel and powered by solar, his fields are fertilized with distilling waste and draff is used as animal feed.

'Because malt provides taste and body in the finished whisky, we didn't want to outsource the delicate process of malting,' he says. That meant building his own 2-tonne (2.2-US-ton) maltings where he can process hull-less and brewing barley, Schlägler rye, wheat, emmer and hull-less oats.

The last is the main grain grown by Oswald **Weidenauer,** a third-generation distiller who has been producing in Kottes since 1997. Unmalted, malted and peated, the oat new makes are matured in new casks. There are also whiskies from corn, einkorn wheat and spelt (*dinkel*), some of which is also malted. Maturation is in French Alliers and Limousin oak as well as local Manhartsberger oak 'for its earthy flavours'. Options galore.

HAIDER

J.H. ORIGINAL RYE 15 YEAR OLD 46%

Delicate rye and sweet baking spice, cinnamon and allspice. In time, rye bread, treacle toffee and cardamom. The palate is very soft and drying with wild grasses.

SPICY/HERBAL

J.H. RYE MALT 6 YEAR OLD 41%

More rye grain and a little new shoe with some black pepper, aniseed and light earthiness moving into dill, caraway and sweet carrot. The palate mixes rye flour and sweet lino.

SPICY/HERBAL

J.H. DARK SINGLE MALT PEATED 10 YEAR OLD 46%

Woodsmoke, some petrichor, sweet mushroom and sweet leather. Palate is curry leaf and fenugreek, lino and then all-enveloping smoke.

SMOKY/PEATY

AUSTRIA, SWITZERLAND & ITALY

OTHER AUSTRIAN DISTILLERIES

Hermann **Rogner** in Roiten near Rappottenstein was another early whisky convert. Having inherited his father's distillery, he now uses locally grown grains (plus some peated malt) with smoked variants and ageing in Austrian oak. Again rye features strongly – his Old John was released at 18 years.

You have to ascend to heights to reach the Warter family's distillery in Radstadt, which takes its name from the nearby **Dachstein** Mountains. Having established an ecotourism business on the farm, in 2009 a distillery was built. Malt comes from the Schladming brewery, and the new make is aged in Burgenland red and sweet wine casks. In 2013, an ice cellar was created in caves on the Dachstein glacier (2,650 metres/8,694 feet) to house casks of Dachstein's Rock Whisky.

Bruno and Eugen **Broger** in Vorarlberg started making whisky as an adjunct to fruit spirits in 2008. German and Scottish barley are used, but wheat, rye and spelt are all made for its Hoamat range. Scottish peated malt is used for the smoked releases Medium Smoked and Burn Out.

The Kuenz family have lived for over 400 years in Dölsach and have been distilling their own fruit for much of that time. Now 12th-generation distillers Johannes and Florian are also making whiskies at the **Kuenz Naturbrennerei**. Günther Mayer uses local peat for a single malt from his **Waldviertler Granitdestillerie** (he also makes a rye), while Harald **Keckeis** uses beechwood smoke for his, as does Valentin Latschen at the **Pfau** distillery in Klagenfurt.

Above *The fruits of labour at Langatun.*

SWITZERLAND

Langatun was established as a brewery and distillery in Langenthal in 1857. Hans Baumberger III took over the business in 1975 and started to produce whisky as soon as the Swiss ban on distilling from grain (and potatoes) was lifted in 1999. The distillery is now housed in the historic Kornhaus in Aarwangen.

The Säntis brand is made by another old brewing/distilling family, **Locher**, who have been working in Appenzell since 1886. Also quick to make whisky post-ban, Karl Locher aged the country's first whisky in 60–130-year-old beer casks. Today, there has been a widening of that into wine and fruit brandy casks. The distillery also runs a Whisky Trek, with 26 different casks being aged in mountain inns, starting in Brauquöll in Appenzell and ending 1,700 metres (5,577 feet) higher at the Berggasthaus Alter Säntis. Each cask also shows the impact of altitude and temperature on the maturing spirit.

LANGATUN

LANGATUN RIOJA CASK MATURED 49.12%

Old oranges, then floor polish, strawberry light camphor, musk in time. Becomes more cedar-like. Light oak. A dry start. Lemony, and chewy in the middle.

FRUITY/SOFT

LANGATUN MONBAZILLAC CASK FINISH 49.12%

Light oils, gentle wood, citrus and marmalade with a scented, honeyed back note. Rounded and concentrated. Seville orange, then the malt.

FRUITY/SOFT

ITALY

The country that 'got' single malt Scotch in the 1960s is another that appears to have been happy with importing rather than producing. It wasn't until 2010 that Italy's first dedicated whisky distillery, **Puni**, fired up its stills in Glorenza in the South Tyrol. The Ebensperger family then embarked on a three-year investigation into how to make a light gentle dram that would suit the local palate. Unusually, hot water coils are used in the stills, allowing minute changes in temperature to pinpoint the trigger points for specific flavour. The initial range has been revamped and now comprises the three-strong Gold, Sole and Vina, with other specials flitting around.

Another 11 distilleries on the Italian mainland are making whisky, with a pair in Sardinia. Most are grappa distillers adding whisky to their skill set. Werner **Psenner** in the village of Tramin in the Adige Valley is the latest member of his family to distil. His eRètico (named after the Adige's Retic people and his 'heretical' thinking) includes some smoked malt from Weyermann in Bavaria as well as local malt. Ferments are ten days with a whisky yeast adding depth and a fruit spirit one for lightness.

Distillation is in two German stills: a bain-marie pot still and a four-plate column still.

The **Poli** distillery has been making grappa since 1898 in Schiavon near Bassano del Grappa, with its Segretario di Stato whisky – made in the same bain-marie stills – a more recent arrival. It's five years old and finished in Amarone casks.

Roner near Bolzano is also an old family-owned enterprise specializing in grappa, fruit spirits and liqueurs. Its TER Lignum is a triple cask-aged (cherry, oak and larch) whisky made with wash from the Forch brewery.

It was in 1800 that Romedio Dolzan began to distil on the **Villa de Varda** estate in the Trentino village of Mezzolombardo. His great-great-grandchildren now run the distillery and have added the InQuota whisky range made from alpine barley and rye, both grown in the Dolomites. Starting in oak, they are using Amarone casks and also spruce wood for finishing.

In Tuscany (Maremma to be precise), the **Nannoni** selection has been boosted by distiller Priscilla Occhipinti's Whisky al Focarile, made in a 200-litre (53-US-gallon) still in a new purpose-built part of the distillery.

The **Exmu** brewery in Sardinia opened its distillery in 2019. The barley is all from local farms and it's double distilled and then aged for three years in virgin oak. Also on the island, **Silvio Carta** near Zeddiani takes a more extravagant approach, distilling in a copper still built to his specifications, then maturing in chestnut casks (dating back to 1852) in which he has matured his Vernaccia di Oristano wine.

PUNI

★ PUNI SOLE 46%

On the delicate and floral side of things. Alpine meadow, peach and patisserie, some strawberry, banana toffee. Mouth-filling and silky. Peach and mango juice. Light acidity.

FRUITY/SOFT

PUNI GOLD 43%

Gentle and estery. Delicate fruits, slightly powdery. Grapefruit, then elderflower. Delicate and fragile but with a lively, bouncy feel. Mouthwatering acidity.

FRAGRANT/FLORAL

PUNI VINA 43%

More presence and cooked red and black fruits. Summer pudding, light oils, roast chestnut, pork char siu and a whiff of gun smoke. Bright.

RICH/DEEP

Above *Puni's remarkable distillery in Italy.*

THE NORDIC COUNTRIES

A combination of quasi-Prohibition and state control of distillation delayed the uptake of whisky distilling in some Nordic countries. With these hurdles now cleared, it has become a busy new whisky-producing zone. Besides looking at developments in Japanese whisky, many distillers have tapped into the New Nordic food movement, whose emphasis on local ingredients can easily be translated into the whisky realm.

FINLAND

Located in a former dairy in Isokyrö, **Kyrö** started in 2012 and makes a week-long fermented, double-pot-still rye that balances the grain's natural spice with an earthy sweetness and, as only malted rye is used, an extra toasty element. Bottled relatively young, the oak is there as support.

'Our philosophy stems from the desire to re-establish the link between our culinary culture and our spirits culture,' says co-founder Miika Lipiäinen. 'We eat a lot of bread, porridge and pies made out of rye. Our approach to distilling, therefore, was to bring the unadulterated taste of rye into the final product and make something that our fellow Finns could drink with pride and note that our national grain has found yet another way to solidify its position at the core of our culinary culture.'

For Miika, this deep attachment to place goes further than geographical location. 'Coming from a new region with a new style of whisky, your location is not just about finding fuel for your marketing stories. You make it based on how good your raw materials are for your whisky and whether there is a willingness for your community to adopt and support you.

'The Nordics are just now taking off! Whisky making is an incredibly varied and energetic endeavour with different approaches to the category. Our insistence on using rye is in no way the only way to go. There are interesting practices, resources and local characters aplenty to choose from.'

KYRÖ

KYRÖ MALT RYE 47.2%

Deep and soft, with toasted oak, pumpernickel, spices, flamed citrus oils and red fruits. On the palate, buttered rye toast, pine and camphor, then apricot.

SPICY/HERBAL

★ KYRÖ MALT RYE OLOROSO 47.2%

Polished and full. There's a purity to the rye, but more muscle from cask. Almost jammy, candied walnut, light allspice, fig and caraway. Prune, rye bread. Complex.

SPICY/HERBAL

Above *All things rye at Kyrö.*

THE NORDIC COUNTRIES

Above *Lahti's Teerenpeli kick-started Finnish whisky.*

The **Helsinki Distilling Company** is going down a similar route. Founded in 2013 by Kai Kilpinen (now the master blender) and Mikko Mykkänen, distillation started a year later with the first whisky appearing in 2017. 'We had been whiskey enthusiasts for 15 years and we were into rye whiskies – and knew that Finnish rye is the best in the world,' as Kai puts it.

To start, he looked at the only ryes that were then available. 'A lot of American, and the few European we could find at the time. We never wanted to imitate, but to learn from them. That's why we chose to go with malted rye.' As well as its signature double-distilled rye, Helsinki is also making single malt, and corn whiskies.

'We believe that the rye and barley we're using have their own "Finnish" taste unlike any other', he adds. 'We also have a project of hyper-local whisky where grain is cultivated and the oak for the casks all come from less than 5km [3 miles] from the distillery.'

Older than both is **Teerenpeli**, whose founder Anssi Pyysing started distilling in Lahti in 2002 'out of a mad desire to find out if we could make whisky here. After all, we have all the basic stuff around us: exceptional water, barley malting, skilful people.'

Water wasn't an issue, as the distillery sits on top of a natural aquifer whose water is filtered through the sand and gravel of the glacial Salpausselkä ridge. The local forest supplies wood pellets for the distillery's boiler, while a local maltings handles the unpeated barley. For Anssi, the Finnish summer plays its part in generating flavour. 'We have long and sunny summer nights, which help to produce an intense flavour in the grain,' he says.

It goes deeper. 'Lahti has a lot of industries based on grain: malting, milling, bakery, brewing as well as distilling. We've formed a "Grain Cluster" and are opening test and pilot facilities with the local university this year. We want Lahti to be a world-leading area of new innovations from grain.' As well as learning from the world, Finland is now also teaching it.

SWEDEN

Winter is cold in the 63rd parallel on the shores of the Baltic. So cold that lakes freeze and hands stick to whisky glasses. Temperature is, however, key to this being the location of the **High Coast** distillery. What began as a way to regenerate the town of Sörviken after its box-making plant closed (the distillery was originally called Box), it was founded in 2007 with production starting three years later.

HELSINKI

★ HELSINKI WHISKEY RYE MALT RELEASE #20 47.5%

Assertive green rye: spruce tip, menthol, then bright spice, caraway and bubblegum. Palate has pickle, gingerbread, dark rye bread, cardamom and spiced pumpkin.

OILY/SPICY

TEERENPELI

TEERENPELI 10 YEAR OLD 43%

Light malt, honey, beeswax, cloudberry. Perfumed with clotted cream, honeycomb and sweet pumpkin. Barley adds grip and texture.

FRUITY/SOFT

★ TEERENPELI SAVU 43%

Sweet smoke, tobacco-like, light apple, green grape, wet stone, then Greek yogurt. Palate has heather blossom, green olive and the thickness of birch water.

SMOKY/PEATY

AGITATOR

AGITATOR SINGLE MALT 43%

Precise and crisp. Dry grass, light cereal, whitecurrants and dusty white florals, rapeseed oil, yellow plum. Silken texture.

FRAGRANT/FLORAL

★ AGITATOR ARGUMENT: LÖNNSIRAPSFAT 46%

Spices. Light red berries, cedar and aniseed, mothballs. Caramelized fruits on the palate, cherry liqueur and Yellow Chartreuse. Spices tempered by deep fruits.

RICH/DEEP

The Ångerman River delivers ice-cold mountain water and, as distiller Roger Melander says, 'the ability to cool the process is extremely important in the pursuit of a characteristic distillate'. He has measured every detail of temperature, plotting each against flavour production.

Understanding this is essential because of extreme seasonal temperature variations. In winter the mercury can hover at -30°C (-22°F), while in summer it can hit 40°C (104°F). 'It means that whisky that matures in large, unheated warehouses is exposed to the enormous forces of nature.'

It's a single malt facility making unpeated, peated and organic from seven washbacks and four pots. The malt used is deliberately low-yielding, the wort is clear, ferments are lengthy and distillation allows a prolonged conversation. 'I had a vision of an unpeated new make with fruity flavours and a pure malty tone, partly inspired by Japanese whisky,' Roger explains. There's a nod to Scotland's west coast in the peated, but the precision of the flavour and the cool clarity of all of his many whiskies is wholly High Coast's.

Distilleries can be found in many different locations, but to the best of my knowledge **Agitator** is the only one inside a military base, in its case in Arboga 93 miles (150km) west of Stockholm. There isn't a visitor centre.

'Our location isn't important,' says distillery manager Oskar Bruno. 'The people and mindset is what creates our specific flavour.' This is a radical approach to whisky making where science rules. 'However,' he adds, 'being Swedish might be the key. That Swedish engineering mindset has most likely helped in developing this plant.'

HIGH COAST

HIGH COAST THE FIRST TEN YEARS 52%

Full and relaxed. Scented, minty pine. Lemon marmalade, some weight to the yellow fruits. Palate has chocolate caramel, coconut and menthol. Ordered.

FRUITY/SOFT

★ HIGH COAST CINCO II 51.1%

Balanced bright oxidized wines, semi-dried fruits. Calm, mineralic, some sultana bread, light shiitake mushroom. Subtle and nuanced. Lemon zest on the finish.

FRAGRANT/FLORAL

★ HIGH COAST THE FESTIVAL 2023 11 YEAR OLD 50.2%

Smoke, varnish, leather oil. More guts. Mixes woodsmoke, rosemary and rowan berry. Precise, delicate and with great balance.

SMOKY/PEATY

Above *The finely tuned High Coast.*

Above *Agitator is Sweden's top-secret producer.*

Above *Mackmyra's vertical distillery.*

OTHER SWEDISH DISTILLERIES

Sweden's seven other distilleries include **Spirit of Hven**, which was founded in 2017 by Henric Molin on the island of Hven and sits between Denmark and Sweden. He approaches distilling in a highly systematic – and effective – manner. Starting with a half-tonne (0.6-US-ton) mash tun, stainless-steel washbacks and two pot stills, he has also installed a wooden-framed Coffey still, modelled on the 1880 Enmore still operating in Guyana's Diamond rum distillery. Hven makes rye and corn whiskies with it. As well as bottling the corn, it's used in a blend – Sweden's first. There's also floor malting using Swedish peat, which on occasion is mixed with seaweed and sea grass.

Tevsjö in Järvsö reinforces the point that that there's nothing that could ever be said to be a standard Nordic approach. Distiller Jonas Larsson makes a column still 'Bourbon-type' from a mash bill of 70% corn, 10% malt, 10% barley, 5% wheat and 5% rye. There's also a double-distilled single malt whisky. All the grains are locally sourced. Considerable expansion is underway.

Smögen started in 2010 with two stills, but in 2018 the style changed when the original shell and tube condensers were replaced by worms, adding depth. This has always been an overtly Scottish-style whisky using imported, heavily peated barley.

There's wet milling, a full lauter tun, stainless-steel washbacks and four Frilli pot stills. The ferments are a minimum of a week using different yeasts (saison for medium-peated, Trappist and distiller's for unpeated) depending on what grain (or result) is wanted. One set of stills is tall, giving high reflux; the second pair are stubby, making a heavier style of spirit. All are run under vacuum 'enabling us to lower the boiling point, and create a spirit with fewer fusel oils'. Malt and fruits are key. 'The only way to improve or to find new exciting flavours is to venture beyond the known production methods,' Oskar says. 'We want to show the world that whisky is more than just Scotch.'

MACKMYRA

MACKMYRA SVENSK EK 46.1%

Typical Mackmyra lightness and precision. Extra layers of sweet spices, some grapefruit peel moving into peppery nuts and light vanilla.

SPICY/HERBAL

MACKMYRA BRUKSWHISKY 41.4%

More wood-influenced, but still very soft and approachable. Vanilla ice cream, raspberry syrup and just a hint of pine. Light with some baked apple.

SWEET/VANILLA

In 1999, when **Mackmyra** started, the idea of a Nordic whisky was a strange concept. It outgrew the original site and in 2012 a new one outside the Baltic port of Gävle was built – a 37-metre- (121-feet-) high vertical distillery, the world's first, with mashing at the top and distillation near to the ground.

Everything Mackmyra did was testing new ground – Swedish peat, smoking over juniper, ageing in Swedish oak and ageing in an old mine whose cool conditions means low cask interaction. Mackmyra remains clean, restrained and distillate-forward.

Core is Svensk Ek (aged in virgin Swedish oak), Svensk Rök (Swedish peat with juniper) and the original, Brukswhisky – now a vintage release. Experiments, including ageing in raspberry and cloudberry wine casks, continue. Once a pioneer...

DENMARK

Can an urban distillery express place? **Copenhagen Distillery** is a test case. Founded in 2014, it runs off a rigorous, scientific examination of flavour and process.

'Nothing is left to chance,' says master distiller Lasse Öznek. 'From selecting Danish organic grains to the way we approach the venture process.'

If location is less of an influence, the aim was still to make it Danish: 'not a take on Scotch made in another country. We don't speak about terroir, but we do speak about a sense of place. We produce local whisky at a single location.'

Single malt is made along with single-grain distillates. A traditional technique for smoking fish and meat is employed, with rehydrated malted barley being cold-smoked over beechwood in a tray oven.

Ferment and distillation is on-grain and run through a hybrid still where Lasse can take spirit off at different points. 'Our process was designed to achieve taste notes that will surprise you,' as he puts it.

Maturation is predominantly in virgin Hungarian oak (*Quercus frainetto*) toasted to different levels to create a range of flavour profiles. As the casks sit in the production space whose temperature ranges between 15°C (59°F) and 36°C (97°F) each week, maturation is speeded up.

The whiskies come in three categories: Raw is organic grain aged in new Hungarian oak, Refined is smoked, while Rare can be unusual grains, cask types or crazy variations like 'Blood and Bone' made by seasoning the cask with a distillate produced from the blood and bones of two deer.

'Denmark is a country where innovation and technology is one of our biggest export success stories,' says Lasse. 'This will also be reflected in the whiskies being produced.' No kidding.

At first glance, **Fary Lochan** is a more conventional single malt operation. Founded by the late Jens-Erik Jørgensen in the Jutland village of Farre (originally Fary) in 2009, it remains a family operation that, in his stepson Thomas Smidt-Kjærby's words, sticks to the vision of 'creating a Danish whisky that was going to stand out'.

The 'standing out' element came from Jens-Erik's memories of a local cheese from his home island of Funen, which was smoked over nettles. So the family collect nettles from the forest, smoke the barley (from a local farmer) and double distil. 'I'm hoping Danish (or Nordic) whisky will be the "new" Japan or Taiwan,' says Thomas. 'Less mass production and more focus on the local. Special smoking methods and terroir, small batches and authenticity. That's how we do it in Denmark.'

COPENHAGEN

COPENHAGEN SCHWEISS 'BLOOD AND BONE' 61.4%

Cherry, za'atar spice, old growth forest. Rich, complex, red and black fruits. Sultry and thick with cigar and heightened spice.

RICH/DEEP

★ COPENHAGEN RAW BATCH NO. 3 58.2%

Layered blue fruits, Ribena and chocolate mint. Higher acid, some strawberry. Soft feel. Balances sweet and sour (rosehip jelly, cumin, sumac). Plum sauce. Layered.

RICH/DEEP

COPENHAGEN SMOKED PX 59%

Very grapey PX, bruised fruits, sloe, blueberry, marzipan. Intense with some light caramel, roasting coffee beans, liqueur chocolate and gentle smoke.

SMOKY/PEATY

Above *Denmark's Fary Lochan is a family business.*

Above *Thomas Smidt-Kjærby at work at Fary Lochan.*

Above *Stauning's stunning new site.*

In the village of **Stauning** in August 2006, nine whisky-loving friends gathered in a butchery to make whisky. Locally grown barley was steeped in a stainless-steel tank, malted on the tiled floor of the cold store, kilned using local peat in a meat smoker, milled in a mincing machine and fermented in a former brining tank. The only thing bought in was a tiny Hoga alembic.

It was such a success that three years later a new distillery was built in a former dairy complete with a self-designed malting machine where steeping and germination both take place on the floor, with the grain being turned by a row of vertical rotating arms. The stills were larger but the same design – and direct-fired. Peat now had heather added, and in recent times seaweed and pine.

What seems experimental was second nature – a combination of the self-reliance of a remote coastal area and the fact that three of the nine were engineers. It was also a conscious decision to challenge the mentality of there only being one way to make whisky. With a significant cash injection from Diageo's Distil Ventures, in 2014 production was increased tenfold in a new distillery with 16 wash stills and 8 spirit stills – all direct-fired.

'We wanted to use local ingredients and that slowly developed into a philosophy that stretched through the whole approach,' says co-founder Alex Højrup Munch. 'One of the big advantages was that we weren't limited by any old, existing Danish traditions of whisky making because there weren't any!' From here, but also of here.

Some 60 miles (97km) north of Stauning is the 202-hectare (500-acre) Gyrup estate, where the **Thy** distillery started. The farm

STAUNING

STAUNING HARDCORE RYE CASK STRENGTH 61.2%

Green aniseed, fennel, horseradish, rye bread with raspberry jam, haskap berry. Solid rye grain with some sweet earthiness and orange peel.

OILY/SPICY

FARY LOCHAN

FARY LOCHAN SINGLE MALT DISTILLERY EDITION #03 47.3%

Slightly bready, then sweet caramel, light earthiness. Intriguing dry leaf elements, nutmeg. The smoke well controlled. Light citrus and clove.

FRUITY/SOFT

FARY LOCHAN SINGLE MALT DISTILLER'S CHOICE #02 (2013) 56.2%

More nuttiness. Sweet. Raisin, light hazelnut, then smoke. The cask adds bulk and balance. Barberry and plum jam. Hint of leather. Rounded and soft with light smoke.

RICH/DEEP

Above *Thy grows all its own grains.*

has been in the Nicolajsen family for eight generations. It's currently run by sisters Marie and Ellen with their husbands Andreas Poulsen and Jakob Stjernholm (who is the distiller). In 2010, the sisters' father Nicolaj started making whisky on the farm as an experiment. The quartet moved back to Thy in 2016 and started an expanded operation in Nordisk Brænderi.

It remains an organic field-to-bottle operation. These self-confessed grain nerds planted old Danish barley varieties (from grains held in deep freeze at the Nordic Gene Bank), added spelt to the rye whisky mash bill and smoked over beechwood.

Malting is in 5-tonne (5.5-US-ton) drums, with sufficient capacity to supply other distilleries and enough flex to produce specialist and smoked malts.

Distillation is in a 1,000-litre (264-US-gallon) pot still built by Müller Brennereianlagen, with a purifier allowing them to distil in a single pass. 'We have always thought that a Danish whisky should be inherently Danish in concept and flavour,' says Jakob.

THY

★ **THY KORNMOD NO. 12** 52.5%

Cereal. Wet canvas, canned pineapple, hay-like, sweet grassiness, then light icing sugar, orchard fruits, cut flowers and sweet barley. Long.

FRAGRANT/FLORAL

'We hope to help Danish whisky in becoming the strong terroir-driven category in whisky that it has the potential to be.' With another 13 distilleries now on stream, the opportunities are huge.

NORWAY

There's an element of playing catch-up in Norway with independent distilling only being permitted in 2014. One of newest on the scene is the 100% women-owned **Feddie** on an island outside Bergen whose first whisky appeared in 2024. It's in for the long term. 'We make whisky,' says sales and product developer Martin Tønder Smith. 'That means making decisions now for 10–20 years in the future and planning for export now because the local market is so small. Without export it makes no sense.'

The location impacts on flavour – and gives headaches. 'We're using surface water that runs through bogs,' says Martin. 'In summer, the water can be 16°C [61°F] and we struggle with condensing. In autumn and winter, there are storms and the water turns salty. The pH changes, which makes it great for fermentation, but it corrodes the equipment.'

All the barley is organic, and after 72 hours of ferment the new make comes from a single pass in a hybrid still. Experiments are underway with the 1,000-year-old Norwegian family of farmhouse yeasts called kveik.

Myken, Norway's other island site, is trickier to get to. The tiny island, 1¼ miles (2km) long by ¼ mile (0.4km) wide, lies 20 miles (32km) off the coast in the Arctic Circle. Housed in a repurposed fish factory, it

OTHER NORWEGIAN DISTILLERIES

Det Norske Brenneri can lay claim to be the first of the country's distillers to launch a whisky when its Audny brand appeared in 2012. What had been a fruit wine producer had switched to whisky production in 2009. A pair of Holstein stills are used.

The former state-owned distiller Arcus (famed for its Linie aquavit) added whisky to its portfolio in 2009 and launched its Gjoleid brand in 2012. In 2022, it merged with Swedish distiller Altia to form the **Anora** Group. Whisky releases have been intermittent, although in 2021 a 10-year-old Mesterens Utvalgte (85% malted barley, 15% malted wheat) appeared.

DET NORSKE BRENNERI

DET NORSKE BRENNERI EIKTYNE 'A QUESTION OF TIME' 51.6%

Deep and oily. Blackcurrant pastille, Ribena, sweet plum sauce, violet chocolate, banana bread, clove, cardamom. Young but bold.

RICH/DEEP

Above *Under the rainbow bridge.*

was founded in 2014 by Roar Larsen and Trude Tokle after they were stormbound there when sailing.

Learning that the population had fallen from 50 to 6, they figured that a distillery could save the community. 'Keeping Myken going is what underpins everything we do,' says Roar. It's not easy making whisky in such a location. The process water is desalinated – thankfully Roar's previous job was as chief scientist specializing in just that.

'Our location is the be-all and end-all of our whisky creation,' he says. Not just the water, but the climate: Arctic storms, sea spray, intense summers and cold winters.' There is an undoubted minerality to the fine-boned spirit.

Myken might be remote, but to find **Aurora Spirit,** the world's most northerly distillery, you have to travel 400 miles (644km) north. Located beside Lyngenfjord, 58 miles (94km) east of Tromsø, it sits on a repurposed NATO base (some of the tunnels and bunkers are used for maturing the whisky as well as one designed like a Viking longhouse).

Built in 2014, it has steadily taken steps to find ways to manifest this location in the spirit. Brewing, originally under contract at the Mack brewery, is now in-house, with ongoing research into yeast strains – brewer's, distiller's, wild local and kveik. Most remarkable has been the sourcing of Arctic barley: six-row Héder, Braga, bere and a Héder x malting barley cross. 'Planting can't start before May or June,' says co-founder and CEO Tor Petter Christensen. 'Harvesting is in September, but the intensity of the sun makes for flavour.' In 2024, a small maltings opened on-site to handle the Arctic crop, which is fermented with a kveik yeast. Distillation is in a hybrid pot and linked column still, with a water bath to stop any burning of the wash. 'It could be said that we've shaped our process to the conditions of our environment,' says Tor, 'rather than trying to force conditions with temperature control or other modern technologies. The climate has an impact on everything we do, but our remote location makes us uniquely resilient.'

The distillery's Bivrost The Nine Worlds of Norse Mythology series showing the work in progress ended in 2024 when the first core release was launched.

MYKEN

★ MYKEN OCEAN HEART 47%

Light. Bread, flowers, then mild cream. A mineral quality alongside jasmine, ambergris, citrus blossom, lemon meringue pie. Warming but delicate and whispering.

FRAGRANT/FLORAL

★ MYKEN SABOTØR 47%

Marine. Whittled stick, ozone, white flour, cold salt water and sel gris. Palate softens and sweetens into elderflower cordial, agave syrup and yellow fruit jam.

FRUITY/SOFT

AURORA

BIVROST ARCTIC BARLEY (CASK SAMPLE) 59%

Crisp, fresh barley, lovage, dried flowers, then lemon. Nutty/earthy/citric, brown apple core and very bright. Still young but hugely promising.

SPICY/HERBAL

BIVROST MIDGARD (BENCH SAMPLE) 46%

This will be the core single malt. Roast chestnut, ripe fruits and a touch of funk and leather. Light smoke, candied walnut and nutmeg. Robust.

RICH/DEEP

EINAR'S

EINAR'S PUFFIN BATCH #10 59.8%

Young, sherried, malty, oxidized. Boot polish, wet sand and dried fruits. Distant smoke, tangerine peel and clove. Mineral.

SPICY/HERBAL

EINAR'S SINGLE MALT CASK STRENGTH 59.8%

Malt-forward. Dry with light resin, smoke, then raisin. An oily distillate, fruit and nut chocolate. Drying smoke at the back. Young but promising.

OILY/SPICY

FAROE ISLANDS

It was in 2016 that Einar Waag, owner of Faroese brewery Føroya Bjór, branched into whisky making with the opening of **Einar's Distillery**. 'We'd dreamed of a distillery for at least 25 years, until 2014 legislation forbade any alcohol stronger than 5.8% [ABV].' His first single malt whisky, made with malt from Simpson's and matured in ex-sherry and ex-Bourbon casks, appeared in 2020.

Three years later, they were joined by **Faer Isles**. There's some familiar Scottish names here – design by Allen Associates, consultant Gordon Steele behind Isle of Harris (see page 93) and many others – and the use of imported (Scottish) barley. 'But,' says founder/distiller Dánial Hoydal, 'the aim is to create a whisky reflecting this place. Scotland has a stable year-round temperature, high humidity, the salty North Atlantic providing optimal conditions for maturation. On the Faroe Islands, we have more of all of it!'

If production is classical, maturation is in traditional Faroese food-curing houses called *opnahjallur*, whose wooden sides have gaps allowing the salt-laden wind to blow through. 'We give our casks maximum exposure to the elements, and we expect the maturation to be an even bigger flavour contributor than with other whiskies. It will create a special terroir.'

ICELAND

Iceland's exploration of locally produced whisky started in 2009 with the building of the **Eimverk** distillery and its whisky Flóki. From the outset, this was a test of self-reliance/sufficiency. That means using geothermal water for mashing and heating the (self-made) stills and local barley.

'We're focused on making Icelandic whisky with all that it entails,' says distillery manager Eva María Sigurbjörnsdóttir. Flóki is made from six-row feed barley (IS-Kría and Philipa), which adds earthiness, grassy notes and an oily texture thanks to it having low starch and high protein. 'The flavour will change depending on the growing season. Some years we can plant in April, others in late May. Harvest can be any time between August and October, but batch variations and the changes in flavour are part of the story.' The smoke for Flóki is another link to food culture in the use of sheep dung – a traditional Icelandic technique for smoking fish and mutton. It's differences like this that makes Nordic whiskies so exciting.

EIMVERK

FLÓKI DOUBLE WOOD RESERVE 45%

Good interactive character. Dry grass and pear, green grape and sweet nut. Clean and spicy palate with cherry tobacco. Dry but with an aromatic lift.

SPICY/HERBAL

FLÓKI SHEEP DUNG SMOKED RESERVE 47%

Smoky and savoury. Mirin. Grassy (rabbit food, cattle cake, silage). Time has made it more smoky and less...dungy. Mineralic and evocative.

SMOKY/PEATY

Above *Eimverk is taking an Icelandic approach to whisky.*

SPAIN

For a period in the late 1980s and early 1990s, Spain was seen as the beacon of light in a world that was turning its back on Scotch, where a new generation was drinking blends liberally dosed with cola. In some ways, this predates the Highball boom that saved Japanese whisky, but oddly at the time, non-Scotch firms attempted to replicate the Spanish phenomenon. Now a new wave of Spanish single malts is beginning to emerge.

Whisky production in Spain actually dates back to 1958 with the building in Segovia of the Molino del Arco distillery by Nicomedes García Gómez. In 1963, DYC (an acronym for his firm **Destilerías y Crianza**) blended whisky was launched. Sales soon outstripped capacity and in 1973 the firm purchased Lochside distillery in Montrose. When DYC was bought by Allied Domecq, the Scottish site was closed. DYC is now owned by Suntory Global Spirits.

This is a considerable operation, making 8 million litres (2.1 million US gallons) per annum of grain (produced in Valverde del Majano) and 2 million litres (528,340 US gallons) per annum of malt. There is now a 100% Spanish as well as a DYC world blend, while the Colección Maestros Destiladores range includes aged single malts.

The modern Spanish whisky industry started in 2001 when Fran Peregrina started his **Liber** distillery in Padul, 12½ miles (20km) south of Granada. Using local barley and water from the Sierra Nevada, his Andalusian take on whisky, Embrujo di Granada, is, not unsurprisingly, aged in ex-PX casks. There's a mix of fermentation times, spanning short (nutty) to long (fruity). Releases are mostly single casks with an average of five years. Liber is also distilling for Spanish bottlers. Madrid's Santamanía has also joined the whisky game, adding it to its existing gin brand.

To find the new Spain, however, you have to head north. **Basque Moonshiners** was the brainchild of Eduardo Martínez de Murguía, Carlos Ortíz de Zárate and José Luis Navarro, who used Carlos's grandfather's still to 'experiment' with distillation – hence the name.

Above *Iberian whiskies are beginning to emerge.*

The operation went legitimate in 2014 and now Basque-grown Quench and Odyssey Pilsner and Munich malts are double distilled in a Holstein still. The locality also means access to ex-Rioja casks (ex-Bourbon and sherry casks are also used).

'Our philosophy is to work with local raw materials and components as much as possible,' says Carlos. 'We don't understand why we need to use malts from I-don't-know-where if we have here some of the most interesting ones. At the same time we contribute with local richness and local economy.

'Spanish whisky is developing quite fast, but there are not many distilleries doing single malt. We're trying to create a group of producers in order to set the basis of a great product. Whisky is serious business. There are no shortcuts in its production and we like to be honest, transparent and true.'

Also in Euskadi (the Basque Country) is the venerable **Destilerías Acha**, which was founded in 1831, specializing in spirits and liqueurs. Whisky is a more recent arrival, with the project starting in 2006 under the guidance of consultant Santiago Bronchales. The releases are under the Haran brand name.

In Cantabria, **Picos de Cabariezo** has added whisky to a portfolio that began in 2000 when six friends decided to revive winemaking in the remote region by planting vineyards. Distilling *orujo* (the Spanish equivalent of grappa) was a logical next step. Now whisky has joined the portfolio. It's malted on site and distilled in six small stills in the Moorish alquitara style – a pot with a short neck and a water-cooled bowl on top acting as a condenser.

Above *Basque Moonshiners are flying the flag in Euskadi.*

BASQUE MOONSHINERS

AGOT PIONEER 4.6 EDITION 46%

Polished brass and old copper coin. Overripe fruits. Slightly steely with a note of cod liver oil. Robust meaty edge. Tamarind, black cherry and dark malt.

OILY/SPICY

BIKKUN AMERICAN EDITION 46%

Here, PX acts as a blanket of sweetness over Puy lentil, brass and meatiness. Some fresh planed wood, then bait box and soft dried fruits. Light char.

RICH/DEEP

PORTUGAL

Portugal's sole whisky distillery was started by Jay **Venakki** in 2012. His Woodwork single malts are made from Spanish barley and are aged in different Portuguese wine casks.

SPAIN

CENTRAL EUROPE

While the number of whisky distilleries thins in Central Europe, this is by no means terra incognita. Influences are coming from a variety of sources – the Czech Republic's vaunted brewing heritage, Romania's significant wine industry and Poland's long experience in distilling and its deepening love of (Scotch) single malt. It's a region that has the potential to develop further.

Above & opposite *Nestville in Slovakia has added whisky to its portfolio.*

CZECH REPUBLIC & SLOVAKIA

The Czech Republic's **Gold Cock** distillery started in Vizovice in 1877, with whisky a more recent arrival. After a hiatus, production has recently restarted. Český Krumlov is home to the Svachovka leisure complex, which is also home to the Glokner brewery and the **Svach** distillery, whose Old Well single malt, made with wash from the brewery, launched in 2019. Future plans include using barley, rye and smoked malt from the Czech Republic, Germany and Scotland.

Slovakia's **Nestville** in Hniezdne (*hniezdo* means 'nest' in Slovak) is based in northern Spiš. Starting as an industrial alcohol producer, in 2010 a purpose-built spirits distillery was being constructed. A seven-column operation produces grain whisky from a corn, triticale and malt mash bill that makes up 90% of its Nestville Blended. There are also individual distillates of corn, wheat and rye, and a single malt using malted barley smoked with beech and cherry wood.

HUNGARY

Hungary's **Gemenc** distillery in Pörböly is the brainchild of Lajos Szöke, who figured whisky would be a point of difference to the country's fruit spirit distillers. He makes two styles from a small pot and linked column set-up: 'grain' aged for two years and whisky aged for a minimum of three. He uses corn, triticale, rye and barley in nine different guises (some malted, some smoked etc.) and combinations. All are aged in Zemplén oak and acacia.

ROMANIA

One of the newest players is Romania's **Carpathian**, which began producing in 2017 in Bucov. Part of the country's largest wine and spirit business, the Alexandrion Group, it's the pet project of chairman Dr Nawaf Salameh, a long-term Scotch aficionado. 'I didn't have any fear,' he says, 'because I knew what extraordinary natural resources Romania has, and I was surrounded by a strong team, led by Allan Anderson who I brought in from Scotland.'

With malt sourced from the local Soufflet maltings (with the exception of peated, which is brought from Belgium), there's a 72-hour ferment before double pots are put into business. Launched in 2022, a mix of small batch and single casks is emerging, some of which have matured in Romanian wine casks such as Fetească Neagră red.

POLAND

Famous for its huge range of high-class vodkas, Poland has recently begun to move into whiskies. Its first comes from the slightly unusual context of honey producer Mazurskie Miody, which added its **Piasecki** distillery in 2012, with an oak-aged honey spirit arriving in 2014. Langlander single malt and the Bee Keeper blend joined in 2017. The former is aged in Polish oak, the latter in a mix of mead and French wine casks. The malt is Pomeranian.

Whisky Jakubiak single malt had a rather more conventional start, being an extension of the **Ciechan** brewery in Ciechanów. Different beers are used for the wash and early releases have been from single casks made from Polish oak. Another brewer/distiller, **Karczma Wrzos**, based in Ustrón, added its Glen Vistula whisky to an already extensive portfolio of beers and spirits in 2019.

Mundivie, which produces Unicum balsam among many other spirits, has also added a whisky to its range. Wild Fields is a 100% rye and distilled in Jasienica in the Silesian Foothills. Debrzno-based **Millside** began as an importer, then became an independent bottler and is now distilling as well, using Golden Promise and a double pot set-up for its single malt.

CARPATHIAN

CARPATHIAN SINGLE MALT FINISHED IN FETEASCĂ NEAGRĂ 46%

Light farmyard. Old wood, drying grass and ginger, then peanut brittle. Some gamey meat, light green pear, then blackcurrant jelly. Needs water.

FRUITY/SOFT

CENTRAL EUROPE

UNITED STATES OF AMERICA

'I am large, I contain multitudes' wrote Walt Whitman, regarded by many as America's greatest poet, allying himself with every fibre of existence in his country. The USA is vast, complex, a weave and web of geography, climate, cultures. It exists but not in a singular sense. The fact that there is no unity is one of its joys. How could there be, given the dimensions? Everything in America is therefore multiplied, multifaceted, never simple.

American literature is native voices and Black experience, as well as white. American music is country and folk, jazz, rock, hip-hop, soul and classical. Many Americas. Its whiskies are the same – impossible to pin down. It's why Whitman's preceding lines are as important. 'Do I contradict myself/Very well then I contradict myself'.

The contradictions within America's approach to whiskey are to be celebrated. It's made in every state in a multiplicity of different ways. It's bent into shape by mindset, but also conditions. It reflects the story of the country and the people who helped to create its contradictory existence.

In other words, it's not easy, but it is compelling.

'How many American distilleries make whiskey?' I asked my friends there. Some said 2,000, others 3,000. So, a lot. A crazy amount.

This isn't new. In the 1820s, there were 20,000 licensed distilleries in the USA. Pennsylvania alone had 3,500, Kentucky 2,000. By the 1880s, Peoria, Illinois was making 18 million US gallons (68 million litres) a year, 3 million US gallons (11.4 million litres) more than the whole of Kentucky. By the 1840s, consolidation started as well as the first stirrings of Temperance, and the total number of US distilleries fell to 9,500, which further reduced by the end of the Civil War. What's happening now is a re-establishing of a lost heritage.

This also extends to the re-emergence of regional expressions. In the 18th century you could choose from Maryland's rye/corn mash bill, the almost exclusively rye mash bill of Pennsylvania's Monongahela rye, Kentucky's Bourbon and Tennessee's variation. There were pots and three-chamber stills, beer columns and thumpers; there were blends and adulterated filth from rectifiers. Sourmashing and sweet, malted rye, oats, different varieties of corn. American whiskey wasn't fixed. The same is happening now, and those who think that Bourbon only comes from Kentucky are being shown that, well, it doesn't.

Right *A new dawn is breaking on American whiskey.*

Above *The stills at St. George help to add to its complexity.*

Above *Checking progress at McCarthy's.*

America's whiskey suffered in the 20th century. Thanks to Prohibition it had shrunk to being a speciality of two states, Kentucky and Tennessee. As the geographical spread shrank, so did variety: pots and the three-chamber stills disappeared, and rye was reduced to a speciality kept alive by stubborn old distillers.

The revival started with Bourbon's pivot to premium in the 1990s; the return of cocktail culture and its fascination with pre-Prohibition drinks gave rye a boost, as did the craft brewing explosion whose lessons, after legislative changes, were soon being applied to distilling.

Then things started to change. St. George (see page 238), Stranahan's (see page 233), McCarthy's (see page 240) and Old Potrero (see page 238) demonstrated that American whiskey could be something other than Bourbon, that single malt was a possibility.

It began to spread and as it did so two camps began to form. On the one hand, there are distillers who want to make a whiskey that takes into

account the conditions where the distillery is located, and experiment with 'new' styles. Farmers, seeing the rise in sales and realizing that their crops are going into someone else's bottle, are taking charge of their destiny. Craft brewers are bringing their knowledge of yeast and malt to create something very different. Heritage varieties are being saved and used, and there's a wider examination of oak, smoking over mesquite and scrub oak or finding local peatlands. Sweet mashes, multiple cereal mash bills, 100% rye whiskey, high-malt Bourbon, blends.

Others have taken a different tack and have conformed to the Kentuckian template because that's what consumers believe American whiskey is. Play safe, or ask questions? What's clear is that there is now greater diversity within American whiskey than at any time in its history. It's great to see developments like Uncle Nearest (see page 215), greater Black representation and many more women distillers. The message that whiskey is not just for white men is getting through, but slowly.

There are other challenges. If you wish to be a local distiller, only selling in your state, then you can, possibly, make a decent living. If you want to go national, you need to get distribution, but there are only two firms with that reach, meaning you are stuck in with everyone else. Retailers' shelves are not infinitely expandable, neither are back bars. Prices are all over the place. It's a joy to see so many whiskies, but are there too many? Undoubtedly there are in terms of the domestic market. America is large, but no distiller I spoke to thinks it can absorb all the whiskey being made. That means a fallout (which will happen) but also a push to export.

The sheer volume of distilleries meant that tough decisions had to be made about who to profile here. The established, the creative, the ones pushing American whiskey forward. A cross section that shows the diversity of the scene. A few of the multitudes.

THE NORTHEAST

Forests and seashores, islands and cities, it's impossible to corral the Northeast into a 'region'. In whiskey, it revels in its differences – rye, single malt, new approaches to Bourbon. In some ways, it's going back to an almost forgotten heritage – New York was a centre of distilling in the 17th century and home to some hundreds of stills pre-Prohibition. The desire to make whiskey wasn't lost but hampered by laws – prohibitively costly permits in NYC and a hangover in Vermont, which went dry in 1853.

VERMONT

Is Vermont rye country? It should be, thanks to **WhistlePig**, the distillery founded by the late distiller/consultant Dave Pickerell, which restarted the whiskey style in an abandoned Shoreham dairy farm in 2007. His aim was a 'rye-velution', to re-establish it as America's whiskey, initially sourcing and blending casks from other distillers in the USA and Canada to create a signature style. Today, its whiskey programme is headed by Liz Rhoades, along with head blender Meghan Ireland.

Starting with aged sourced whiskeys, Liz has been steadily increasing the amount of whiskey made at the distillery from grains grown on the 740-acre (300-hectare) farm.

Denko rye was selected after field trials to assess yield and viability and then how it performed in the distillery. 'Sustainability isn't just utilizing resources in the most efficient way,' she says. 'The agronomics also have to be correct.' There was also the not insignificant matter of flavour, which she puts down to levels of 4G-hydroxycinnamic acids in rye's cell walls – the precursor for rye's baking spices.

'Rye can be bold and in your face, but there's a lot of different sides to it driven by field, climate and variety. It can go in any direction,' she says. 'There's Kentucky rye, Empire rye, Monongahela, Maryland or three-chamber – that's what makes it cool.'

WhistlePig runs a secret mashing regime to transform these precursors with the aid of a specific yeast (also secret). Ferments (in one of the old barns where a snoozing pig made the sound of a whistle) are short, and distillation is in an eight-tray hybrid.

It leads to a range celebrating rye in many guises: a flagship 10-year-old, a (sourced) 21-year-old single malt, a wheat and rye whiskey blend and the FarmStock Beyond Bonded, using 100% own-grown grains.

'None of this is plug and play,' she says. 'We're always innovating. Rye is how we started. Now it's wider thinking.' A single malt will come, and there are trials with new strains of barley and rye, grown not just for yield but new strains that can cope with climate change.

Above *WhistlePig helped to usher in a new era for American rye.*

For Liz, it all comes back to the field. 'It's all about agriculture,' she says. 'Whiskey should be about the taste of place. If distillers don't pay attention to it...well...'

WHISTLEPIG

WHISTLEPIG LIBRARY SERIES 11 YEAR OLD #2407 55.3%
Robust. Dry tea leaf, varnish, acetone, cold butter. Coffee, then aniseed/caraway, manuka honey. Complex with orange peel and crackling spice.
SPICY/HERBAL

★ WHISTLEPIG LIBRARY SERIES 11 YEAR OLD #20832 59.8%
Crème brûlée, caramel toffee, then allspice and oak. Thickly textured, slow start, then sudden spice (menthol, peppercorn, caraway). Drying at the end. Complex.
SPICY/HERBAL

★ WHISTLEPIG OLD WORLD 12 YEAR OLD 43%
Divided between and finished in Madeira (63%), Sauternes (30%) and port (7%) barrels before blending. Sauternes to start: apricot jam, honey, overripe peach, then blossom, red fruits. Elegant and complex. Light lavender, aniseed and blue fruits.
SPICY/HERBAL

THE NORTHEAST

NEW YORK STATE

There is a stark difference between the bucolic farmland setting of WhistlePig and the middle of Brooklyn. Although urban whiskey distilling is not unusual in America, what is normal in Louisville raised more than a few eyebrows in the New York borough when **Kings County** started in 2009.

'Brooklyn is a melting pot of cultures and a hotbed of creativity, things that are a part of our approach to whiskey making,' says Kings County's founder Colin Spoelman. New York's distilling boomed during the 19th century, then ground to a halt during Prohibition. It took a change in law in 2004 for legal distilling to restart.

While Colin's initial aim was to revive this lost tradition, the focus has shifted slightly. 'As we've moved to older and age-statement

KINGS COUNTY

KINGS COUNTY STRAIGHT BOURBON BOTTLED-IN-BOND 7 YEAR OLD 50%

Honey, mace, dried red fruits, light nut, cinnamon and rosewater, then peppermint, toffee. Balanced with some nuts and pul biber.

SWEET/VANILLA

KINGS COUNTY EMPIRE RYE 51%

Low spice and more on the crushed fruit side. Deep and slightly oily. Cranberry sauce, black fruits, low spice. Concentrated and mildly medicinal with thick plum jam, raisin and tingling spice.

RICH/DEEP

Above *The imposing entrance to Kings County.*

whiskeys, we've adopted a more international outlook, making whiskeys that borrow the best of distilling cultures around the world. It's unusual, but more interesting than trying to replicate Kentucky Bourbon in Brooklyn.'

Kings County's whiskies are all 'uncompromising, old-tradition pot still whiskeys. Because pots make better Bourbon', with a mash bill utilizing a higher than average ratio of malted barley. There are also open-top, off-grain ferments, sweet mashing and a huge array of casks.

'We're a part of a group of determined craft makers that have focused on provenance,' he says. 'We were initially dismissed as young and expensive, but the market has really moved to us.'

For him, small is not only beautiful but essential. 'Whiskey is best made at a small scale, where the farmer and distiller know each other, the farms are close to the distillery and the consumer can come see every part of the making process.

'Globalization and consolidation destroyed a lot of authenticity in American whiskey, and small distillers are just now clawing it back, creating idiosyncrasy and sublime spirits that for 100 years have been value-engineered into being completely undifferentiated and homogeneous. There's an opportunity to create variety and differentiation.' Here is the voice of a resistance, an overt critic of the 'monolithic' distillers who he feels have 'usurped' American whiskey.

Also Brooklyn-based, **New York Distilling Company** was started in 2008 by Tom Potter, one of the founders of Brooklyn Brewery. There's Bourbon, rye, single malt and Ragtime Rye from New York State-grown rye.

In 2003, Ralph Erenzo was looking to establish a climbing school in upstate New York. When his plans were thwarted, he decided to

EMPIRE RYE

In 2015, a group of six distillers (**Black Button**, **Coppersea**, **Finger Lakes**, **Kings County**, **New York Distilling** and **Tuthilltown**) agreed on a set of rules to define a rye whiskey made in New York State. Under the agreement, Empire Rye must be mashed, fermented, distilled, barrelled and aged at a single New York State distillery; contain a minimum of 75% New York State-grown rye (raw, malted or a combination); the spirit distilled to no more than 160° proof (80% ABV); aged for a minimum of two years in charred, new oak barrels at not more than 115° proof (57.5% ABV). The low barrel proof was the norm for the pre-Prohibition ryes.

SAGAMORE

SAGAMORE SPIRIT RYE 41.5%

Long pepper spice mixes with a herbaceous element. Sweet start and gently drying before the spices (clove, allspice) emerge quietly with a zing of orange peel. Overall gentle.

SPICY/HERBAL

SAGAMORE SPIRIT DOUBLE OAK RYE 48.3%

More oak and boiled sweets. Higher concentration of sweet fruits, adding some hazelnut butter, coconut and cacao, given lift by bubblegum esters.

SPICY/HERBAL

★ SAGAMORE SPIRIT CASK STRENGTH RYE 56.1%

Heavy and thick with some boiled sweets, dark marmalade, treacle, then allspice and tangerine. The herbal element is retained. Has elegance.

SPICY/HERBAL

renovate an old grist mill and start a distillery instead. So was born **Tuthilltown**. He also lobbied successfully to pass the Farm Distillery Act, allowing the state's farms to establish distilleries – and retail on site.

Now owned by William Grant & Sons, the range has been tightened with Hudson Whiskey Bright Lights, Big Bourbon as the three-year-old core grain-forward straight Bourbon made from 95% corn and 3% malted barley, while Hudson Whiskey Do The Rye Thing is boldly rye-forward (95% rye and 5% malt barley). The same ratios are used for the considerably more gentle corn expression.

MARYLAND

In pre-Prohibition days, Maryland was famed for its ryes. In 1910, there were 44 distilleries in the state, half of which were in Baltimore. Based on a rye/corn mash bill, Maryland rye was a softer style to the punchy ryes from Pennsylvania. It wasn't until 2015, however, with the arrival in Baltimore of **Sagamore Spirit** that the state began to produce once more. All of its grain comes from 100 farmers in the state. Some sourced whiskey continues to be used, though the aim is to become wholly self-sufficient.

Above *Local sourcing is important to Sagamore Spirit.*

MASSACHUSETTS

It was in 2000 that Randy Hudson began to make The Notch single malt whiskey at Nantucket's **Triple Eight Distillery**. The decision to take that path rather than Bourbon or rye was a simple one. 'We're a brewery and had no facility for distilling grain, so it had to be single malt.'

What that means is using the old brewing variety Maris Otter: 'The orange kind of fruit it gives is one of my favourite flavours in beer and whiskey – that's the beauty of the grain. I was told from Day 1 that we should find a different variety because of XYZ – Z being it would be discontinued because of its low yield, but we've done our part to let farmers know we find what they do is beautiful and there's a reason to continue to grow it.'

Coming at whiskey from a brewer's mindset ran counter to conventional thinking. 'The distilling folks I spoke to early on spoke of yeast being yeast, grain being grain and yield being king. I said, "I don't think you're right. I'm just going with my gut"', he laughs. 'I think using Maris Otter and a combination of brewer's yeasts makes things taste better with a broader range of flavours.'

The wort isn't pasteurized: 'I know some people do, but that misses out on possibilities of flavour', and ferments are between five and ten days to get acidification using yeasts that drive up esters.

A stripping column and Holstein pot gives a spirit between 73 and 80% ABV that's barrelled at 55%. 'A cue we took was from brandy distillers (Triple Eight started by distilling brandy from its winery) was using "faibles" (whiskey diluted with water) to slowly lower proof in barrel. I think we get less astringency and tannins. It makes it softer.'

Some speciality malts have been trialled in recent years and there's peated (made from Scottish barley) in cask.

While Nantucket's sand and oceanic climate means cereal growing isn't possible, Randy feels that the environment has a direct effect on the style. 'The feeling of the air is different here – invigorating. When you come over by boat or stepping off a plane, the difference is stark. Those puffs of island air make a difference to the whiskey because it's bathing in it for years.

'There's not another whiskey like The Notch because there's not another place like here. That's super-important. We make great decisions on making whiskey up until we put it in barrels. Then it is really a product of this place.' Those barrels are a mix of ex-Bourbon and only a little new wood: 'Single malt doesn't seem to behave as well in new American oak.' He also shaves and toasts some of the old wine barrels. 'It's important to have a broad range of flavour,' he says. 'One of my passions is oil painting. I couldn't imagine painting without one colour or another. Blending is our strong point.'

TRIPLE EIGHT

★ THE NOTCH 8 YEAR OLD 48%
Rich and sweet. Fermenting oranges, ras el hanout, toasted raisin, brambles with structured oak giving grip. Mulled spice, hint of malt, black grape and caramel. Elegant.
FRUITY/SOFT

THE NOTCH 12 YEAR OLD 48%
Fragrant with a mild grassy side, tomato leaf, Assam tea and more acidity. Hoppy and crisp moving into sumac and supple oak.
FRAGRANT/FLORAL

★ THE NOTCH 15 YEAR OLD 48%
Fatter fruit. Complex toasted oak. Ripe and powerful with sweet berry-filled chocolate, but still the balancing lift of dried grass, citrus and generous malt.
RICH/DEEP

RHODE ISLAND

After a career in finance, in 2009 Mike Reppucci founded **Sons of Liberty** in South Kingstown, thereby restarting distilling in the state for the first time since Prohibition, though in a different fashion, as he was making single malt. It started, he says, with the question that Randy Hudson had also asked: 'Why is no one distilling single malt from the beers we love?' In tandem with Dave Pickerell, he applied brewing thinking to whiskey making, starting with the understanding that there's a difference between a distiller's wash – a crude ferment of barley – and what you would drink as, well, beer. So he widened the grains and used different kilnings, expanded the yeast recipe and distilled in a pot with rectifying plates.

It's first release, an oak-infused distilled stout called Uprising, set the tone. A move into flavoured whiskey followed with Pumpkin Spice, hopped whiskey and Battle Cry. Now the portfolio is split between single malt releases and a four-strong flavoured range.

Uprising is no longer infused but aged in a mix of new American oak and toasted French oak. There's also a 5-year-old single barrel bottled-in-bond version (see box, page 135), while Battle Cry uses malted rye and honey malt and a Trappist yeast. Beer thinking applied to whiskey.

CONNECTICUT

In the northwest of the state is the small town of **Litchfield** where the Baker family started to make their whiskeys in 2014. Like so many of the new players, a main consideration was finding connections between location and the character of their new creations.

'The importance of location, place and terroir in shaping our whiskey is paramount,' says Robert Baker. 'The mineral-rich terroir of New England is the secret sauce in our whiskey.' The mash bill of 70% non-GMO corn, 25% rye and 5% malted barley is distilled slowly and released as a 4-year-old straight Bourbon at entry level, with the flagship being a year older and double-barrelled. There's also an arrangement with a local maple syrup house that uses ex-whiskey barrels for ageing and then returns them to the distillery to be filled with Bourbon. Other used barrels are sent to coffee roasters and breweries.

It's not easy setting up any new distillery, especially in a part of the country not known for whiskey making. Robert remains optimistic, however. 'We see a bright future for whiskey. The Bourbon category will grow steadily, and we're enthusiastic about the future of American single malt.' This is Litchfield's latest avenue of exploration, a classic example of long-term thinking and an appreciation of the need for hard work.

LITCHFIELD

LITCHFIELD STRAIGHT BOURBON 46%

Big, sweet, focused. Honey, light toasted oak, greenhouse/tomato, mulch, tobacco with top notes of bright citrus. Good length. Roasted spice with a zing of grapefruit.

SWEET/VANILLA

LITCHFIELD STRAIGHT BOURBON CASK STRENGTH 61%

Autumnal, boiled sweets, fruit syrups, wood oils. Dynamic. Hefty, the depth comes from the distillate, not wood; petrichor, chewy feel.

FRUITY/SOFT

LITCHFIELD DOUBLE BARRELED BOURBON 5 YEAR OLD 45%

Has presence. Big, dry, oak notes to start, then cocoa powder, burned caramel, light fir tree. Dries slightly on the palate, hint of roasted corn.

OAKY/DRY

★ LITCHFIELD PORT CASK FINISH BOURBON 46%

Bittersweet. The port nudges things into Griottines cherries, tomato leaf, sloe and grilled pepper with rye adding mischief. A release of fruit at the end.

FRUITY/SOFT

Above *Terroir-based thinking drives Litchfield.*

Above *Tamworth takes a 'hyperlocal' approach to distilling.*

NEW HAMPSHIRE

According to founder Stephen Grasse, **Tamworth Distilling** (founded 2015), which sits in the foothills of the White Mountains in New Hampshire, takes a cue from the 19th century Transcendentalists of New England, who believed in the need to reject society in preference for a life of self-reliance and spending time in nature.

'We draw inspiration from the surrounding atmosphere,' Stephen says. 'Our recipes are born from the crops we grow, and the histories buried in the soil below. Terroir is an irrefutable part of our distillery ethos.'

Above *Tamworth's distillery is home to some transcendental ideas.*

A wide range of spirits is made using 'hyperlocal' flora, all farmed or foraged within 150 miles (240km) of the distillery. 'We've set up our distillery in accordance to the land around us, not against it. A single 750ml bottle results in some 6.5lb [3kg] of carbon dioxide being released into the atmosphere.'

His aim is to show that there is an alternative to the standard approach of whiskey making. 'There's good Bourbon being made outside of Kentucky, the interest in rye and the calling out of specific varietals of grain.

'My goal was to go where no other distillery had gone before, to explore everything that is possible in the process of transmuting the life force of plants and grains into spirits.' That has resulted in whiskeys such as The Deer Slayer (whiskey infused with venison) and Crab Trapper, based on its 4-year-old Old Man of the Mountain Bourbon (82.4% organic corn, 11% organic Maine rye, 6.6% malted barley) blended with distilled green crab stock and a distilled spice mix.

If you're confused about the apparent contradiction between talking about sustainability and then using 90lb (41kg) of crabs for each batch, green crabs are an invasive species that is destroying New England's coastal ecosystem.

TAMWORTH

TAMWORTH CRAB TRAPPER 46%

Aroma of old usquebaugh, sweet saffron, coriander seed. Crabbiness comes on the palate. Marine, sour citrus. Does what it says on the label.

SPICY/HERBAL

KENTUCKY & TENNESSEE

Imagine for a moment that Prohibition hadn't happened in the USA and that whiskey making in Kentucky and Tennessee had continued to evolve in the same way as it did around the world: larger plants, consolidation, branding. It would have meant whiskey being produced throughout the 1920s. There would have been no need or desire for bootlegged Scotch or Canadian whiskies, meaning that neither country would have been able to establish a strong foothold.

Post-World War II, Scotch would not have been able to draw on its established reputation to rebuild itself – and the UK economy. And Canada would not have had a thirsty outlet for its whiskies. The whisky world would have been a very different place.

At least Kentucky and Tennessee survived, unlike whiskey distilleries in the rest of the country, even if the full rebuild of their whiskey industries was a long process as Prohibition shut down most of the states' producers. Those who reopened then saw plants closing in the war, effectively meaning there was hardly any stock being laid down between 1919 and 1946.

By the time Bourbon was up and running again it was fighting for shelf space with super-confident Scotch and Canadian industries, both of which made whiskies that were of a different style to that traditionally made in Kentucky and Tennessee.

The response of some distillers to go light to try and appeal to this changed palate was a misstep. It would not be until the 1980s that the

Above *Rickhouses (here at Heaven Hill) are flavour-generating machines.*

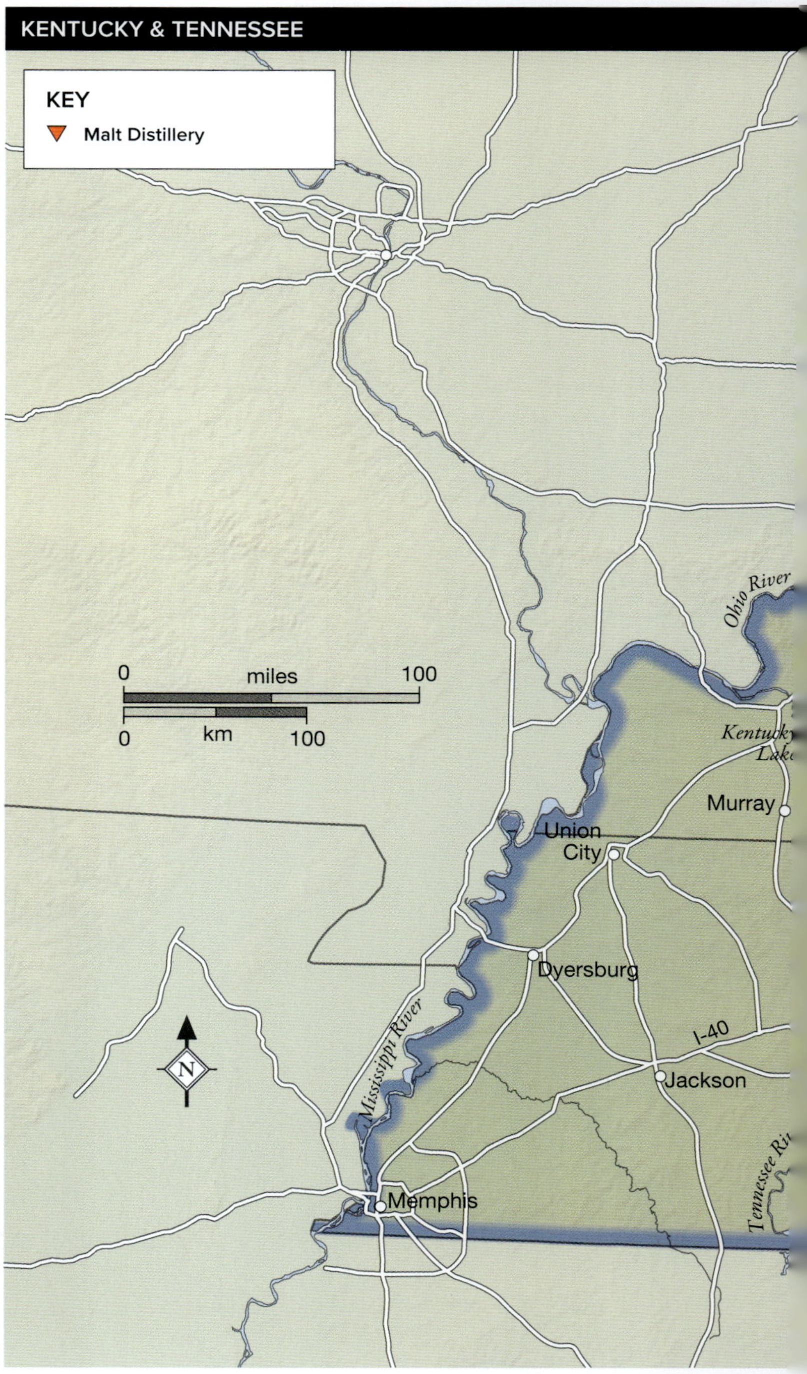

idea that bolder, richer and indeed more 'Bourbon-y' flavours was the way forward took hold.

Now distilling is booming in both states with that boldness in many cases now being turned up to 11. Another era has begun.

KENTUCKY

Kentucky is Bourbon's heartland, earning the state $9 billion a year. In 2022, according to the Kentucky Distillers' Association [KDA], its 100 distilleries employed 23,000 people and made more than 2.7 million barrels of Bourbon, while 12.6 million barrels were lying in warehouses. Bourbon has also boosted the state's agricultural base, which has a seen a 300% increase in production of corn since 2009. Overall, distillation has risen by 475% since 1999. Good times – or a reckless expansion into over-capacity?

LOUISVILLE

Although **Heaven Hill** is based in Bardstown, its whiskeys are distilled in Shively, Louisville because of what happened on 7 November 1996, when a rickhouse on the top of the hill overlooking the firm's distillery on the outskirts of Bardstown caught fire. High winds spread the flames from one warehouse to another and soon a river of flaming whiskey headed downhill to the distillery, which erupted. The firm had lost 92,000 barrels and the place to make its Bourbon.

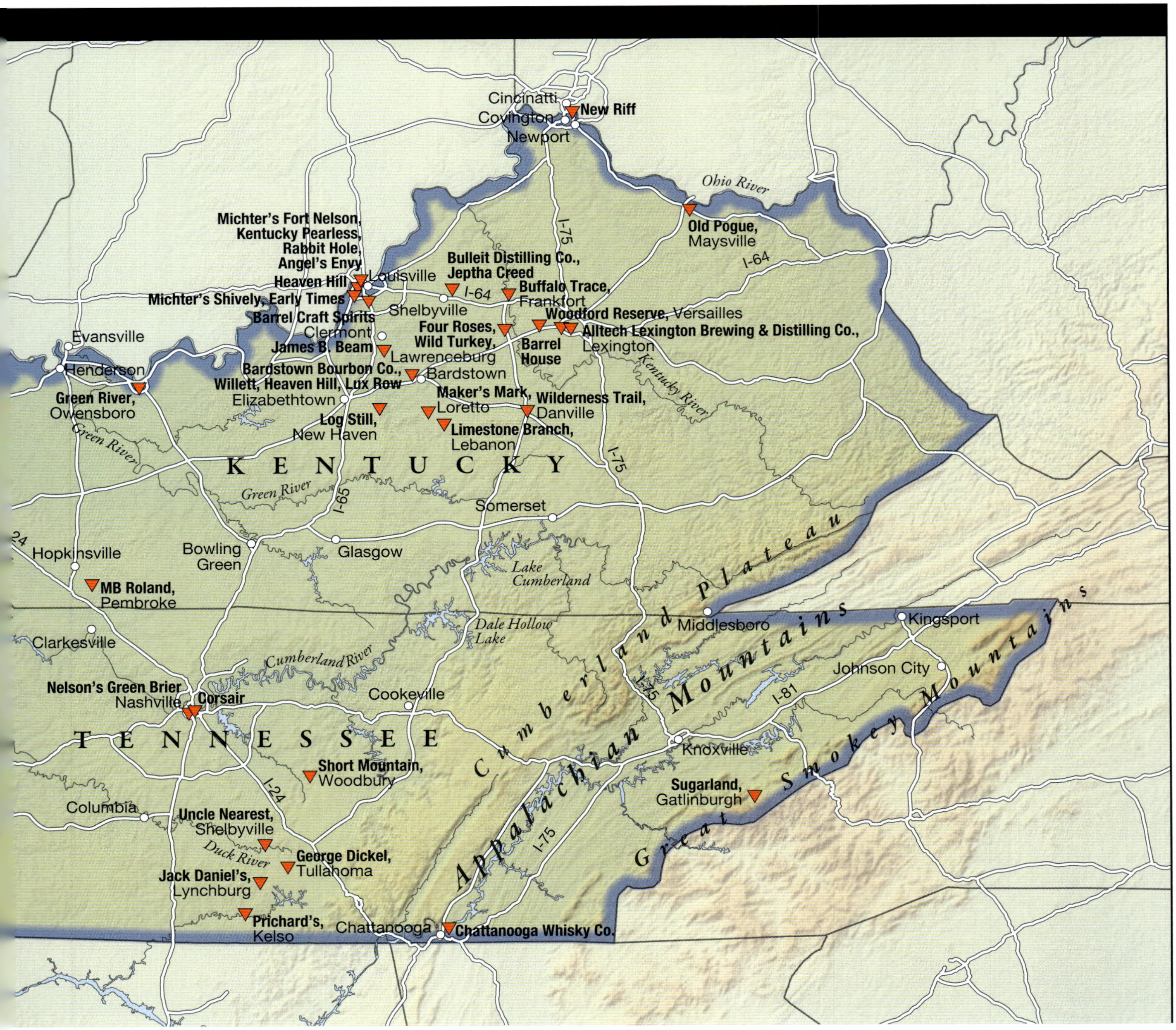

Above *Distillers at Heaven Hill have a flavour map of each floor of a rickhouse.*

After a period sharing space with other firms, in 1999 Heaven Hill bought the Bernheim distillery in Louisville, which Diageo had closed down. Every distillery has its quirks and personality. Bernheim was highly automated, Heaven Hill was more hands-on. Bernheim had less copper in its still, Heaven Hill's was rich in it. So, readjustment was needed. Out went much of the automation, in went copper.

There was expansion in 2007 and 2014 and the site is now making 300,000 barrels a year. In 2013, the Evan Williams Bourbon Experience opened with a small distillery making one barrel a day.

Heaven Hill's strategy has always been about seeing the gaps, and taking the risks that other firms might have been leery about. It has pioneered wheat whiskey with Bernheim, reintroduced corn (Mellow Corn) and kept faith with rye (Rittenhouse), while under former master distiller Parker Beam in the late 1980s it was a pioneer of premium. Where was the Bourbon equivalent to single malt Scotch?

The answer was Evan Williams Single Barrel, Larceny, Henry McKenna and Parker's Heritage; and bottled-in-bond variants (distilled in a single season in one distillery, aged for four years, then bottled at 100° proof/50% ABV).

Within the rickhouses there are now temperature sensors that feed back data to better understand how climate affects the barrels inside and the flavours created by the cooler, thicker air at the bottom where proof drops, compared to the hotter air at the top where temperatures can reach 145°F (63°C) and strength rises.

Flavour isn't just mash bill but airflow, temperature, orientation. That understanding allows blending between different levels or pinpointing sweet spots for specific flavours. A complex mix.

No matter what angle you take when you're writing about an American distillery, at some point you'll hit Prohibition and its effects. It heralded the ending of one era, while Repeal saw the stumbling start of another. It wiped away styles and distilleries, and wrote off family businesses and thinking about how to make whiskey. Consolidation and trademarks meant brands shifted ownership and distilleries moved state. Telling its story is trying to restitch a ragged tapestry.

Michter's is a case in point. Its convoluted story doesn't start in Kentucky but Schaefferstown, Pennsylvania in 1753 with a man called John Shenk. By 1860, the distillery was owned by Abraham Bomberger. Among the brands it made was A.H. Hirsch. It reopened under new owner Louis Forman after Repeal, who renamed it Michter's after his sons MICHael and peTER. The distiller was Charles Everett Beam who created an 'old style' pot still whiskey called Michter's Original Sour Mash. In 1989, bankrupted, the distillery closed. But in American whiskey there's often a second or third act.

In the 1990s, Joseph Magliocco bought the trademark and relocated the business to Kentucky, restarting the brand with sourced whiskey and then cuckooing in other facilities, before opening its own site in Shively, Louisville in 2012. There are two pots plus a 32-inch (81cm), 46-foot (14-metre) beer column with doubler. A smaller distillery – where the Schaefferstown distillery's two pots are – with a visitor centre operates in Louisville's Fort Nelson district.

The question facing the new site was what direction to take? Michter's is secretive when it comes to details, but Original Sour Mash has returned with a 'fair amount' of rye and corn, suggesting that the mash bill is neither classic straight rye nor Bourbon. An American whiskey uses refill barrels (but no neutral spirit).

MICHTER'S

★ MICHTER'S US*1 SMALL BATCH BOURBON 45.7%

Direct and sweet, a touch of boiled sweet, raisin, sultana and dried black cherry. Moves to aniseed, cinnamon bark and a little peppery rye. Thick and sweet on the palate with lush fruits.

SWEET/VANILLA

★ MICHTER'S 10 YEAR OLD RYE 46.4%

An almost feral mix of dusty spice, sour cherry and polished oak. Soft tannins with the rye fizzing away. Cherry tunes, cigar leaf, dried oregano and mint chocolate. Refined.

SPICY/HERBAL

HEAVEN HILL

ELIJAH CRAIG SMALL BATCH 47%

A classic Kentucky Bourbon. Green garden twine, bottled cherries, thick sweetness, vanilla, nutmeg and cooling mint. Hits all the right notes.

SWEET/VANILLA

★ HEAVEN HILL BOTTLED-IN-BOND 7 YEAR OLD 50%

Big-boned. An intriguing floral element alongside black cherry, bramble briar. Danish oil, then pitch pine and spice. Bitter chocolate, Brazil nut, allspice and cherry jam.

FRUITY/SOFT

OLD FITZGERALD BOTTLED-IN-BOND 13 YEAR OLD 50%

Classically framed. Cherry spice, citrus, oak, dusty/earthy rye with balanced sweet, sticky fruits. Moves to peach tea, velvety tannins and burned pastry. Balanced.

SWEET/VANILLA

Above *The golden glow of a fermenter at Peerless.*

The majority of Michter's is barrelled at 103° proof (51.5% ABV), significantly lower than the norm (the rye is at 125° proof/62.5% ABV), marking a return to the standards of the 1950s–70s. Low proof means more water, more barrels and more space, but it also changes the way in which spirit and oak interact, lowering astringency and oak tannins, and raising wood sugars and esters. Heated warehouses give year-round interaction, rather than the high summer and low winter activity typical of much of Kentucky ageing profile. Old thinking, new understanding.

The idea of revival is inextricably linked to the world of new American whiskey – styles, approaches but also names and family histories. **Kentucky Peerless** is one of those. Originally made in the E.W. Worsham distillery in Henderson, its reputation grew in the 1890s under the ownership of Henry Kraver. Though it produced 'medicinal whiskey' during Prohibition, it foundered afterwards. Then, in 2014, Kraver's great-grandson, Corky Taylor and his son Carson bought the lapsed trademark and restored an old building in downtown Louisville to house their new, revived distillery.

Unlike many start-ups, they decided not to use sourced whiskey or follow the sourmash route. 'Sweet mash is unconventional,' says COO Cordell Lawrence, 'but there's no risk of cross-contamination – just fresh water, grain and yeast. It's more labour-intensive, as there is more cleaning and the grain cost is 20% more, but I believe you get a more complex result.'

It's another that barrels at low strength – here it's 107° proof (53.5% ABV). 'When you go up in proof, you'll pull out harsher tannins,' Cordell says. 'Herbs, fruits, botanical notes and freshness are easier to capture at lower proof.' There's also no chill-filtering to preserve mouthfeel.

Distillation has been overseen by Caleb Kilburn who, after tinkering with home distilling in the family barn, worked at Barton, Beam, Four Roses and Wild Turkey. 'Essentially, he picked the best points of difference from each and put them in place at one site,' says Cordell. 'He's shown us things that we thought impossible. He's pushed the envelope.'

Above *The long and short of it at Peerless.*

KENTUCKY PEERLESS

PEERLESS HIGH RYE BOURBON 54.3%

Rich, with dark fruits, cigar and wet canvas. Textured with grippy wood. Bold rye (on the clove side), roasted spice, burned sugar. Boisterous, sweet, plummy/damson.

SPICY/HERBAL

PEERLESS DOUBLE OAK BOURBON BATCH 8 54.35%

Highly concentrated. Pigskin leather, charred red pepper, high cacao, dark fruits, cherry cola, hickory. Resinous. Well-textured.

OAKY/DRY

PEERLESS STRAIGHT RYE 54.7%

Red fruits, clove, green cardamom. Dense and grippy. Concentrated sweetness, blackberry. Slightly oily, loosening the grip.

SPICY/HERBAL

While mash bills are secret, the ferments are three to five days and the white dog comes off at 130–131° (65–65.5% ABV). 'We're doing that to concentrate flavour and complexity by adding water at the front end, not diluting the final product,' he explains. 'Ageing in #3 char with heavy toast in a single storey rickhouse also contributes, as there's less temperature fluctuation. The key is building complexity at every stage.' It's not a shy style, and takes time for the dense flavours to open and reveal a dark honeyed sweetness.

'American whiskey, maybe all whisky, reflects the issues facing the world,' says Cordell. 'Growth at all cost or being condition-dependent/resource-aware. If that restricts capacity, so be it. Do you want to get bigger to make money, or make better quality?' It's clear which side of that line Peerless is on.

Shively is also home to Brown-Forman's **Early Times** distillery where that brand and Old Forester are made using two distinct recipes. The 'old-fashioned country style' found in Early Times starts with a mash bill that is 79% corn, 11% rye and 10% barley malt, and uses the IA yeast strain from the 1920s. On the other hand, Old Forester's mash bill ups the rye to 18%, which adds an extra dusting of spice. It too has its own yeast.

CLERMONT

What does a major distiller that has helped set the tone, dictated the pace, defined the styles do when young whippersnappers emerge not just across the country but in their own backyard? Ignore them and keep on the same course that has worked for decades...or adapt to the new landscape? You just have to wait to see what emerges in bottle.

A case in point is the **James B. Beam Distilling Co.**, Bourbon's largest producer whose main site, the Fred B. Noe Distillery, is in Clermont. Opened in 2021, it's where the firm's ever-growing portfolio of super-premium and small-batch whiskeys is made, as well as Clermont Steep, the 5-year-old, column-distilled, aged in toasted/#1 char single malt, Beam's first foray into the category. That range alone shows that Beam isn't sitting back and relying on White and Black Label to do all of the heavy lifting.

Maybe that's no great surprise. The Beam story is one of resilience and a willingness to change. Jim Beam and his son Jeremiah restarted the business after Repeal in 1935 and the family is still there. Booker Noe – Jim's grandson – was master distiller, and his grandson Freddie is now at work.

JIM BEAM

JIM BEAM WHITE LABEL 40%
Dry, peppery/papery. Lemon with a dusty back note, then corn. Overall prickly with some tart red fruits. Light and made to be mixed.
SPICY/HERBAL

JIM BEAM BLACK LABEL EXTRA-AGED 43%
More time and more oak-driven character. Some nutmeg, toffee and caramel, then the lemony Beam bite and dusty rye.
FRUITY/SOFT

CLERMONT STEEP SINGLE MALT (JAMES B. BEAM) 47%
Rich in the American Single Malt (ASM) style (see box, page 240). New leather shoes, dried pear and apple, malty, celery. Firm oak. Starts light, then highly spiced elements emerge.
SWEET/VANILLA

Above *Whiskey-making at scale: Jim Beam.*

The premium Bourbon movement in the 1980s wasn't started by marketing departments but by distillers like Booker Noe, Jimmy Russell, Parker Beam, Elmer T. Lee, Bill Samuels and Jim Rutledge. Booker's, Knob Creek and Basil Hayden showed an alternative route. Flavours dictated not by mash bill but by distillation and barrel proof, position in rickhouse and time.

Now, there's Hardin's Creek Kentucky Series, three 17-year-old Bourbons made from the same mash bill and aged for the same time but at different 'campuses': Clermont, Frankfort and Boston. Each shows different flavour characteristics. Even with these and Freddie's fascinating explorations into blending with the annual Little Book releases, it's single malt that is exciting them the most.

'We foresee American single malt whiskey as the new frontier,' says the firm's Kate Nelson. Even more surprisingly she adds, 'and we see it even surpassing Bourbon in the next five years.' Stay still and you get eaten.

Above *It all starts with the grain.*

BARDSTOWN

It's hard to keep track of brands in the complex, ever-shifting world of American whiskey. Consolidation means brands can get passed around like swap cards. Then there's sourced whiskeys with a myriad of different names alluding to distilleries that don't always exist. Many of those draw liquid from MGP. A lot also get it from **Bardstown Bourbon Company** (BBC), the giant distiller you might not have heard of.

New distilleries need whiskey to tide them over until their own matures. Since 2014, BBC has answered that need. Starting with a single column still, by 2024 there will be three, and production of 10 million US gallons (38 million litres) a year.

'One of our pillars is to promote collaboration and transparency within the industry,' says brand manager Brandon Smith. BBC's 'collaborative distilling program' allows brand owners/distillers to develop their own mash bills, ageing techniques and flavour profiles. They can also draw from 50 different mash bills. There's also custom distilling, giving distillers access to data points within production to analyse the process in real time. The software is the first of its kind to be used in spirits. Brands using it include Jefferson's, High West, Kentucky Owl, Blue Run, Hirsch, Calumet, James E. Pepper and Cyrus Noble.

There's more to BBC than this. It also makes its own under the watchful eyes of master distiller Steve Nally, who worked previously at Maker's Mark and Wyoming Whiskey. In an industry where mash bills and yeasts and production details are often frustratingly secret (even though no one could copy them), BBC is refreshingly open – as seen on the labels of its own products.

Its Origin Series comprises three 6-year-old whiskies: a Kentucky straight Bourbon, a wheated Bottled-in-Bond and a Straight Rye. Fusion blends younger Bourbon from its own mash bills with old, sourced Bourbon. Each release in the series increases the amount of the BBC distillates. The Discovery Series blends whiskies from around the world, and there's also a Collaborative Series of finishes.

It's an interesting balance. 'Contract distilling has been the foundation and will continue to be key,' says Brandon. 'After all, rising tides lift all ships. It injected cash flow and allowed expansion, allowing us to wait to release Origin. As we continue to grow, and our barrels age, we will continue to integrate our house-distilled mash bills.'

There's ambition at work here. The aim is to be a top ten super-premium whiskey brand worldwide. By 2025, full global expansion will take place. Be prepared, and be aware.

The **Willett** family's association with Bourbon starts in the 1860s when John David Willett became part owner of the Moore, Willett & Frenke Distillery. It wasn't until 1936, however, that Thompson Willett broke ground in Bardstown to build the family's first wholly owned distillery, which ran until the early 1980s.

In 1984, it was renamed Kentucky Bourbon Distillers (KBD) and changed from distiller to independent bottler/blender. As well as drawing from old stock, new brands such as Rowan's

WILLETT

★ WILLETT FAMILY ESTATE BOTTLED BOURBON
ABV VARIES

Fruity. Sultana, dried pear, clover honey, beeswax. Vinous feel. Green grape, sweet sawdust and precise citric burst, then macadamia and coriander seed. Subtle.

FRUITY/SOFT

BARDSTOWN

★ BARDSTOWN WHEATED BOTTLED-IN-BOND BOURBON 50%

Complex and layered with juicy peach cobbler, currant, haskap berry, blackcurrant with a lemony lift and a long caramel toffee finish.

SWEET/VANILLA

BARDSTOWN DISCOVERY SERIES #10
(BLEND OF DISTILLERY'S 6-YEAR-OLD HIGH RYE AND 18-YEAR-OLD FROM TENNESSEE) 57.12%

Gentle, supple and fruity. Candied orange, light sagebrush, then some dried fruits.

RICH/DEEP

Above *Diageo's Bulleit is steadily expanding.*

BULLEIT

BULLEIT 10 YEAR OLD STRAIGHT BOURBON 45.6%

Sweet spice. Slightly creamy oak, with red fruits, barbecued corn, light rye. Starts softly, then pickling spices, scented sandalwood and lemon.

SWEET/VANILLA

BULLEIT STRAIGHT RYE 45%

In the Bulleit style. Lifted and light, quite aromatic with fresh citrus and mint. Palate is fruitier than you expect, the rye kick coming late on.

SPICY/HERBAL

Creek, Noah's Mill, Kentucky Vintage and Pure Kentucky emerged. In 2008, the Willett name re-emerged with the launch of the (sourced) Willett Family Estate Bottled rye and Bourbon. Four years later, the family distillery with its two pot stills restarted. In 2014, a 2-year-old rye was released, joined in 2016 by a Bourbon.

Bardstown is also the spiritual home of **Heaven Hill** (see page 199), which was founded in 1934 by a group of investors and the Shapira brothers (Mose, George, Gary, David and Ed). In 1937, the brothers took full control and the firm remains in the family. After the fire in 1996, production switched to Louisville, but in 2022 ground was broken in Bardstown and the building of the new $200 million Heaven Hill Springs Distillery began. It will ultimately produce 450,000 barrels a year. The two distilleries will run in tandem.

MAKER'S MARK

★ MAKER'S MARK 45%

Gentle. Orange peel, light oak, cinnamon toast, ripe red cherry and fudge. Freesia notes lift free from soft creaminess. Finishes on roasted spices and Snickers bar.

SWEET/VANILLA

MAKER'S MARK NO. 46 FRENCH OAKED 47%

Much darker. Roasting coffee bean, black banana, black cardamom and geranium. More oaky grip, char and tannin, but the gentleness is there.

RICH/DEEP

LORETTO

A recurring leitmotif in American whiskey is how, on Repeal, a distiller picked himself up, dusted off the old recipe and got down to business once more. Bill Samuels Sr took a different tack. In 1953 he ripped up his family's old recipe and started afresh with the aim of making a premium Bourbon. It was a bold decision, given that the family had been distilling for 170 years and prices were being driven down.

Star Hill Farm in Loretto was picked as the site and the recipe switched away from rye to red winter wheat. It remains 70% corn, 16% wheat and 14% malted barley.

The Samuels' **Maker's Mark** distillery, now owned by Suntory Global Spirits, is larger these days, but each expansion has involved building a replica of the original rather than simply putting in larger stills because that could change the character. The same house jug yeast is still used. Distillation is to 130° proof (65% ABV), then aged in air-dried #3 char barrels that, unusually, are moved between different temperature zones in the rickhouses. Every label is hand-cut, each bottle is hand-dipped in wax.

Samuels' vision may be unaltered, but things have changed, such as the cellar hewn out of limestone rock where whiskies are finished and also the Cellar Aged is matured. The surrounding 1,000-acre (405-hectare) farm has been given over to regenerative farming, which encompasses growing grains for the still. Trialling this on site, the argument goes, means a greater chance of suppliers to apply the same sustainable practices. It has also established the world's repository of American white oak trees for research purposes, including genome mapping research. Whisky starts in the soil.

SHELBYVILLE

It was in 1987 when Tom Bulleit left the legal world to start the **Bulleit Distilling Co.** Bought by Seagram in 1997, its whiskey was produced by MGP, then Bulleit became part of Diageo when the Canadian giant pulled out of spirits. Having used sourced whiskey made to its specifics for its first years, in 2019 its distillery in Shelbyville opened (though the rye continues to be made by MGP), followed by a second site in Lebanon, Kentucky that is powered by 100% renewable electricity and has a capacity of 10 million US gallons (38 million litres) a year. There are two (secret) high rye mash bills, which are fermented for three days using five proprietary yeasts, giving blenders Eboni Major and Andrew Mackay ten streams for blending. Distillation is in a beer and doubler set-up and put into barrel at 125° proof (62.5% ABV). Will there be a Bulleit single malt? 'We're always exploring new frontiers and are constantly experimenting with new mash ideas of different grains and recipes,' says distiller Ramsay Borthwick. 'So, stay tuned!'

Shelbyville is also home to the farm-based, family-owned, woman-led **Jeptha Creed**. Joyce Nethery was a process engineer in industrial-scale distillation, while husband Bruce is an agronomist with five generations of family experience farming in the state. In 2016, they started the first legal distillery in the county since before Prohibition.

The approach is substantially different to the rest of the state by starting in the field – the family grow all their own requirements from 1,900 acres (769 hectares) and use Bloody Butcher corn rather than the standard Yellow Dent. The slightly lurid name comes from the kernels that are flecked with red, like blood on a butcher's apron.

'My mother had been growing it for food,' says co-owner Autumn Nethery. 'It was Dad who said, if we're making all this effort to build a distillery, then he'd rather use Bloody Butcher. He'd noticed that when yellow corn and Bloody Butcher were in the same field, the wildlife preferred to browse on the latter. It also means we have to anticipate the critters eating some of the harvest every year!'

Distillation trials proved his hunch to be correct. 'It's tall, so prone to wind damage and the yield is lower,' she says, 'but the flavour difference is great. It's also non-GMO and open-pollinated, meaning that we save seeds year after year to replant.' There are now 450 acres (182 hectares) planted.

It makes up 70% of the mash bill, which also breaks with standard practice by having three other grains (15% rye, 10% wheat and 5% barley), all being malted 'because malted grain, especially rye, just works better with Bloody Butcher'. The corn adds its own sweetness, but also an earthy undertone. It's Bourbon, just not as we know it.

JEPTHA CREED

★ **JEPTHA CREED FOUR GRAIN BOURBON** 49%

Sweet, slightly grassy with a malty background. Chocolate digestive, florist shop, lilac. Some powdery cereal, cedar, then varnish and agave syrup. Sensual and complex.

FRUITY/SOFT

FRANKFORT

The **Buffalo Trace** distillery has established itself as an industry leader in the new American whiskey world. What started as a whiskey warehouse on the banks of the Kentucky River in 1811 had become a distillery by 1858. Its story can be charted through the names of many of its brands. Colonel E.H. Taylor bought the distillery in 1869, renaming it OFC (Old Fire Copper), then George T. Stagg took over in 1878, renaming it after himself in 1900. Three years earlier, 16-year-old Albert Blanton joined as an office boy. By 1921 he was president.

Elmer T. Lee arrived in 1949 and began a decades-long process of expansion and modernization. In 1884, his madcap idea of launching a single barrel Bourbon, Blanton's, appeared. Since 1992, the distillery has been part of Sazerac (who in turn bought it from Japanese Takara Shuzo), becoming Buffalo Trace in 1999.

Above *Jeptha Creed is a family affair.*

Above *The return of the Bloody Butcher.*

Above *Buffalo Trace, on the banks of the Kentucky River, is home to a huge range of innovative whiskeys.*

JAMES CROW

Hailed as the man who brought rigour to 19th-century distilling in Kentucky, Scotsman James Crow emigrated to the USA, arriving in Philadelphia in 1823, and became what we would now call a consultant distiller. In 1826, he moved to Kentucky, and although he worked at a number of sites, his main base was the Oscar Pepper distillery where he worked for 15 years.

Wrongly attributed as the inventor of sourmashing, what Crow did do was apply science to distillation through the use of hydrometers, and refining what at the time were somewhat chaotic sourmashing techniques.

Through measurement and analysis, he worked out the appropriate temperature parameters and volumes of backset required to reduce bacterial spoilage. This in turn allowed him to ascertain the right levels of acidity to provide yeast with the optimum environment in which to grow and create a consistent product.

He also introduced rummagers in the copper pots. Cleanliness was key to quality across the distillery. He died in 1856.

As well as the Blanton's and Elmer T. Lee brands (still owned by Takara), the distillery also makes the Van Winkle range as well as owning Virginia's A. Smith Bowman and Kentucky's Glenmore and Barton 1792.

Today, Buffalo Trace is not only large but at the forefront of innovation and experimentation. 'The past decade has been a defining and exciting chapter,' says distiller Harlen Wheatley. 'In 2013, the Distillery announced a $1.2 billion expansion that saw capacity increase by 50%.'

Throughout, he has overseen continual, industry-leading investigations into what American whiskey can be. 'One thing you can count on is continued experimentation,' he says. 'We've explored variables such as cereals (amaranth, emmer), mash bills, barrel char levels, temperature control and different types of wood. We firmly believe the whiskey category is only bound by the ideas we have not thought of yet.'

'Innovation is in our blood,' adds master blender Drew Mayville. 'We're experimental by nature. At Sazerac we don't sit around waiting. We don't just want to make good whiskies today, but great whiskies for tomorrow.'

Stay still and you run the risk of being left behind. What were innovations become calcified; tradition rather than being fluid becomes sluggish and moribund. 'What people like today they won't necessarily like tomorrow,' he continues. 'If you sit on your laurels you'll end up being left behind.'

Whiskey isn't fixed – it's dynamic. The key is finding those points of flex: variety, mash bill, barrelling strength, coopering, warehouse location. The joy is that all of what has happened here is production rather than marketing led.

'Bourbon wasn't big when I started at Buffalo Trace,' Drew continues. 'In 2004, we were wondering if Eagle Rare would ever sell or whether we'd discontinue it.' The question is, what's next? His answer was

BUFFALO TRACE

★ BUFFALO TRACE BOURBON 40%

Spiced honey with subtle citrus, menthol, orange blossom and lemongrass. Corn adds roundness and rye, baking spice. Effervescent. A Bourbon for the Scotch drinker.

SWEET/VANILLA

★ EAGLE RARE BOURBON 45%

Aniseed, red cherry liqueur/cough syrup and sweet oak. Deep corn base and firm oak. Bold dried fruits, crème de violette. Becomes peppery at the end with hibiscus tea and acorn coffee.

SPICY/HERBAL

1792 SINGLE BARREL BOURBON 49.3%

Dry and autumnal with some dusty wood, new cigar. Oak phenols more obvious. Wood and toffee, oozing caramel texture, then comes allspice. Full-bodied.

OAKY/DRY

★ BLANTON'S GOLD EDITION #656 51.5%

The higher strength adds another layer of complexity. More oak, vanilla and coconut, cigar. A light charred note, then a mix of nutmeg, cinnamon, aniseed, peach juice and citrus.

SWEET/VANILLA

BLANTON'S ORIGINAL SINGLE BARREL #386 46.5%

Sweet and floral with a little touch of mandarin peel and oak. The palate is generous, soft with light tobacco, butterscotch and a touch of allspice and clove and cherry stone.

SWEET/VANILLA

one that was repeated across the country. 'Blending. There's so much potential, in different oak types, heritage varieties and then mingling them. You get unlimited flavour profiles. It's a blender's dream. I'm not convinced people see it yet, but it's our job to convince them, to educate them that the sum is better than the individual parts.'

All of that means looking further afield. 'We're seeing an appetite for different tastes in the US, but this market will be saturated. The global market has expansion room. I'm not sure we'll conquer Scotch or single malt tomorrow, but our aim is to have American whiskey everywhere.'

WOODFORD

WOODFORD RESERVE STRAIGHT MALT
45.2%
Grainy, slightly nutty with new oak poking through. Barley stalks, dried beans, antique shop, V8 vegetable juice. Sweeter than you think. Clearing garden in autumn. Short.
MALTY/NUTTY

WOODFORD RESERVE BACCARAT EDITION
45.2%
Mix of toasted oak, coffee, raspberry jam, crème brûlée, bright citrus and parma violet. On the lighter side. Spice and some grip.
SWEET/VANILLA

VERSAILLES

It's strange to think that when, in 1996, the **Woodford Reserve** distillery started on the site of the old Oscar Pepper distillery that it was considered a slightly crazy experiment. Making Bourbon with three pot stills? Madness! The mash would be too thick, the grains would burn, the yield would be low. Maybe most of all it would taste different. Now, though pots might not be standard, they are increasingly common. Different flavour is prized. It was appropriate that it started here at the distillery where James Crow helped to refine Bourbon production (see box).

Woodford Reserve is based on the Old Forester mash bill of 72% corn, 18% rye and 10% malted barley. Ferments are a week long with a specific house yeast. Distiller's Select continues to be a blend of the distillery's pot still spirit with column still from Brown-Forman's Early Times distillery. The less 'efficient' pots build in more flavour than a spirit of the same strength out of a column. Capacity doubled at the start of the 2020s, and the range has increased. Now there's a wheat variant (52% wheat, 20% malt, 20% corn, 8% rye), a 53% rye and a straight malt (51% malted barley rather than single malt). In 2023, as part of its experimental Distillery Series, a blend of its Bourbon, rye, wheat and malt appeared, the first release from master distiller Elizabeth McCall.

Above *Woodford Reserve distillery, both modern and ancient.*

LAWRENCEBURG

Kentucky is home to a number of whiskey dynasties: the Beams/Noes, Samuels, Shapiras and, at **Wild Turkey**, the Russells. In the 1960s, blended Scotch became lighter to suit a changing American palate. Faced with declining sales, some Bourbon brands followed suit. Others, however, held firm to the ways in which they had been raised. This too helps to tie whiskey to its location. It's more than just where it's made – it's how a whiskey can become the heartbeat of a place. Often that pulse is manifested in a person.

Jimmy Russell was one of the men who ensured it survived. He started working at Wild Turkey in 1954. At the time of writing he is still there. Wild Turkey continues to be made his way, the way he was taught: corn in the low 70s percentage-wise, allowing the small grains more say. Collected off the still at 124–126° proof (62–63% ABV), barrelled at 110° proof (55% ABV), bottling at higher strength. Jimmy's son Eddie has been there for 45 years and now Eddie's son Bruce has joined. It may be owned by Campari, but the Russells are in charge.

The 8-year-old 101° was the old flagship – a glimpse into the big-boned Bourbons of the past. Its place is now taken by Russell's Reserve. Little in the process has changed, though in 2023, ground was broken for a new distillery that will take capacity from 9 million to 14 million US gallons (34–53 million litres). More seven-storey rickhouses will be built, creating an ever more complex map of flavour depending on which floor the barrel sits or whether it gets more or less sun – musky soft oak in the bottom, spicy and sweet in the hotter upper reaches. These nuances in flavour according to position have been highlighted in the annual Single Rickhouse series.

'Whiskey is made by people' can be a lazy cliché. Sometimes, however, the soul of the people who steer it manifests itself in flavour. In days of cynical marketing and bad history, it's something to be cherished. Only Jimmy, his son and grandson could have made this.

It's time to talk yeast. Kentucky's distillers are never shy of speaking of how special their strain (or strains) are. Guarded by families, locked away in refrigerators and cellars, it's one of the keys to each distillery's character. Some will use one strain per whiskey, others a combination. **Four Roses** takes it to the extreme thanks to its purchase in 1943 by Seagram, a firm that was obsessed with yeast. There were in excess of 300 strains in Canada and each of its American sites had their own selection, all of which were held at Four Roses.

It was the only positive thing about the Canadian giant's ownership. Four Roses became export-only, then a blend (straight whiskey blended with neutral spirit). In 2002, it was taken over by Japanese distiller Kirin (see page 282), who saw the potential and how that lay in the yeast.

Another theme repeats itself – the faith of one man in his distillery. In Four Roses that was former

WILD TURKEY

RUSSELL'S RESERVE 10 YEAR OLD 45%

Rich but balanced. Sumptuous oak and rich distillate, baked vanilla pod and earthy rye. Sweet and mouth-filling, black cherry, gentian, clove-studded orange.

FRUITY/SOFT

★ RUSSELL'S RESERVE SINGLE BARREL 55%

Sandalwood, dried lime, milk chocolate. Zest, zing and swing. A touch of coffee, the rye used as an accent. Dried orchard fruits. Great balance.

FRUITY/SOFT

WILD TURKEY RARE BREED 58.4%

Gutsy. A beefed-up distillate with balancing oak, strawberry, chestnut honey and blackcurrant. A suggestion of weight in the middle, then wintergreen and lots of rye accents.

RICH/DEEP

Above *Barrelling at low strength adds to Wild Turkey's character.*

Above *The yeast pioneer: Four Roses.*

master distiller Jim Rutledge. He created a complex matrix of blending possibilities by using two mash bills (E and B). The former is 75% corn, 20% rye and 5% malted barley; the latter 60% corn 35% rye and 5% malt. Each are fermented with five different yeasts: V (delicate fruits), K (spice), O (bold fruits), Q (floral/fruity) and F (herbal).

All of the spirit is then aged separately, giving ten whiskeys that are used in different percentages for each release. When Kirin took over, its master blender Jota Tanaka suggested that the flagship Yellow Label needed more weight, so the average age was upped to five years and all ten variants used.

In the days of precious little straight rye, Four Roses' zesty energy set it apart from Bourbon's usual seductive weight and oak. That spice is there, but handled in such a way that there's a seamless integration, rather than the often abrupt step change when rye comes crashing through the door like a drunk being thrown out of a saloon. Blending, yeast...and keeping the faith. In 2019, capacity was doubled. The bad old days were over.

NATIVE AMERICAN CORN

Most histories of whiskey making in Kentucky and Tennessee speak of how settlers came over the Appalachians into the new territories and began to cultivate, then distil the corn that was growing there. There was nothing, ever, about the people who grew that corn for millennia. Bourbon's roots come from the Cherokee. They grew corn (*se-lu*), specifically southern white dent ('she-corn' in Cherokee), which they crossed with flint corn varieties, giving varieties such as Cherokee White Eagle and many more.

Distillation was unknown, but fermented drinks, such as a corn drink (*kanâhe'na*), were used for celebrations and as medicine. The settlers arrived in a land that was in the process of being cleared, not of vegetation but people. In 1838, the Cherokee were forcibly moved to Oklahoma along what they called 'The Trail of Tears'.

Plant breeders such as Robert Reed began to build on Cherokee strains, his son, also Robert, crossing one of those with Northern Big Yellow flint and creating the genetic parent of the bulk of the non-GMO corn used for today's whiskey.

Did old whiskeys taste different? Yes, because prior to Reid's Yellow Dent takeover there was a wider range of corn to choose from. Now though, heirloom varieties such as Bloody Butcher and Jimmy Red (see page 229) are returning. Cherokee White Eagle is also still available...

FOUR ROSES

FOUR ROSES ORIGINAL 40%

On the camphor side along with fresh red berries, ripe pear, a touch of aniseed, then preserved lemon. Restrained and gently floral. Sweet oak.

SPICY/HERBAL

FOUR ROSES SMALL BATCH 45%

Slightly tighter with light vanilla spice, more of the wild rose, then dark chocolate, allspice. Thick texture, cranberry and caramelized wood. Refined.

FRUITY/SOFT

★ FOUR ROSES SINGLE BARREL 50%

Higher rye accents but with balancing sweetness. Plummy fruits, toasted oak. Lightly nutty, then zesty baking spices and light fruits. Maple syrup and vanilla on the finish.

FRUITY/SOFT

Above & right *New Riff are finding new tunes to play on an old recipe.*

NEWPORT

Newport sits on the banks of the Ohio River opposite Cincinnati. It's home to **New Riff**, founded in 2014 by Ken Lewis and Jay Erisman, former owners of liquor outlet The Party Source. As the name suggests, they wanted a new riff on an old tradition.

'In the early 2000s, we saw the rebirth of Bourbon and rye, and realized that we could shift from selling other people's Bourbon to making the stuff ourselves. We sold the store (to the employees) and put the distillery in the parking lot. And it worked!'

NEW RIFF

NEW RIFF SINGLE MALT 8 YEAR OLD 56.9%

Airy and only lightly wood-accented. Freshly baked sourdough, then cleavers. Mid-palate is malt extract, bitter chocolate and togarashi spice on the finish.

FRUITY/SOFT

★ NEW RIFF MALTED RYE 6 YEAR OLD 50%

The chocolate edge of malted rye comes through, but the distillery's lifted character adds brightness and acidity. Tayberry, roasted squash and a hint of meatiness. Great balance.

SPICY/HERBAL

Thankfully, underneath the car park was a limestone water aquifer, giving what Jay calls, 'big-tasting, mineral-rich water that becomes big-tasting Bourbon and rye'. Non-GMO corn is sourced from a single farm, as well as heirloom corn, rye and wheat for limited-edition releases. 'These heirloom projects serve as a preservation of old flavours in whiskey, sustaining and uplifting an agriculture that might otherwise be lost,' he adds. 'We're entirely committed to what we term the "Kentucky Regimen": sourmash, slow, open fermentation with native microflora; on-grain distillation in a continuous column beer still, then doubler.

'There's no chill-filtering, and everything is bottled as bottled-in-bond (bar cask-strength whiskeys and Sour Mash Single Malt).' Now, that might seem that this is a pretty old riff, but then he adds, 'We grant ourselves, shall we say, *considerable leeway* in our recipes and mash bills. We have made over 40 different mash bills, including the use of heirloom corn, rye and wheat grains; 100% malted rye; a rare "oated" Bourbon mash bill evoking a chocolate oatmeal stout ale; and *six* different malt whiskey mash bills that are vatted together to make New Riff Sour Mash Single Malt.' That's more riffs than Keith Richards at his best.

'Our mission is to become, someday, one of the world's great small distilleries, and to do so with Kentucky whiskey. We cannot accomplish this by half measures.' That knowledge of the market that only a retailer has is bearing fruit.

SOURMASH/SWEET MASH

A defining element in many Bourbons is the use of sourmash (aka backset or stillage), the acidic residue left at the base of a still after distillation. Adding this to a fermenter will acidify the wash, allowing a better environment for yeast and lactobacillus. This is particularly useful when the water used is alkaline, as it is in Kentucky and Tennessee.

It isn't, however, an American invention. The use of dunder (aka sourmashing) was common in rum production in the late 17th century. In 1743, the first 'improver' of Scotch whisky, Henricus Van Wyngaerden, encouraged the use of 'sourmashing'.

In American whiskey it appears to have started on the East Coast – perhaps following on from the American rum industry that predated it. It spread from there, first to Pennsylvania and then to Tennessee where its use in the 1820s was widespread, before heading into Kentucky where sweet mash (i.e. no backset) had been preferred.

DANVILLE

Wilderness Trail came from nowhere: 'from zero to a hero', as co-founder Dr Pat Heist puts it. 'Ten years ago we were just firing up our still. We've gone from pot to column distillation, which didn't really change our product quality or flavour, but was a major process change. Plus we went from one barrel a day to 220 barrels a day.' The other thing was the $420 million investment by Campari in 2014.

There was also a huge degree of knowledge. Both Pat and co-founder Shane Baker also work at Ferm Solutions, providing yeast and technical support to distilleries and breweries, a troubleshooting background that was invaluable when it came to building their own state-of-the-art distillery. 'We look at every step of the process very scientifically and ask, "What are we trying to accomplish here?"', says Pat. 'We're using modern science to optimize old-world distilling traditions.'

There's a wheated Bourbon (64% corn, 24% wheat, 12% malted barley), a high rye Bourbon (64% corn, 24% rye, 12% malted barley) and a rye whiskey (56% rye, 33% corn, 11% malted barley), but also 4-grain and other mash recipes. All are sweet mashes: 'It lends to a softness that makes the whiskey very smooth and drinkable even at higher proofs,' Pat says, cooking it at what he claims is the lowest temperatures in the industry. 'Converting starch to fermentable sugars happens at 185°F [85°C], so we cook at that temp rather than over 200°F [93.3°C] like most other distilleries.' It's fermented to 85% alcohol, distilled in two copper columns to around 137° proof (68.5% ABV) and barrelled into #4 char casks at 110° proof (55% ABV).

It might seem surprising that he then speaks of location as being 'hugely important' for barrel-aged spirits. 'You have the terroir of the grains from the soil and climate, the conditions with which they were dried/cured. You can make great whiskey anywhere in the world providing you have great ingredients and know-how. Then you're at the mercy of the climate. Fluctuating temperatures help with extraction of wood chemicals and also drive chemical reactions such as esterification, creating many of the classic flavours of Bourbon and rye whiskies.' Then there's rickhouse type, height, orientation and rotation, or not, of barrels. 'It gets complicated real quick,' he laughs. Complicated, it would seem, is exactly what he wants.

TENNESSEE

For the first decade of the 21st century Tennessee had only three distilleries. The boom in distilleries around America appeared to have passed it by. The reason for this wasn't a lack of desire to revive the state's distilling heritage but the fact that whiskey production was only legal in three of its counties (Moore, Coffee and Lincoln). When the law changed in 2009 permitting production in 41 counties, new operations began to spring up across the state.

LYNCHBURG

Jack Daniel's is the second biggest-selling whisky in the world – just behind Johnnie Walker – and with that comes responsibility and also pressure. Any mega-brand needs to maintain momentum and give the people what they expect. It also has to decide whether it's happy to transcend its category – think Bacardi – or help build it. What is Jack? On one hand it's a brand made at a distillery that just keeps getting

Above *The fires blaze at Jack Daniel's.*

Above *Hidden among the trees, Jack Daniel's' barrels slowly mature.*

bigger. In 2023, a seventh set of stills and the 97th barrelhouse were built. It's also, proudly, a Tennessee whiskey. The question then is how to put a spin on that at a time when so many other distilleries are pushing American whiskey in new and exciting directions.

Two mash bills operate. Tennessee Whiskey is 80% corn, 12% malt and 8% rye; its Rye is a punchy 70% rye, 12% malt and 18% corn.

Both are soured, mashed and cooked with the cave spring water, and fermented for seven days before beer column and doubler distillation. Then comes the extra dimension that distinguishes Tennessee Whiskey – the mellowing through maple charcoal (see box). 'It removes grainy character (mostly corn aroma) and allows esters to become more prominent, giving the sweet and fruity Jack Daniel's' character,' says master distiller Chris Fletcher.

JACK DANIEL'S

JACK DANIEL'S TRIPLE MASH 50%

Jack's lean quality comes through. Dusty rye, a touch of soot, some dried mint, boiled sweet, light oak and pekoe tea. Becomes buttery as it reaches the middle, then tightens.

SPICY/HERBAL

'For our barrels, we toast for approximately 12 minutes before charring,' he adds. 'Then the whiskies are aged for between four and seven years, with some being much longer.' Age statements come in at 10- and 12-years-old.

'Innovation has become a focus,' Chris says, speaking about the age-statement releases, double-oaked expressions, single barrels, bottled-in-bond expressions that now include a rye and the Tennessee-exclusive Distillery Series, all variations on the two mash bills.

The most radical arrival came in 2022 with the launch of its first single malt. 'Our approach is slow and steady. Our processes and systems are very different, with keeping the whole grain through to distillation, column, then pot distillation and of course the new barrel first and then finishing in oloroso. It offers a modern twist on single malt.'

The malt also appears in a triple mash bottle-in-bond blend of Tennessee whiskey, rye and single malt where the last acts as a bridge, adding coherence to the mix.

'It's exciting to see Tennessee whiskey continue to carve out its place among world whiskies,' he says. 'I think it's anyone's guess on whether we've reached saturation point, but we sell more outside the US than in, so export is a priority for us. There remains opportunities to bring Tennessee whiskey to the world.'

TULLAHOMA

It wasn't that long ago that **George Dickel** was the forgotten member of the Diageo whisky empire. Now, under distiller Nicole Austin, it's not only expanding the idea of Tennessee whiskey but whiskey in general.

Stylistically, it is totally different to Jack. Dickel is all about lush fruits and sweetness. Even though its rye (launched in 2012) has 95% of the grain in its mash bill and is aged for a relatively short 4 years, this mild-mannered fruity character prevails. Its classic mash bill is 84% corn, 8% rye and 8% malted barley.

There's precision at work here. Dickel's is all about the balance of the oak with citrus, fruits and spice. Even the Bottled in Bond speaks more about maturation than becoming a showcase for oak. The range shows deep understanding of more than one dimension.

One of the most exciting developments has been the inter-distillery blend of Todd Leopold's Three Chamber Rye (see page 232) and Dickel (see tasting note). 'That led to us digging in the archive,' says Nicole, 'and we uncovered evidence to suggest that a three-chamber still was used for Dickel's Bottled in Bond Pennsylvania Straight Rye from 1938.' Although she says nothing fundamental has changed in production, what has happened is a greater understanding and appreciation of the brand, which has taken them into an interesting space. 'Location is becoming more important across the industry, but for Dickel specifically, I think the opposite, actually. We spoke SO much about the location historically that consumers lost some of the greater context – that Tennessee whiskey is a regional identity of Bourbon, that Bourbon lovers can appreciate it. We want that challenge and believe we can compete against the best American whiskey producers, not only those in Tennessee.'

So if Tennessee is where you are from, but American whiskey is where you want to be, then where are things going? 'We're nowhere near oversaturation,' she says. 'If you look historically at the share spirits have occupied, and the incredible growth of small, regional producers that beer and wine have experienced, there is still plenty of room for spirits producers.'

GEORGE DICKEL

GEORGE DICKEL BOURBON 8 YEAR OLD 45%

Scented, high-toned. Attic dust, grassy rye, candied peel and candy floss. Light medicinal note. Palate has gentle tangerine, drying spice. Poised and precise.

SWEET/VANILLA

★ GEORGE DICKEL BOTTLED IN BOND TENNESSEE WHISKEY 50%

Rounded, soft. Chocolate honeycomb, dried fruits, bitter chocolate, slightly herbaceous. Polished fruits, some caramel.

FRUITY/SOFT

★ GEORGE DICKEL X LEOPOLD BROS COLLABORATION BLEND 50%

A blend of column and three-chamber still. Sweet. Orange blossom water, lavender, Turkish delight. Elegant and perfumed. Pink peppercorn, strawberry, pancetta and Sichuan pepper.

OILY/SPICY

LINCOLN COUNTY PROCESS

The properties of charcoal as a filtering agent have been known for millennia. Home water filters appeared in 1750 and its use in spirits started in Russia in 1785 when charcoal filtration was used to purify vodka. By the start of the 19th century, it was being used in whiskey production in Kentucky before being adopted by distillers in Tennessee. Now it's the key method that defines Tennessee whiskey.

Also called the Lincoln County Process (though there's no evidence it was started there), it involves new make (white dog) being run through vats of sugar maple charcoal in order to remove fusel oils, sulphur compounds and harshness. It doesn't add flavour, but some argue it might reduce mouthfeel and strip out some esters.

Above *George Dickel is now making up for lost time.*

CHATTANOOGA

Distilling has boomed in Tennessee, though it took Tim Piersant and Joe Ledbetter two years to lobby for a change in law to allow them to distil in Chattanooga. Tennessee applied state-wide Prohibition in 1917 (Chattanooga had gone dry two years earlier) and even after Repeal many counties remained dry. With the passing of House Bill 102 (Senate Bill 129), aka 'The Whiskey Bill', 50 counties were allowed to start (or restart) distilling.

Located in the Tennessee River Valley near the foot of the Cumberland Plateau, **Chattanooga Whiskey Co.** began in 2011 with sourced whiskey from MGP. Distilling started in 2015 and now everything bar its Founder's brand is from the distillery.

It's a member of the TNSS (Tennessee Sustainable Spirits) initiative, which requires the distillery to meet a minimum 25 criteria in practices related to water, air, waste, energy, sourcing, community and maintenance. For production manager Grant McCracken, it demonstrates how American whiskey is becoming a more locally focused product.

'We believe that characteristic will only become more important. All the corn comes from a local grower,' he adds, 'and we're working with regional maltsters to create recipes like Appalachian Grist that celebrate the terroir.

'We also believe that stylistic exploration will become increasingly important as the number of craft distilleries increases. We decided to blaze our own trail, and create our own style of whiskey – we call Tennessee High Malt Barrel 91.'

This takes the high rye principle and ups the quantity of malted barley to 25%, with speciality malts also in the mash bill – the application of brewing/single malt principles wrapped up in Tennessee. 'The three-malt 99 Rye is our malt-forward approach to straight rye malt, while there's also a finishing series, Bottled in Bond, with a multi-mash bill blend.'

White dog comes off the still at less than 132° proof (66% ABV) and barrelling entry varies between 113° and 117° proof (56.5% and 58.5% ABV) depending on the whiskey being made. Further flavour options are fed in with the use of a combination of charred and custom-toasted barrels, and then a solera-style system to build consistency.

UNCLE NEAREST

UNCLE NEAREST 1884 SMALL BATCH 46.5%

Laurel and dried raspberry. On the powdery side, but there's balancing hawthorn jelly in the middle. With water, some sweet sawdust, then peach.

SWEET/VANILLA

UNCLE NEAREST 1856 PREMIUM 50%

Richer. Light dustiness, boiled sweets, white pepper, coconut, pink peppercorn, then mint cream. Some caramel, orange bitters, clove and hazelnut butter.

RICH/DEEP

UNCLE NEAREST SINGLE BARREL #61 60.5%

Chewy wood. Haskap berry, black cherry and stewed rhubarb. More oak and a light touch of char. Has power and a degree of spicy succulence.

RICH/DEEP

Above *The single barrel warehouse at Uncle Nearest.*

Above *Filtering at Nelson's Green Brier.*

SHELBYVILLE

In the first edition of the Atlas (2010), when writing about Jack Daniel's I mused about how appropriate it would be if Nearest Green was given some recognition. This old slave, after all, was the man who taught young Jack how to make whiskey, one of many Black distillers in America whose names have been excised from American whiskey's story.

Now it's changing. In 2016, Fawn Weaver saw a photo on the cover of the *New York Times* International Edition. 'It showed Jack Daniel surrounded by his entire team, who were all white. In the centre was a Black man, George Green, son of "Uncle" Nearest Green. The headline said, "Jack Daniels embraces a secret ingredient, help from a slave". Nearest Green's story was finally being told.'

Fawn enlisted a team of historians, archivists, archeologists, genealogists and conservators to dig ever deeper and in 2017 she launched **Uncle Nearest**. To complete the story, in 2019 Victoria Eady Butler, the great-great-granddaughter of Nearest Green, was appointed master blender.

This, then, is a story that's more than about whiskey. This is about correcting the record. It's also an example of whiskey's ability to be a lens to examine a culture or a society. 'There's a level of hope in this story,' says Fawn. 'We're doing this for Nearest's legacy, and for those who are at the beginning of their journey.'

The distillery is built and the whiskey – previously sourced – is appearing and utilizes four mash bills. The 1884 Small Batch (the date commemorates the year Uncle Nearest retired) uses 80% corn, 10% rye and 10% malted barley, and is aged for four years. The 1856 Premium (the date when Uncle Nearest was working on the Dan Call Farm and met Jack) blends the 80:10:10 with the distillery's 84% corn, 77% corn and 70% corn mash bills. It too is four years old. A Bourbon using malted corn is ageing with an aim to release small amounts from 2027.

The ryes are still sourced because, as Fawn explains, 'rye does not grow well in Tennessee. Wild onions nearly always ruin the crops. Also, we are basically at capacity and we wanted to concentrate on what we know.'

A number of vital elements in whiskey's evolution meet here. Recognition of Black contribution, the start of greater representation not just in production but in whiskey clubs and a rise in the number of women making, owning and drinking whiskey. Uncle Nearest is a 100% female-led company. 'In 2023, a study found that 37% of whiskey drinkers are women,' Fawn says, 'a number that's doubled since 1990. This could be due to a rise in female representation in the whiskey industry, something we are immensely proud of here at Uncle Nearest.'

NASHVILLE

Charles Nelson was an enterprising young man. A German émigré, he was head of the family at a young age after his father drowned in a tragic shipwreck. He ended up in Nashville, where he opened a grocery. In time, realizing that whiskey was one of the biggest sellers, he bought the distillery that was supplying him and renamed it Nelson's Green Brier. The first to be in a clear glass bottle, at its peak it was selling two million bottles a year. After his death, his wife Louisa ran the business until Prohibition. It never reopened.

In 2006, Andy and Charlie Nelson and their father were driving to Greenbrier to pick up half a cow. They stopped at the gas station and saw a plaque telling of the old distillery that was once opposite. It transpired that they were Charles's great-great-great-grandsons. It was too good a story to pass on. In 2014, the revived **Nelson's Green Brier** opened. In 2019, Constellation Brands took a majority stake.

Charles was so keen to speak of his whiskey's quality that he published the mash bill in newspaper ads. The bottle is the same, as is the label. 'Wheated whiskey seems to have been more common in Tennessee,' says brand manager Michelle Jones, 'and its more gentle character speaks to what Charles was trying to do.

'All grains are as local as possible,' she says. 'We're part of an organization that is helping with white oak, so rather than harvesting and saying thanks for the barrels, we're putting back into those communities as well as to farmers. Without that we have no business. When you're creating product that relies on your city and your community, you better make sure that you're giving back to them in a big way.

'Some of our best feedback is people telling us how approachable the whiskey is. Maybe in part that's because the majority of departments at the distillery are run by women, so we're always aware of making sure that the approachability extends through all types of whiskey drinkers.'

NELSON'S GREEN BRIER

NELSON'S GREEN BRIER TENNESSEE WHISKEY 45.5%

Gentle, with TN sweetness alongside tobacco and a thick, slightly honeyed drop in the middle. Cornbread, light cocoa, menthol, soft fruits. Silky and balanced.

SWEET/VANILLA

★ **NELSON'S GREEN BRIER TENNESSEE WHISKEY CASK STRENGTH** 60.7%

Sweet oak, nicely restrained. Boiled sweets, vanilla and coconut, cardamom, leaf mulch, bacon and cigar shop. Light grip with clean acidity cutting through. Broad and deep.

RICH/DEEP

THE SOUTH & TEXAS

The number of new distilleries whose story starts with the claim 'we were the first in the city/town/county/state since 1919' demonstrates how devastating an impact Prohibition had on distilling across America. The state of Texas is a prime example of this. Now, however, its distillers are looking deep into their location and coming out with whiskeys that are without any doubt wholly Texan in flavour.

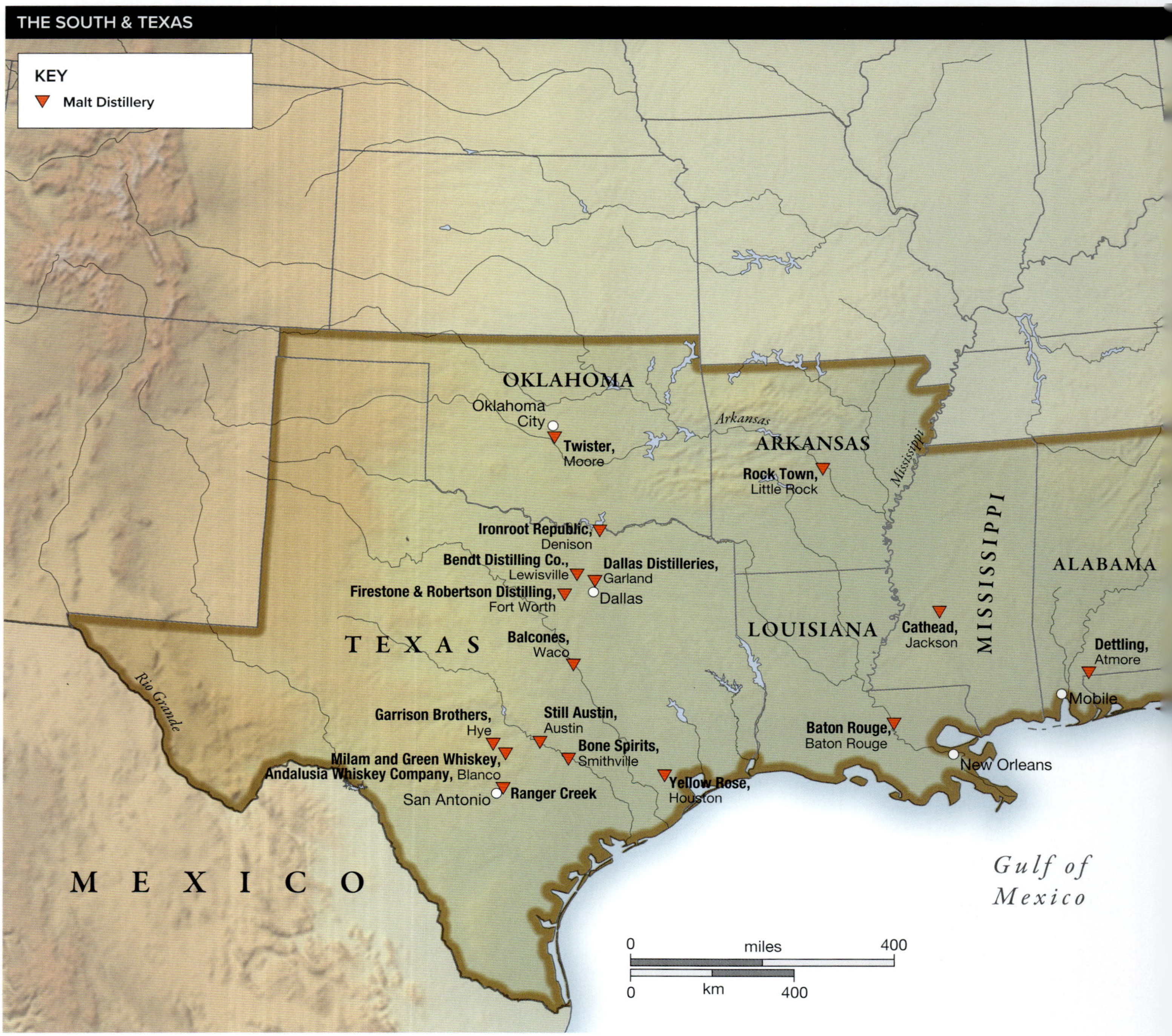

WEST VIRGINIA

One of the most fascinating developments in whisky's recent global evolution is how distilleries are being established as a way to reinvigorate communities. 'West Virginia has been economically depressed. There is no manufacturing,' says John Little, founder/whiskey maker at Maxwelton's **Smooth Ambler**. 'We've been left behind in the last 120 years, but we have great people, climate, water, air. The idea was to create something world class in a small town and try to change people's lives.' Distilling wasn't top of the list, but when his father-in-law read an article about craft distilling, the germ of an idea started. In 2010, Smooth Ambler opened for business. Now it employs 24 people, the local hotels are full, the shops busy.

It wasn't all plain sailing, particularly when they launched their Bourbon. 'I get the "only Kentucky makes Bourbon" argument regularly,' John says, 'but we fight our fight and we believe the products are different. That takes time for people to figure out, but we're finally seeing results.' It's a central issue in the development of 21st-century American whiskey. Why should Bourbon or rye conform to a single template? A Bourbon from the Appalachians *shouldn't* be the same as one from Louisville.

Community extends beyond the distillery and town. 'We wanted to source as close to the distillery as we could for environmental reasons,' says John. 'We also wanted to use non-GMO corn, so needed to work with farmers to grow it. But, as we don't grow a lot of grain here, finding them was a project.'

In 2017, Smooth Ambler was bought by Pernod Ricard and along with investment came distiller Travis Hammond. 'He began by changing every variable,' says John. 'Everything to do with flavour was analysed to get the flavours we wanted. All in all there were 32 changes. You have to be obsessive about everything.'

The result was different yeast, a lower cooking temperature for the rye, extending ferments to 80 hours, moving from pot to column and dropping barreling strength from 115° to 110° proof (57.5 to 55% ABV), all with the aim 'to make a gentle, fruit-forward whiskey suited to our conditions'.

Those conditions also mean understanding the maturation profile. 'We're milder than Kentucky, so there's a different ageing profile. We have a single-storey rickhouse, so things age more evenly and slowly. It's a good thing, but the whiskey needs more time.'

His belief that the barrel isn't 'a magic box' is also linked to the locality. In the same year as the takeover, the $40 million West Virginia Great Barrel Company stave mill and cooperage, 'the most technically advanced in the world', was built 10 miles (16km) south in White Sulphur Springs, again driven as a response to the need for investment in community. The distillery can now draw on casks made from trees grown locally, and fill them with whiskey made from grain from surrounding fields. 'People thought that we were crazy,' he laughs. 'Even now, folks say we're the best-kept secret in West Virginia. We're trying to be the worst-kept one!'

SMOOTH AMBLER

★ SMOOTH AMBLER FOUNDERS' CASK STRENGTH SERIES BOURBON BATCH 1 58.5%

Lifted and energetic. Bruised fruit and spice, jasmine, rose, then a lemony spike. Punchy, with sweet oak, toasted corn, leather, intense citrus, cough syrup and spices.

FRUITY/SOFT

SMOOTH AMBLER CONTRADICTION BOURBON 46%

More citric giving a clean, lightly zesty start. Perfumed with heightened aromatics, boiled sweet, banana. A softer flow and ripe fruit.

FRUITY/SOFT

SMOOTH AMBLER FOUNDERS' RYE CASK STRENGTH SERIES 61.5% BUT VARIES

Crackling rye. Floral/green note. Padrón pepper, coffee, peanut brittle, cumin, sumac and Sichuan pepper. Low oak, mint sauce, sour cherry and a nutty back note.

SPICY/HERBAL

VIRGINIA

If Smooth Ambler was trying to show that Bourbon could be made anywhere in America, then the **Virginia Distillery Co.** (VDC) in Nelson County went even further back into whiskey's history for its inspiration. 'Ideally James Anderson (see below) would have made single malt,' says VDC's lead blender Amanda Beckwith, 'but rye grew better in the rocky ground, then corn became cheaper. I'd still argue that America should always have been making single malt!'

It was that belief which persuaded VDC's founder George Moore to start distilling with help from Scotch luminaries Dr Jim Swan and Harry Cockburn. By 2015, Virginia was, finally, making single malt.

The barley comes from 600 farmers, is malted in three maltings and fermented with two yeast strains. 'One is the workhorse doing the conversion job,' Amanda explains. 'The other [it's a secret developed over three years, unique to the distillery] we rely on to develop esters over 72 hours. I love walking through the tanks, smelling banana, nut bread, apple skin, pear. It's crucial that we have that to develop the orchard fruit, which comes later on.' Distillation is in copper pots.

It's when the climate kicks in that the signature Virginia style begins to develop fully. 'You can have one day at 75°F [24°C] and the next it's snowing. All of that means we see quick development in cask. The downside was a 12% angel's share, which they have brought down to 3–6% by switching to palletizing and changing the entry proof to 125° (62.5% ABV).

She has also created a flavour map from data analysis of the orientation of the metal-sided warehouses, their internal heat and humidity pockets. 'Jim always said that you shouldn't focus on changing the environment to change the whisky, but change the whisky to suit the environment. We work with what we have and embrace it.' While most is in used ex-Bourbon barrels, there are also fino and manzanilla sherry and STR casks.

'Bourbon has become the great American comfort blanket, so it's exciting to see people fall in love with single malt, though the biggest preconception is that for many people "single malt" means "peated Scotch". We have to fight that battle.'

Above *Virginia Distillery Co. is a pioneer of American single malt.*

The state is also home to the museum distillery at George Washington's **Mount Vernon** estate. It was here in 1797 that the manager, a Scot called James Anderson, persuaded America's first president that distilling made economic sense. He and his son ran it with six slaves – Hanson, Peter, Nat, Daniel, James and Timothy – until a fire destroyed it in 1814. An archaeological dig in 1995 revealed the old distillery. By 2005, it had been restored and began to make whiskey the (very) old way.

Mount Vernon now makes Bourbon, rye and, in a nod to Anderson, single malt. The grains are barrel-fermented and transferred by bucket to an old-style (wood-fired) alembic with removable top.

Fellow Virginia distillery **Copper Fox** takes an equally individualistic approach to its whisky making (there's no 'e'). The first American distillery to restart on-site floor malting since Prohibition, it smokes six-row Thoroughbred barley over fruitwood. Apple wood chips are then added to the maturing casks.

A further variation on the theme of whiskey is taken by Purcellville's **Catoctin Creek**. Founded in 2009 by distiller Becky Harris and husband Scott, it only makes rye, though from four varieties – three from Virginia and one from Amish Country in Pennsylvania. Each gives different variations on grass, herb and spice, allowing a more complex range of flavours. They are distilled, slowly, in two small hybrid pots. In a reversal of the current trend, in 2023 the Harrises bought back the minority shareholding held by Constellation Brands.

Washington DC's **District Made**, formerly One Eight Distilling, is making Bourbon, rye and single malt, with the first using Hickory King Corn, a variety grown in Virginia since the 19th century. Two mash bills are used – one wheated and one high rye – which, unusually, are aged individually and then blended.

SOUTH CAROLINA

It's fascinating to see how a brewer or winemaker applies their knowledge to distilling. What happens, though, when a baker becomes a distiller? This was the case with Scott and Ann Blackwell at Charleston's **High Wire**. 'It was Ann who saw the trend in craft distilling and thought we should pursue it,' Scott says. 'We sold the bakery, took the plunge.' There are no regrets. 'Distilling is amazing and flavour, if anything, is intensified. It's the ultimate form of preservation and exploration of flavour.'

Their culinary background came in useful when deciding what to make. Charleston is a centre of the New South culinary movement, which has gone about reclaiming often lost local ingredients, and revitalizing a food culture. Bread was part of that, as was corn. In 2014, they heard about Jimmy Red, a corn variety on the brink of extinction. A red dent corn, it was grown by local farmers and bootleggers on James Island, a practice that ended in 2000.

VIRGINIA DISTILLERY CO.

VIRGINIA DISTILLERY CO. COURAGE & CONVICTION 46%

Cereal to the fore. Ripe stone fruits, then subtle musky sherry. Oxidative, fresh and bready. Texturally rich. Light hazelnut crunch, tayberry and bright acidity.

FRUITY/SOFT

Above *High Wire is home to Jimmy Red.*

The last two corn cobs were taken from the last moonshiner's plot and given to farmer Ted Chewning, a guardian of heritage crops. In time, it was being used (and grown) by Charleston's top chef Sean Brock. The Blackwells decided to see if it made as good a whiskey as Brock's grits and flapjacks.

'We work with whole grain, mill and mash it, but there's no souring. Ferment with an ale yeast is held at the low 80s [c.28°C] for about a week and then we run it through a hybrid Christian Carl still in a single pass.'

Barrel entry ranges from 103–110° proof (51.5–55% ABV) and it's matured in new 53-US-gallon (200-litre), 36-month air-dried #3 char barrels.

'All of these make a positive difference, but the biggest in flavour terms is the variety of grain and how it's grown.'

Here's a 100% corn whiskey that has the complexity of a four-grain mash – sweet and oily, almonds and red fruits, estery yet with a drying earthiness and a tickle of spice.

Jimmy Red took corn whiskey into a new area, showing that whiskey's workhorse could be a thoroughbred. It's one of the most important whiskeys in America, not just due to its flavour but because it has demonstrated what distilling can do to help heritage grains.

The Blackwells aren't finished. 'We're constantly playing with other varieties of corn, ryes and wheats among other things,' Scott says. 'There are several new things in the pipeline.'

GEORGIA

Founding stories are fascinating. What was the epiphany, the motivation, the touchpoint, the inspiration? Why here? And if I am here, what do (or can) I make? In the case of Jim Chasteen and Charlie Thompson of **ASW Distillery**, it was making a clear whiskey out of the Terressentia facility in Charleston.

They then hooked up with Justin Manglitz, who owned a homebrew supply store in downtown Athens, Georgia and who had been bitten by the whisky bug when he was given a bottle of Macallan 18-year-old. Maybe that's one reason why

HIGH WIRE

★ HIGH WIRE JIMMY RED CLASSIC 47.5%

Richly fruited. Overripe melon, drying into mace, cinnamon nachos, fried nut, then red fruits, floor polish. Complex, thick, sweet and oily. Just corn. Revelatory.

FRUITY/SOFT

★ HIGH WIRE JIMMY RED SINGLE BARREL 54.9% BUT VARIES

Vanilla, also sumac, red lemonade, comb honey moving to molasses. A fresh pine top note. Massive concentration of fruit. Unctuous feel, coffee, raspberry and chocolate on the finish.

FRUITY/SOFT

ASW's whiskey making (it makes a wide range of spirits) is a mash-up of Scottish and American approaches.

'In effect, we're combining traditional, Scottish-style (Vendome) double-copper-pot distillation with the innovative, New World technique of grain-in distillation,' explains 'chief acorn officer' Chad Ralston.

'This allows us to pull the maximum amount of flavour from our grains, for a very robust flavour profile.' Mash bills here give a sense of the restless and constant innovation, and asking those important questions 'What if?' and 'Why not?'

'A typical large Scottish single malt distillery or large American Bourbon distillery has just two to four mash bills,' Chad explains. 'We have at least a dozen if you count all the different expressions of Ameireaganach American single malt.'

Different barley varieties are also used. Optic Promise is a 50:50 blend of, you guessed it, Optic and Golden Promise. There's also a 100% Maris Otter, and Burns Night single malt, 'which uses the feints from our heavily peated single malt along with two-row Munich malt and others. There's a lot!' All of the range, bar the Fiddler brand, are made at the distillery. 'With that one we forage the US for different whiskies of repute and "fiddle with" them.'

All mature quickly in a wide range of new, charred and toasted casks plus wine, sherry, rum and refill. The answer to 'What can I make?' in ASW's case seems to be...anything and everything.

ARKANSAS

In 2010, Little Rock native Phil Brandon decided to start up **Rock Town**. 'I wanted to make Arkansas whiskey,' says Phil, 'because it hadn't been done before.' That meant being resolute in only sourcing local (within 125 miles/200km) grains. 'I think it's important to use Arkansas grains to make Arkansas whiskey,' he adds. A wide range is made, with a corn-rich, wheated Bourbon as the signature. Distilled in a Vendome pot, it's barrelled at 100° proof (50% ABV) into 25-US-gallon (95-litre), 24-month air-dried, #4 char barrels, which are coopered in the state by Gibbs Brothers Cooperage. There's a bottled-in-bond version, a straight rye (56% rye) and a four-grain with wheat added to the mix.

Pushing things further out, he also makes a peated straight Bourbon using 9% peated and 9% unpeated malt in an 82% corn mash bill. The idea of a Bourbon/malt hybrid continues with an 82% corn, 9% chocolate malt and 9% malt, and another that replaces the chocolate

DEEP SOUTH SMALL-BATCH DISTILLERS

Across the border in Alabama is the **Dettling** distillery, founded in 2014 by Seth and Vicki Dettling, descendants of Franz-Xaver (*sic*) Dettling, a 19th-century Swiss distiller whose firm is still noted for its kirsch. Based in the Big Escambia region, it decided to go its own way rather than use sourced whiskey, with an aim to become 100% farm-produced. Using non-GMO, locally grown corn, any small grains are sourced from further afield – including Germany.

The small-barrel option chosen by many start-ups to give rapid maturation has also been spurned with only full-sized (53-US gallon/200-litre) air-dried barrels used. Taking a cue from brewing, there are five grain types in the mash bill, including roasted rye and oats.

It's hot here, and the house Bourbon is bottled at two years, Select at three and Cask Strength a year more. The experimental side appears under the Single Barrel, Bespoke Collection label and has included Four Wheat, White Wheat and Double Oak. Production remains small and distribution restricted, but its reputation is growing.

Even smaller is **Baton Rouge Distilling,** the first post-Prohibition distillery in the Louisiana capital. Armed with a 52-US-gallon (197-litre) still, husband and wife team Ricci Hull and Natacha Krzesaj have concentrated on small-batch releases, each of which uses a slightly different mash bill, and often yeast type. Rye first appeared in 2019, a 100% corn a year later and since then there have been explorations into Bourbon and wheat. On average there's only 50 x 75cl bottles per batch.

Above *ASW is expanding the remit of American whiskey.*

malt with Golden Promise. The joy of there never being Arkansas whiskey before is that you are not restricted by history or expectation. Rice Bourbon anyone? You'll find it here.

In 2020, Phil also started trucking the Arkansas grains to Bardstown Bourbon and distilling there. The results are bottled under Rock Town Column Still Collection.

If the Bourbon is the signature, he's most excited about the potential of Arkansas single malt. Again, there's an American approach to this with the use of two mash bills: 72.5% Brewer's pale malt, 5.7% Crisp Brown malt, 10.4% Briess 2-row malt and 11.4% Simpsons Peated malt, the other 83% Brewer's pale malt, 5.2% Crisp Brown malt and 11.8% Briess 2-row malt. This is then double matured first for two years in new oak, then in ex-Cognac puncheons.

'I think the future of whiskey is very bright with the emergence of American single malts,' he says. 'I believe that our attention to detail, use of quality grains and barrels and passion set us apart from the average distillery. I want to make Arkansas one of the top whiskey regions in the USA.'

Above *Rock Town is putting Arkansas whiskey on the map.*

TEXAS

The statement, 'I don't want to make whiskey in Texas, I want to make Texan whiskey,' has become a principle for distillers echoed with local variations across America. Place matters. It came out of the **Balcones** distillery in Waco, which was bought by Diageo in 2022. 'This is the heart of Texas, Blackland Prairie bioregion,' says master distiller Jared Himstedt, who has been here since the start. 'When we embarked on our journey into whisky making, we founded the distillery where we were.'

Balcones makes single malt, Texan style. There's no 'e' in their spelling. 'Our production is pretty straightforward,' he insists. 'Malted barley through a roller mill, into a lauter tun. Mashing at around 148–150°F [64.5–65.5°C], continuous sparging. We pitch with M1 yeast, which is usually done fermenting at day three or so, but we continue on to six-plus days to let the lactobacillus continue to work. Our fermenters are not open top, but are not sterilized and are openly vented to the air outside, allowing for some amount of house microflora influence.'

Jared contrasts making whisky as 'an engineer, manipulating all the variables to such a degree that you could make your whisky anywhere', with the Balcones approach, centring around local sourcing and understanding the bioregion.

'We're part of the provenance, and we like what we like...so how can we get great texture? How can we accentuate the fruit character we love? How do we adjust the distillation process, cask choices, entry proof etc. to give us whisky that is truly a result of all the inputs, including the place?'

Part of that 'engineering' is distillation in Forsyth pots with upward-angled lyne arms, and distillation rates are varied to increase oils, add in some sulphur or remove it and accentuate esters. Spirit runs can be 12 hours.

The soul of Balcones lies in the thinking behind that 'straightforward' process (which is anything but). 'I think your philosophical approach should have an impact on which levers you decide to pull,' Jared says. 'We choose to approach getting to know how the microflora contributes, how the climate contributes, what kind of character that barley grown in this hot and dry place will have, and then adjusting the process to steer things in a given direction. If we were to fight the place or resist its expression in our making, then why be here?'

That is different to just sitting back and letting the whisky do what it wants. The craft of distilling is about a deep understanding and appreciation of your components – be they microflora, climate or ingredients.

It's now 16 years into the Balcones story. Is the Texan style now established? 'Whisky is very diverse here,' he says, 'so I'd hesitate to say it has an established style. One of the commonalities we do find consistently, though, is an overall density of flavour. With our climate, whether you make Bourbon or malt, use virgin or used casks, there seems to be a compactness to Texas whisky, almost like a reduction, where the same character is there, but packed into a smaller package, with all the levels turned up.' Even within that concentration, Balcones' whiskies capture tropical fruit notes (especially in refill). Rich yet graceful.

For Jared it's only the start. 'We are just beginning to understand how the place wants to partner with us,' he says. 'There's an inherently different relationship to a thing when you make it. You come to see yourself and the creation differently, and the boundary between you

ROCK TOWN

ROCK TOWN SINGLE BARREL CASK STRENGTH
(MGP AND DISTILLERY) 56–57.5%

Oak. Worcestershire sauce, tamarind. Intense, concentrated. Softens a little in time. Smoky wood chipotle, new leather. Supercharged. Weight and grunt.

OAKY/DRY

Above *Not a whiskey from Texas but a Texan whiskey: Balcones.*

becomes porous, loaded with self-discovery and meaning. It started as a fledgling relationship, and with each step we're still wandering and still arriving at some unknown destination.'

Out there watching and listening to the land, the stills, the vision.

If Jared Himstedt is seeking the Texan pathway for his whiskies, then Dan Garrison of **Garrison Brothers** in Hye is looking to see what a Texan Bourbon could be. After ten years in a high-flying advertising career, Dan was looking to take early retirement before being financially beaten up by the 2001 Enron crash. As he tells it, he 'decided to go on the KY Bourbon trail and get drunk for a week'. When he returned, he was ready to start distilling. In Texas. The land was bought in 2003, the still went in in 2005 and the first bottles appeared five years later.

That in itself was a struggle – less so in understanding process, as he was Dave Pickerell's first client, but rather in public perception. 'I'd written to the TTB [alcohol governing body] asking can I distil Bourbon in Texas and call it Bourbon? Of course they said yes.

'I'd carry their letter with me when visiting bars or retailers and show 'em it if they questioned whether I could make Bourbon in Texas. Got a lot of hate mail from Kentuckians as well!'

What he made was Bourbon with the volume turned up to 11. Big, oaked, concentrated, but with depth and complexity. The mash bill is 74% corn, 15% winter wheat and 11% malted barley. Wheat rather than rye? 'My favourite Bourbon when I was 13 was Weller 12-year-old, then Blanton's. I've always loved wheated.'

Both of those were made by Elmer T. Lee, who it transpired had a connection to the original Garrison Bros Vendome still. Unable to afford a top-end new model, Dan rented an old pot with rectifying column that he then discovered was the experimental still on which Elmer had first made Blanton's. 'I've been paying it off at $500 a quarter. In 2032 it'll be mine!' Now it's been joined by two (new) 500-US-gallon (1,893-litre) and a 2,000-US-gallon (7,570-litre) model.

A single pass through the stills helps to capture the oils from the corn, wheat and barley. 'It's the oil you'd fry chicken with. It adds to the thickness and the mouthfeel if you distil low and slow. Dave taught me that.'

Oils are also there in maturation, thanks to the Texan climate. 'We're ageing in 100° [F/38°C],' he says. 'That's forcing that Bourbon into the wood every day for nine months a year. It brings out texture, wood oils and those grain oils.' It too was an education. Completely filling the barrels in this dynamic climate meant the first ones exploded. Now they're only filled to 70%.

He's also worked with small cooperages to create barrels with thicker staves. 'When we started, there were three cooperages in the US providing all the barrels. I knew if we were going to start craft distilling, then we needed to infuse capital into smaller cooperages, so we did with two in Minnesota, one in South Carolina and another in Kentucky. The thicker staves they give us bleed more of the oils into the product. It's helped to differentiate us from Kentucky. George T. Stagg [from Buffalo Trace] was a revelation. It comes from the top of the rickhouse. I said, I'm going to an exaggerated version.'

Dan thinks big. 'If your vision is regional or local, then you'll build a small business. My vision has always been to sell across the US and internationally. We're now in every state and six countries. It's been a great ride.'

The diversity of Texan whiskey is demonstrated with the approach taken by the Likarish family in their **Ironroot Republic** distillery in Denison, close to

GARRISON BROTHERS

GARRISON BROTHERS BALMORHEA 57.5%

Bold and extractive. Thick corn, leather, bruised blueberries, iron, savoury. Dry oak, roasted, tobacco. Heading towards burned. Of the earth. It's Texas baby!

OAKY/DRY

BALCONES

★ BALCONES LINEAGE SINGLE MALT

(REFILL CASKS) 47%

Light pear drop. Juicy Fruit chewing gum, yellow plum, semi-dried fruits. Fried banana, toasted cereal, peach, persimmon. Red fruit density. Oily and robust.

FRUITY/SOFT

BALCONES TEXAS 1 SINGLE MALT

(VIRGIN OAK) 53%

Some resinous oak, dried cherry, hawthorn and a hint of leather. Chewy caramelized peach. Light grip and nicely bitter touches. Barbecued meat.

RICH/DEEP

BALCONES PILGRIMAGE SINGLE MALT

(REFILL CASKS, THEN SAUTERNES FINISH) 65%

Lots of chocolate and a touch of cigar smoke. Big pepper, positive thick Golden Promise barley feel, then redcurrant, apricot and light nuts.

FRUITY/SOFT

Above *Big thinking in a big state: Garrison Brothers.*

the Oklahoma border. Jonathan is head distiller, brother Robert does sales and their mother Marcia keeps it all in order.

Ironroot is named in tribute to Thomas Volney Munson, the Denison viticulturist who developed the rootstock that saved the French wine industry from phylloxera in the 19th century. It also saved Cognac, which coincidentally is where a lot of Ironroot's approaches come from. 'We use traditional Cognac methodologies with a bit of a Texas twist,' as Robert puts it.

Rather than grape varieties, Ironroot uses different corns. Its four mash bills are all based on Yellow Dent and Bloody Butcher. Peruvian Purple is added to mash bill #2, while Floriani flint is in #3 and #4, along with rye in the last.

'Corn has nearly become a monocrop in the US,' he says. 'These corns are essential to be able to maintain local agriculture throughout the US. They've developed over generations to be grown in different climates and different soils. Maintaining the genetic diversity of the crop is essential to its survival, especially as climate change continues to impact our world.

'Modern Corn can be very tough on soil. One of the beautiful things about some heirloom varieties is that they don't take as much of a toll on the soils and often use different nutrients than the standard yellow corns.'

After long ferments, there's double distillation in a 1,057-US-gallon (4,000-litre) pot with a boil bulb ('onion') and steeply downward-angled lyne arm. 'We've adjusted the shape of the still to create more copper contact and reflux in the neck above the onion because corn is inherently quite a bit oilier than grapes,' he explains.

The most significant Cognaçais element comes in maturation, where a modified version of élevage is used. (Élevage is the method of interventionist maturation: physically moving the spirit from fresh to used wood, blending during ageing or shifting from dry to damp warehousing.)

After a year in barrel, the whiskey is tasted every three months to determine if any adjustment to the barrel needs to be made – changing proof or location, or in the case of non-straight Bourbon or rye maybe decanting to a different barrel.

With proof rising between 1° to 4° a year, they will add water slowly to reduce the rate of ingress of alcohol-soluble compounds in the wood, such as tannin, and increase the rate of extraction of water-soluble compounds. 'We may end up dropping the proof of a barrel from 135° down to 112° [67.5–56% ABV] and then allow it to raise back up over the next several years.'

All this, as well as moving barrels from hot to humid conditions, adds extra flavour options for the final blend, though it doesn't stop there. Once the blend ratios have been worked out, the casks are slowly reduced to bottling strength – which is also a Cognaçais technique to eliminate saponification.

It's hugely time-consuming, but as Robert says, 'It gives us the beauty of mature grain, esters and wood brought together in ways that couldn't be achieved otherwise. In intense climates, the extractive parts of maturation and the high evaporation can lead to extremes very quickly – these techniques allow us to harness those things and use them to express that sense of place without being run over by them.'

IRONROOT REPUBLIC

IRONROOT HARBINGER STRAIGHT BOURBON 57.5%

Earthy. Caramel, dark chocolate and dried cranberry, roasted corn, smoked nuts, red velvet cake. Woodpile, dry leather. Firm and robust. Water opens up fruits.

OAKY/DRY

MIDWEST

The Midwest is a fascinatingly mixed bag, which is maybe no surprise when you consider the huge amount of territory it encompasses, but that shows the ways whereby whisky is repopulating the USA. Here you'll find the whiskey well from which so many draw their liquid, in addition to new start-ups in former garages and corset factories, farms and purpose-built facilities. The following are some of the best.

OHIO

'There wasn't really any craft distilling in the US in 2005,' says Ryan Lang. 'But I'd learned it from my grandfather and wanted to do something that was in my heritage. Then I saw there were guys on the West Coast, so I travelled out to St George and then ping-ponged around looking at stuff. There were probably only 50 of us in the whole country.'

Ryan and his business partner then headed back to Columbus, Ohio and started **Middle West Spirits**. 'We realized quickly that we were under-resourced in terms of volume and space,' he says. 'We looked at sourcing and decided it wasn't our route. We'd have to focus on scale.' In 2010, they opened a bigger facility, then larger ones in 2015 and 2023. There are now four separate still systems.

'The focus has always been on making spirits from the region,' he explains. 'The water has always been the same. The yeast has been fixed. We're hyper-focused on grains.

'Ohio's got 11 million acres [4.5 million hectares] of farmland, so there are lots of opportunities, and we have farmers who are ridiculously knowledgeable. We've been working to bring rye back to Ohio, and then with specific varietals. How they all grow is interesting, as is where they're grown – the heavy clays of northern Ohio or the sandy soils in

Above *There's no shortage of raw materials in the Midwest.*

MIDWEST

the south also affect quality. The same with corn, so we're constantly moving material around the state.'

The first release was a (95%) wheat whiskey made with soft winter wheat that's used as a baker's variety. The corn they use is an open-pollinated, non-GMO yellow, while the search for the right rye ended up with them landing on the Pumpernickel variety. 'It created a crazily different rye profile.'

'We took those three and increased or decreased the percentages in the wash to see what impact it had on flavour, ending up with the right ratio in relation to 5–15% malted barley. The bourbon is 63% corn, 19% wheat, 13% Pumpernickel rye and 5% malt, while the rye is the same grains but in a 80% rye, 10% corn, 5% wheat and 5% barley mash bill.'

Its scale has meant that Middle West has begun to distil for other producers. 'We felt we could be an outlet for guys who needed material customization and unique grain profiles. I love it. We have guys who want oats, or red corn or spelt in their Bourbon. Those heritage corns give you more oil and less yield, but crazy flavour, which is very cool.'

Above *Middle West Spirits has become a significant producer.*

Above *No corsets or whips are made at Journeyman anymore.*

For their own, he says, the key is balance and 'making it as soft as possible off the still. We also got lucky when Speyside Cooperage put a cooperage here in Ohio. We focus a lot on toasting, then match that to a very different char range and then barrel age them differently, giving a spread of flavours that you can then back-blend. The barrel's 50% of the flavour, so why stick with the status quo?'

The biggest lesson has been patience. 'It can take five years to find out if something's worked. In whiskey you have to know it takes a long time to make a buck. You might not hit the sales goals because you've to age these products for better quality, but you have to accept that and continually improve. We should never be satisfied of where we are.'

MICHIGAN

In 2010, Michigan's **Journeyman** distillery started in an unlikely location, a 19th-century corset and (buggy) whip factory in Three Oaks, which, to be clear, was fully renovated. In 2023, a second location in Valparaiso, Indiana opened, also in another renovated historic factory.

The choice was deliberate. 'We believe that the old factory buildings have wonderful stories,' says founder Bill Welter. It is, as he points out, another element of sustainability. 'In Three Oaks, Journeyman has helped spark a revitalization of a village of just 1,000 people. In Valparaiso, the conversion has had a significant impact on the town. The ultimate in sustainability is taking something that exists and revitalizing it versus building new.'

His journey into distilling started when working in hospitality in Scotland where he met Aussie Greg Ramsay (see page 341), who then headed off to start a distillery in Tasmania. Bill followed soon after. 'I returned to the US with nothing more than enough courage and experience to believe I could open a distillery.'

It had to be one that reflected place. One hundred per cent of Journeyman's whiskey is made in-house. The distillery grows some of its requirements on a 50-acre (20-hectare) plot next to the distillery as well as on the family farm in Indiana. The decision was taken to concentrate on rye and wheat, the former

JOURNEYMAN

JOURNEYMAN CORSETS, WHIPS AND WHISKEY (100% WHEAT) 58.5%

Light oak, citrus. Big vanilla, minty toffee, millionaire's shortbread, chocolate pudding, dried red berries and cherry tobacco. Peaches and cream and kumquat.

SWEET/VANILLA

★ JOURNEYMAN CASK STRENGTH SINGLE BARREL LAST FEATHER RYE c.60%

A sweeter, floral rye. Baked rye bread, new leaves, bean sprouts and lemon and honey sweets. Becomes medicinal. A celebration of the grain.

SPICY/HERBAL

because it was simply his favourite style, even though at the time rye only accounted for 1% of the market. The mash bill remains 60% rye, 35% wheat and 5% malted barley. The wheat came from its use at Buffalo Trace. 'We loved that so much that we also made a 100% wheat barrel-strength whiskey called Corsets, Whips and Whiskey.' Again, there's a sense of backing the underdog.

'The wheat category has been a bit overlooked, but we feel like we're bringing awareness and in the process owning and building the category. We feel we've shown that wheat can take on and even beat some of the best Bourbon and rye in the world.'

He's happy to admit to learning from mistakes. 'We started off knowing very little about distilling. There was limited know-how in terms of resources, but also limited people, so effectively we were forced to learn ourselves. I was the first distiller at Journeyman and I admit we didn't know shit, but neither would we have done the things we did had we followed a more traditional route.'

It's this willingness to be different and true to themselves that sets the best new American distilleries apart. 'In Scotland I gained an understanding of whisky as a multigenerational fabric of their society. We're still in the infancy of American whiskey. I'll be interested in seeing the long-term impact of craft distilling on the American whiskey landscape, and the realization that not all roads lead to Kentucky.'

INDIANA

Like many of America's distillers, the Huber family has German ancestry, although it wouldn't be until the 1990s that Carl Huber made the switch from winemaking in **Starlight**, Indiana into distilling. As the largest fruit farm in the state, it started with grape brandies. Though a natural fit, it required a change in legislation and the creation of the Farm Brandy Distillation permit for them to start the business. It then took until 2013 for Carl's sons Christian and Blake (under the tutelage of Dave Pickerell) to persuade the authorities to change the law permitting them to distil grain as well.

It's not just grain though. They cultivate Sweet Lancaster white corn, Bloody Butcher, Hopi blue corn and five yellow strains. Barley and wheat come from their farms. A sweet mash leads to a week-long ferment, then distilling in a Vendome hybrid.

Carl T is a blend of a three-grain (60% corn, 20% rye, 20% malted barley) and a four-grain (55% corn, 20% rye, 20% malted barley, 5% wheat). There's also a rye from an 85% rye, 15% malted barley mash bill. Unsurprisingly, coming from a winemaking background, a wide range of casks is used for an extensive finishing programme.

MGP/Ross & Squibb is actually the same distillery. I'll explain. MGP is the home of the bulk of America's sourced whiskeys (including significant brands such as Bulleit rye). At the time of writing, over 50 distilleries/brands source either all or some of their whiskey from the massive site in Lawrenceburg, Indiana. If it says 'distilled in Indiana' on the label, the likelihood is that it's from here.

MGP (Midwest Grain Producers) was founded as an industrial alcohol producer in Kansas by Cloud L. Cray in 1941, but the Lawrenceburg site is much older, starting life as the Rossville distillery in 1847. It was bought in 1933 by Seagram as part of its post-Prohibition expansion. With the fire sale of the Seagram estate in 1999, it briefly became part of Pernod Ricard, operating under the name LDI. It was then bought by MGP in 2011. Today, it's one of the largest producers in the USA and the largest producer of rye whiskey by some distance. It's also run entirely on solar power.

In 2021, MGP bought Kentucky-based Luxco, which had both distilleries and, importantly, a distribution network, allowing MGP to sell its own whiskies: Rossville Union rye, George Remus Bourbon and Eight & Sand blended Bourbon. All these say they're made at Ross & Squibb – another name for the Lawrenceburg site.

With Luxco on board, the portfolio also includes Bardstown's Lux Row distillery (Blood Oath, Ezra Brooks and others), and Limestone Branch distillery in Lebanon, Kentucky (Yellowstone, Minor Case).

MGP's clients can choose from five Bourbon mash bills: 99% corn, high (36%) rye, lower (21%) rye, high barley (49%) and a 45% wheat. There are also six whiskey mash bills: 95% wheat, 95% rye, 51% rye, 100% malted barley, a light (99%) corn and corn with 15% rye. All are distilled with yeasts drawn from the extensive Seagram depository.

Ageing takes place in concrete-floored warehouses where the strength in barrel decreases. MGP allows new distilleries a chance to craft their own styles, age profile and flavour profile – and at volume. Without it, American whiskey would be a very different place.

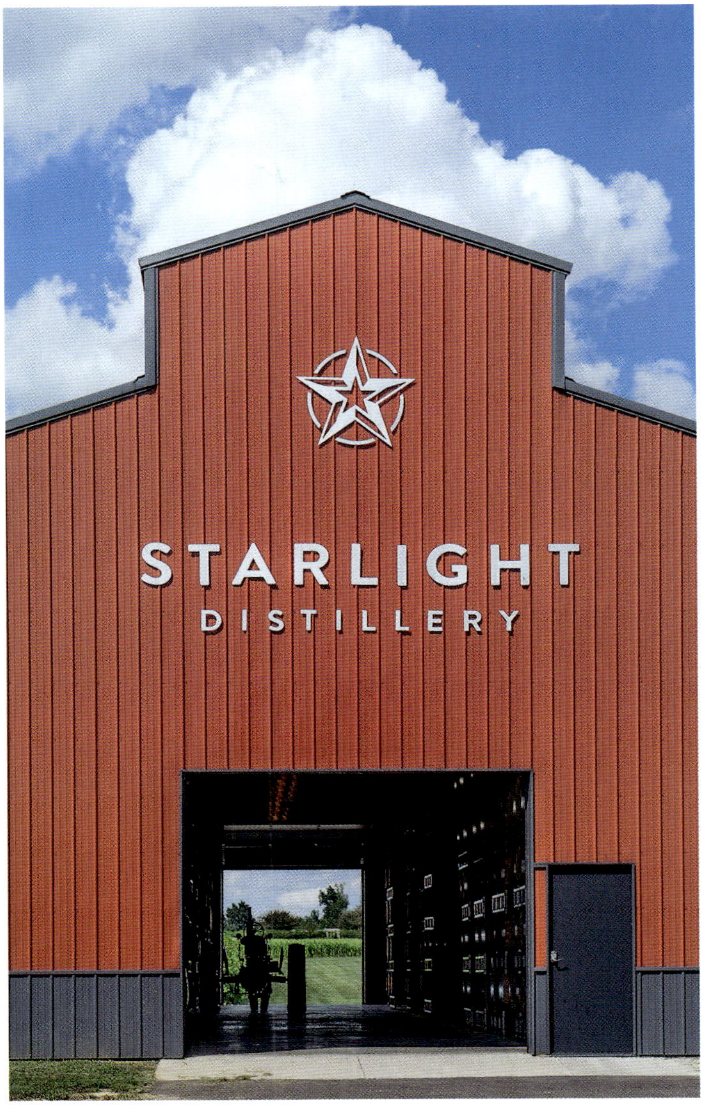

Above *A big red barn and big thinking at Starlight.*

Above *The man who could: Paul Hletko at FEW.*

Above *Cedar Ridge is an American single malt pioneer.*

ILLINOIS

Another recurring theme of the new American whiskey saga is battles with authorities. Prohibition may have ended in 1933, but each state, indeed each county, district or town, could still impose a state of dryness on its population. This was what Paul Hletko faced when he wanted to open a distillery in the north Chicago suburb of Evanston. This wasn't just any old spot. Evanston had been the headquarters of the Women's Christian Temperance Union run by Frances Elizabeth Willard, and old beliefs died hard.

It took time to convince the authorities to change the law, but in 2011 **FEW Spirits** (named after guess who) was born.

As Paul explains, there was a personal element to his decision to distil. 'With the Nazi invasion of Czechoslovakia, my grandfather lost the family brewery and his entire family to the camps. When he passed, it occurred to me that the family history and legacy was gone forever if I didn't do something about it.'

Maybe that central European heritage is why there's a distinct rye-high punch to FEW's whiskeys, rather than a polite sprinkling of spice on top of the creamy vanilla.

FEW

FEW BOURBON 46.5%

Generous and warm. Lightly perfumed. Maraschino, baked squash, turmeric and candied fruits. Light mid-palate, drying into citric-accented coriander seed.

FRUITY/SOFT

FEW STRAIGHT RYE 46.5%

Sweet rye with some camphor adding lift. Turmeric again, sweet carrot. A chewy, lightly dry start. Good grip and drive. Ginger, chocolate and baking spices. Clean acidity.

SPICY/HERBAL

The use of a Belgian saison yeast to drive up fruit is another nod to Europe. The Straight whiskey is a blend of FEW's Bourbon and rye with a cherrywood-smoked malt whiskey, while the rye is 70% rye, 20% corn and 10% malt.

Here again, flavour could come down to location. 'We proudly make Chicago-style whiskey,' says Paul, 'and our location plays a role. We're in the rye belt, not the corn belt. Everything plays a role, from the water to the weather, from the grains to the gumption of the people. It all matters.'

The first of Chicago's new distilleries also has connections to European distilling. Austrian-born Robert Birnecker comes from generations of distillers and was taught the craft by his grandfather. Maybe it was inevitable that he and his wife Sonat should give up careers in academia to follow the urge. In 2008, **Koval** (which means 'someone who forges ahead' in Yiddish and was also Sonat's grandfather's nickname) started. Robert also remains a sought-after consultant for new distillers.

Like FEW, it follows a grain-to-bottle ethos. All the grains are organic and grown in the Midwest, and the water is from Lake Michigan (charcoal-filtered before use). With a wide range of spirits, distillation is in a pot with two linked rectifying columns with plates that can be disengaged depending on what is being made. The exploration into flavour has produced some radically different approaches to whiskey.

Koval's Bourbon uses 49% millet to balance the 51% corn, while Four Grain's mash bill comprises oats, malted barley, rye and wheat. There is also a 100% oat and a 100% millet whiskey. The middle European approach of capturing the essence of the ingredient runs through the entire range.

IOWA

Corn farmer Jeff Quint and his wife Laurie started **Cedar Ridge** in Winthrop in 2005, another distillery that was the first in its state since

Prohibition. The first of their whiskies arrived five years later. Now the range includes straight Bourbon, a Reserve variant, Bottled-in-Bond, a 100% white wheated whiskey, malted rye and a single malt, all distilled by son Murphy.

With corn coming from the family farm (it makes up 74% of the Bourbon mash bill) and the small grains sourced from local growers, Murphy uses malted rye as well as malted barley. Grains are mashed and lautered before fermenting for between three and four days, then double-distilled in one of a number of Christian Carl stills – the smallest is 35 US gallons (132 litres), the largest 800 US gallons (3,028 litres).

Leaving it to naturally mature in full-sized, #3 char casks for between three and five years is a bit of a heartbreak for Murphy, as the angels drink up 18% of the volume every year, though the subsequent concentration of what's left in the barrel adds depth.

While the Bourbon and rye can cope with the attentions of new oak, all manner of ageing techniques are used for the QuintEssential (geddit?) single malt, with initial ageing in refill before being split into a selection of other casks for finishing: port, sherry, brandy and rum. It's then blended and married in a solera vat. Layered and complex, it's become one of America's leading single malts.

KANSAS

In 2010, **Union Horse** started in Kansas City after the Garcia family, brothers Damian, Eric and Patrick and sister Mary, visited a craft distillery and switched from thinking of starting a brewhouse to whiskey making. Originally called Dark Horse, they changed the name after a trademark dispute with Gallo. It's yet another that was the first in its city post-Repeal. Wheat comes from the state, small grains from other Midwest states and barrels from Missouri. Two (secret) yeasts are used.

The white dog emerges from a Vendome 500-US-gallon (1,893-litre) hybrid pot and is barrelled at 110° proof (55% ABV) in #3 char barrels before resting in a non-climate-controlled warehouse, allowing the temperature extremes – hot summers, cold winters – to get to work.

Starting with a Bourbon from a straightforward mash bill of 80% corn and 20% rye, and a 100% rye (enzymes are used instead of malted grains), the range is steadily expanding into new variants. The 4-year-old Rivalist (*sic*) single malt uses Odyssey, chocolate and caramel malts. This is also used in the Rolling Standard where it's blended with the wheated Bourbon. In many ways it's a microcosm of the main trends in American whiskey.

GRAIN INITIATIVES

An increasing number of America's distillers are beginning to explore the flavour possibilities offered by heirloom varieties of grain. Corns such as Jimmy Red, Bloody Butcher, Boone County White and Japonica (the last two trialled by Buffalo Trace), as well as older barley varieties and rye strains being grown. All the distillers are aware that yields will be low, but are willing to accept that for the perceived benefits in flavour.

The issue, as seen with barley in Scotland, is that they are working against what is a commoditized system that prizes yield over flavour or nutrition. There are few small mills, or independent smaller-scale maltings.

The only way in which farmers can continue to increase yields is by also increasing inputs (fertilizers, nitrates, phosphates and so on). They have little option. The problem with that is the nitrogen and phosphate run-off goes into the watershed. Not only is soil fertility decreasing, so is the life of rivers.

In addition, more than 57 billion US tons (51.7 billion tonnes) of soil have eroded in the Midwest, making agriculture unsustainable. The National Resources Conservation Service estimates that cropland in Iowa has lost an average of 6.8 inches (17.3cm) of topsoil since 1850. With that, fertility decreases, biodiversity is reduced, productivity falls and once again inputs increase. A vicious circle.

Possible solutions are the development of regenerative (or organic) farming practices, no-till farming, crop rotation and the planting of cover crops, all of which protect the soil from erosion, add nitrogen and increase fertility. Western Kentucky is leading the way in some of these no-till initiatives.

By establishing closer links with farmers, plant breeders, increased diversity and helping to build rather than reduce soil fertility, maybe the whiskey industry can show that there is an alternative. It has to start with the farmer, which in turn means paying by acre not yield. Looking at the figures, it's clear that time is running out.

THE WEST

This vast, cereal-growing area has always had the potential of having whiskey as a logical adjunct to its rich agricultural heritage while the Pacific Northwest's position was the epicentre of the craft brewing movement. The result is brewers turned distillers, farmers making whiskey from their own crops. What is happening now in America isn't a reset, or a revival. It's the birth of something whose potential was never achieved.

WYOMING

In the town of Kirby (population 76) in Hot Springs county is the state's first distillery, **Wyoming Whiskey**. It was the brainchild of Brad and Kate Mead who run the Gulf Springs ranches in the foothills of the Grant Tetons and Kirby. Their initial idea of a winery was kicked into the long prairie grass. After all, there was everything needed for whiskey growing nearby – and the area had a history of bootlegging and moonshining.

The distillery arrived in 2009, another using local resources to give added value to crops (all grown on the Rageth Farm in Byron) and to benefit the community.

'The philosophy is to celebrate the cowboy culture and heritage of the Wild West in Wyoming,' says Brad. 'The distillery tailors its whiskey production to fit this harsh environment. Being at 4,000-plus feet [1,219 metres] means we have a lower boiling point than at sea level. Here, water boils at about 204°F [95.5°C]. We have to monitor our boiling points in the still much more closely. Barometric pressure changes dramatically, which also affects the boiling points that then need adjustment.'

WYOMING

★ WYOMING WHISKEY 10 YEAR OLD 51.7%
Sweet. A deep, forested element, leaf mulch, shiitake, ryegrass. Well controlled mixing floral elements, gunpowder tea and boiled sweets. Honeyed. Cocoa, floor polish, mint humbugs. Complex.
RICH/DEEP

WYOMING DOUBLE CASK 50%
Soft with dark fruits, hot gorse, heavy lily/orchid, then thickening caramel. Mulberry on the palate, then zippy rye. Green tea and soft fruits on the end.
RICH/DEEP

Above *Cowboy culture meets whiskey at Wyoming.*

THE WEST

A wheated Bourbon (68% corn, 20% winter wheat, 12% malted barley) is the flagship and has changed from early iterations, the grain-forward style of the early days becoming more fruit-accented, the result of a second (wine) yeast being used. Now there's more citrus, orchard fruits and floral top notes.

The whiskey is barrelled at 114° proof (57% ABV) and 'harvested' at five years of age.

'We're talking extreme temperatures here,' says Brad. 'Sometimes in a day, the temperature swings more than 40 or 50 degrees [4.5–10°C]. There are short, intense, hot summers that drive the whiskey deep into the barrel. The winters are long and bone-chillingly cold, leaving the barrels to hibernate and mellow out.'

The whiskey caught the eye of Macallan owner Edrington, which took a majority shareholding in 2023. With consultant master blender Nancy Fraley (Ironroot, Still Austin, Virginia Distillery, J. Henry) on board, longer maturation is planned.

'We're just at the beginning of the craft whiskey boom in the US,' says Brad. 'The consumer is starting to taste, understand and believe in the quality of the Bourbon industry, and production quality is only getting better each day.'

COLORADO

Leopold Bros. was started by Todd and Scott Leopold in 1999 when they opened a brewery and micro-distillery in Ann Arbor, Michigan. In 2008, they upped sticks to the outskirts of Denver. The wide spaces of the high desert seem to suit Todd's ideas better. Trained in Germany as a brewer, he is a visionary who can then put those ideas into practice (with Scott keeping an eye on the finances).

One of those is opening the windows, allowing Colorado into the open fermenters. 'I want the native yeasts to get to work,' he explains. 'The building has been designed to pull air in. The ferment is 72 hours, but then there are another 48 hours to get the lactic to do its thing until the pH is below 3. I have a longer view when it comes to ferments; it's the German training in me. Be patient, but watch it. It's spontaneous – that's why people think I'm crazy.'

It also comes back to the grain itself. 'Lactobacillus depends on where it's been farmed, then what variety it is, where it's malted. If you change any of those, you change the whiskey. It's obvious to me that educated distillers will look at older varieties and methods – because it's about flavour. We're in Colorado. Our whiskey is different. It's peach, pear, apple versus the stone fruit in Kentucky Bourbon. That's why we do what we do.'

The spontaneity is also controlled – that paradox of craft at work. 'Modern [commercial] malt is too "hot", too many enzymes,' Todd explains. 'So when we mash in, it's almost impossible to hit the original gravity. That's why I want control.' It has resulted in the brothers building the largest floor maltings in America on site.

'We're trying to move the industry forward, but at all times what we need to do is keep being Leopold Bros. Very few people get creative freedom in this world, so we have an obligation not to squander it.'

The greatest example of that freedom is the three-chamber still (see box). 'To learn that 100% of rye producers used these, I thought, what the hell? Why did no one know this? One ran in Baltimore 'til the '60s, but there were no details. None of the big guys knew.' He found a flow chart of the Hiram Walker plant in Peoria from the 1930s, took it to Vendome and worked with them to recreate something that no one actually knew how to run. It took a while to learn.

Then came what rye to use. 'The downside was that the commodity grains used by the industry have had their flavour bred out, as the lust for yield has dominated the thinking. In 1898, a rye grain used would have contained 60–65% starch. Today it's 80%.

LEOPOLD BROS.

★ LEOPOLD BROS. THREE CHAMBER RYE 50%

Voluptuously oily. Damask rose, raisin, Swedish cinnamon buns, cookie dough/suet, then dried lavender. Halfway through a release of citrus, flowers, ryegrass and oils. A classic.

OILY/SPICY

LEOPOLD BROS. MARYLAND STYLE RYE 5 YEAR OLD 50%

A sweeter, fruitier style. Banana, light clove, then into some pepper and old books. Lightly floral mid-palate with lower but still tingling spice in the background.

FRUITY/SOFT

Above *Stranahan's: one of American single malt's pioneers.*

THREE-CHAMBER STILL

Wooden chambered stills began being used in American whiskey in the mid-19th century and became the go-to for many distillers, particularly those working with thick rye mashes that burned in direct-fired pots. They were also cheaper, as copper was taxed. Copper versions began to appear towards the end of the 19th century.

The still is made up of three chambers sitting above each other, with a preheater chamber at the top. Chamber 3 (at the bottom) contains the exhausted stillage from the last batch. This is pumped out.

Opening the valve to chamber 2 lets the wash from there flood chamber 3. Once that is filled, chamber 2 is filled with the wash that has been in chamber 1, while it, in turn, is filled with fresh wash from the preheater. That is then filled with wash from a beer well.

Once each is charged, live steam is injected into chamber 3, passing through the wash and rising, via a central column, to permeate through chamber 2, where it strips off alcohol and congeners before rising again to chamber 1.

The alcohol stream then passes through a pipe in the preheater, warming the wash before being carried through a thumper, and then a condenser, where it is separated into heads, hearts and tails.

The live steam passing through the wash in chamber 3 increases in both pressure and temperature, hitting 208–220°F (98–104°C), compared to 198°F (92°C) in a column. This pulls out oils and aromas that you won't normally find in rye. The result is bigger in every dimension – thicker-textured and intensely flavoured. Rye like you have never tried before.

Starch gives alcohol, but doesn't contribute to flavour, so there will have been a commensurate loss of that element over the past century.'

Thankfully, he found a local farmer growing the old Abruzzi rye variety, which has 62% starch. It might take 30% more grain to make the same volume of spirit, but that extra cost delivered extra flavour.

The results were a revelation. Rye became floral (roses and lavender) and oily, the spices hanging back – a mouth-coating, uproarious explosion of flavour. The taste of the past becoming the taste of a future. New American whiskey has to be about opening up new avenues. It's not just a great whiskey, it's a vitally important one.

Why do it, Todd? 'Stupidity!' he laughs. 'More distilleries were popping up and being able to stand out was harder. Also I just wanted to know what those old whiskeys were like. Why did the big companies of the time put in an inefficient still like this? Flavour. I'm worried that these days change is being driven by finance, not flavour.'

For his whiskeys, flavour also means controlling oak. 'I don't like oak, oak, oak and more oak. It's the zeitgeist and I hate it! It's homogenization.' His whiskeys mature in humid, sand-floored dunnage warehouses. Even in the Colorado heat, the angel's share drops. 'The entry is 100° [50% ABV] and at eight years it's not moved, so we're bottled in bond without any late dilution.' He looks around the distillery. 'There's a lot of science here, but most of what we do is magic.'

Denver is a whiskey city and also one of the birthplaces of American single malt. In 1998, firefighter Jess Graber attended a call-out at a barn owned by brewer George Stranahan. A lover of all things barley-related, Jess was soon distilling George's beer. In 2004 the first whiskey was launched, and in 2009 **Stranahan's** moved to the outskirts of Denver. It was bought by Proximo Spirits in 2010. Jess stepped away and started his Tin Cup brand (see page 234).

It's a rambling site with the brewing antecedents clear – a mash tun followed by a boil kettle with whirlpool. It's processing 60,000lb (27,216kg) of malt, sourced in the San Luis Valley, a week. After a three-day ferment, it's distilled in a wash still whose base looks like a squashed bagel. A short neck has four bubble plates. The spirit stills are the same design. The white dog is collected at 140° proof (70% ABV), then barrelled at 110° proof (55% ABV).

High altitude and dry heat mean losses of between 11 and 15% a year in #3 char and a rise in strength. The single barrel, for example, comes out at 119.8° proof (59.9% ABV). Malty and nutty with some citric elements, it's one of the benchmarks of American single malt.

STRANAHAN'S

STRANAHAN'S ORIGINAL SINGLE MALT 47%

Full-bodied. Some oak, cinnamon, nutmeg and butterscotch, some light banana and orange. Palate is soft with sweet oak, light malt, thyme, paraffin, leather and white pepper.

FRUITY/SOFT

STRANAHAN'S DIAMOND PEAK
(FINISHED IN AÑEJO TEQUILA) 45%

Deeper toffee, more herbal palate (thyme and sage). Full fruits. Ripe pear, sweet vanilla from the oak.

FRUITY/SOFT

Above *Denver's Laws Whiskey House takes a patient approach to its whiskey-making.*

Tin Cup is bottled here and blends 3% of a 4-year-old Stranahan's malt into 97% of an MGP Bourbon – 64% corn, 32% rye and 4% malted barley.

The third of Denver's distilleries is **Laws Whiskey House**, which was started in 2010 by former financier Al Laws. It's also deep into its location, only using low-yielding, heritage and heirloom varieties of wheat, rye and barley, grown by the Cody family in San Luis Valley. At 7,500-feet (2,286-metre) altitude, the high day temperatures and cold nights means many plants will struggle – but not winter rye.

The Bourbon is a four-grain mash bill (60% corn, 20% wheat, 10% rye and 10% malted barley), while the rye is 95% heirloom rye with 5% malted barley. There's also a single malt and a wheat.

After souring, the mash is fermented for four days in open-topped fermenters before being distilled in a stripping column, then a pot with four plates. The Bourbon has an openness and lift, but it's the rye that seems to capture the wildness of these high plains.

New distillers face a dilemma. There is a need to say you are different, but that doesn't always balance with a commercially driven impulse to cleave to the norm. You might excite the geeks with the newness, but if you don't conform to a familiar style dictated by the established players, you might not sell. There again, they have the commercial heft to blow you out of the water. Who would be a distiller?

This is why these Denver distillers are fascinating. All have looked at 'whiskey' and tackled it in different ways, which speak of where they are from and challenge the norm. 'Take us on our terms,' they say. 'Trust us. Come on the ride.' Every state has people with this approach. The ones who lead are the ones to follow.

LAWS WHISKEY HOUSE

★ LAWS SAN LUIS VALLEY RYE 47.5%

Rich, intense, earthy with a menthol herbal/ ryegrass nose. Lemon, buttery toffee, then bergamot. Light orange oils, banana, drying spice and green peppercorn.

SPICY/HERBAL

LAWS SAN LUIS VALLEY CASK RYE

60% BUT VARIES

More oomph. Light mint chocolate, floor polish, tobacco leaf, toffee. Serious, dense and earthy. A sense of the soil. Nougat. Pushes back against the oak.

SPICY/HERBAL

LAWS FOUR GRAIN CASK BOURBON

60% BUT VARIES

More vanilla. Rye coming through: caraway prominent. Honeyed with water, corn nuts. More toffee, red and black fruits. Countess Grey tea, Danish pastries.

FRUITY/SOFT

NEW MEXICO

Heading south, **Santa Fe Spirits** was the brainchild of English expat (and former architect) Colin Keegan and started up in 2010. Whiskey forms just part of what is now a huge portfolio. Distiller Stephen Julander (who previously worked at Laws) uses two-row barley for a white whiskey and six-row for the signature Colkegan, which is kilned over mesquite, giving as clear a link to environment as you can imagine. It's then run through a beer/doubler set-up before ageing in a mix of used and new American oak. Being at 7,000 feet (2,134 metres) helps with energy and gives pretty stable conditions.

Above *Whiskey is just one part of Santa Fe's range.*

ARIZONA

It was in 2006 on a Tucson night that the smell of mesquite smoke triggered something in Elaine Paul's mind. Rather than peating a whiskey, what about using the native cooking wood? She and husband Stephen weren't exactly unfamiliar with mesquite, given they ran a furniture design firm specializing in it. After an interim period of home experimentation with floor malting and distillation, in 2011 Hamilton Distillers Group was started with the whiskeys under the name **Whiskey Del Bac.**

Dorado uses two-row barley that's kilned over mesquite, distilled in pot and aged in new American oak. Classic is unsmoked. There's also an MGP-distilled rye, Sentinel, which is finished in Tucson in refill Del Bac casks and filtered over mesquite charcoal.

In the language of the local Tohono O'odham people, Whiskey Del Bac means 'Whiskey of the Place Where the River Reappears in the Sand.' A spring. A start. A southwestern single malt, from the desert.

UTAH

In 2004, biochemist Dave Perkins and his wife Jane moved to Park City, Utah with a plan to open a distillery. Three years later, **High West** started distilling, another that was the first in its state. Dave left in 2016 after selling to Constellation and, under its new owner, a new, larger facility has opened 20 miles (32km) north in Wanship.

'The inspiration was the great Scottish Highland distilleries, which are removed from the hustle and bustle of the city,' says master distiller Brendan Coyle. Maybe that's one reason the Wanship site has Forsyth pots instead of the Holsteins in Park City.

A lot is made of the 7,000-foot (2,134-metre) altitude, but Brendan believes it's of less significance than people think. 'Yes, it gives us a lower boiling point, but it doesn't play a significant role in the flavour profile. It impacts on maturation, though, because of the dryness of the air. We'll lose water in the cask, which means we barrel at a low proof to get better solvent extraction as the proof rises. We're getting robust maturation profiles from eight years onwards.'

High West has always been based around blending. 'We've always sourced from MGP and Bardstown,' he says, 'and those whiskeys make

DEL BAC

DEL BAC DORADO SINGLE MALT 46%

Mango chews, melon and banana. Light caramelization, heavy florals, then the mesquite campfire smoke: turpentine, woodsmoke. Padrón peppers on the barbecue. Savoury smoky wildness.

FRUITY/SOFT

DEL BAC CLASSIC SINGLE MALT 46%

Banana, pear drop and lemon sponge. Palate-coating. Both softly, tropically fruited, dustily dry. Some caramel and cream at the end. Of its place.

FRUITY/SOFT

DEL BAC SENTINEL RYE 46%

Adobe, mint, wormwood, privet blossom and a little caraway. Subtle mesquite smoke. More sweetness: Froot Loops/Love Hearts sweets, raspberry leaf yet still somehow rye.

SPICY/HERBAL

up the foundation of blends like Double Rye, which is [sourced] column still forward with some of our own. Rendezvous Rye has more of our own pot still rye with older column.'

It's not unusual to hear criticisms of brands using sourced whiskeys, the inference being that they are somehow cheating, but Brendan sees it as opening up flavour possibilities. 'People ask: "When will you fully transition?" I say "Never!" Blending is fun; it allows an opportunity to say how we make the products, and why we believe it drives complexity. We've always been open about it. We don't see any shame in what we are doing.'

More mature stock does mean that more wholly distilled whiskeys are appearing, in particular a rye that uses 80% raw and 20% malted from southern Idaho and western Colorado. There's also a high-rye Bourbon, which started being blended in at the end of 2023.

Single malt production began in 2015 based on three recipes, including one using pale, C69 caramel and chocolate malt. 'We take our blending approach to this as well,' Brendan explains. 'Three recipes and ageing in three different cask types: new, used and fortified wine.'

HIGH WEST

HIGH WEST RENDEZVOUS RYE 46%
Huge clove and caraway, sweet carrot and dill (leaf and seed). Has solidity and presence. Palate is straight ahead. More herbs and subtle spice. Hugely quaffable.
SPICY/HERBAL

HIGH WEST BOURBON 46%
Quieter. Charred red pepper, coriander leaf/parsley stalk. Black banana, then into foam banana, sweet corn. Easy-going.
FRUITY/SOFT

Above *White dog flows at High West.*

He's excited about the potential. 'I think the most significant advancements towards regional profiles is coming from American single malt. It doesn't have the history of rye and Bourbon, which can curtail regional profiles.'

Does that constraint apply to High West? 'Rye and Bourbon are more challenging,' he admits. 'Competing with long-established brands is challenging in terms of distribution and for consumers. Malt whiskey's like the Wild Wild West! What's driving awareness is that interesting innovative whiskeys can come from other places aligned with different values.'

Experimentation with yeast is also underway. 'If you want to show what an eccentric Belgian yeast can do, then show it as a single whiskey. A painter can make a more beautiful picture with more colours – and that's what yeasts, mash bills and casks are. The more colours I have, the more beautiful the picture.'

Above *High West – high in altitude and reputation.*

Above *Frey Ranch: from field to still to glass.*

NEVADA

The Frey family has been farming in Fallon to the east of the Sierra Nevada mountains since 1854. It wasn't until 2006, however, that Colby Frey and his wife Ashley decided to add whiskey distilling to the mix. 'It made sense for the future of the farm,' Colby says. 'By producing spirits from rotation crops like wheat, rye, barley and corn, we could create something long-lasting and meaningful with the crops we've always grown. Parlaying our passion for whiskey into distilling felt like a great way to showcase our crops and share them with the world. Without the farm, we wouldn't have a distillery – and vice versa.'

It wasn't until 2013 that the law was passed permitting the commercial operation of distilleries. Since then, there have been investigations into grains, mash bills, yeast and distilling. The grains used are 100% from **Frey Ranch**. 'It means we factor in northern Nevada's climate, terroir, water source and topography to ensure we are yielding the best crops. We have the driest climate in the US, and we have access to snow-melt run-off from the Sierra Nevadas.'

The draff is given to the dairy cows and the muck is used to fertilize the fields. They also grow alfalfa to boost nitrogen levels naturally, and planting is on a seven-year rotation. 'We have a saying that with better inputs you yield better, and get more desirable outputs.

'Given that we control every aspect of the production process, there are fewer variables and unknowns in our operation. We aren't bound by the same supply chain issues and unpredictability as other distillers.'

The flagship Four Grain Bourbon is 66.6% corn, 11.4% winter cereal rye, 10% winter wheat and 12% 2-row barley (malted on site), aged in 53-US-gallon (200-litre) new casks for five years. 'It's long enough to accentuate the grain, but we don't need to age any longer because we aren't masking or covering up a flawed white dog.

'In the wine industry, there's a lot of focus on where the grapes come from and how they're grown. In the world of whiskey, grains are rarely discussed – even through grain is the main ingredient. By growing the whiskey grains ourselves, we grow the grain for a highly specific purpose – which rarely happens in the whiskey world.'

FREY RANCH

★ **FREY RANCH BOURBON** 45%

Sweet with a light vegetal note. Honeyed and unhindered by oak, allowing the grain to speak. Complex, corn oil, sweet spice, ice cream, apple, blackcurrant and hot pastry.

FRUITY/SOFT

Above *Still ahead of the game: St. George Spirits.*

SAN FRANCISCO AREA

Old Potrero (originally Anchor Distilling Company, which became Hotaling & Co. in 2018) was started in 1993 by Fritz Maytag, one of the godfathers of craft brewing in the USA. What he did was look at rye and make it as ornery as he could. It was the only pot still-distilled whiskey in the USA for years. Using 100% rye, it remains fascinatingly uncompromising.

Sonoma County Distilling Co. was started by Adam Spiegel in 2010, a bold entry into wine country. It's making a mix of different styles. A Bourbon uses corn, wheat, rye and malted barley, distilled in direct-fired heated copper stills, adding that identifiable weight.

CALIFORNIA

If you want to discover the starting point of new American whiskey, you need to come to a former naval base in Alameda, a short ferry ride from downtown San Francisco. There you'll find the **St. George Spirits** distillery. Its origins stretch back 41 years to when Jörg Rupf started distilling eau-de-vie 'at a time when no one wanted eau-de-vie', says CEO Lance Winters. The same could apply to his decision to make single malt, or to be precise *Californian* single malt.

Why single malt? 'Because I'm an elitist,' Lance says. 'There's a cultural connection – my family came from Scotland. But also, I hadn't tried a Bourbon that blew my mind – it seemed to be more about the cooper than the distillery. Rye I like, because there's a sense of grain.'

With a knowledge of brewing, he started to distil in his garage, then turned up at Jörg's with a malt whiskey as his CV.

There's a sense of restlessness about St. George, a willingness to try anything. 'We're an attention deficit distillery!' says Lance, 'but we are also our own worst critics.' That questing nature has led to whiskeys such as Baller blend, an idea that started in a sushi bar and a need for a good Highball whiskey. 'We'd made shochu,' says master distiller Dave Smith, 'and it was plum season, so we made umeshu, aged that and finished a maple charcoal-filtered two-row pale/Munich malt whiskey in it.' That kind of sums the place up.

Within that open-minded thinking lies their idea of place. 'The Bay Area is about who ended up here, and why,' says Lance. 'That influences the mindset and what they eat and drink.' The constant waves of new arrivals run through the St. George approach.

'People here are willing to try something different,' says Dave. 'Tradition is important; it can be meaningful when there's something to carry forward. It can also become "traditional" because the boundaries haven't been pushed. We want to create longevity, challenge, but also respect.'

Does that make things more difficult? 'We're not avoiding difficulty,' says Lance. 'That makes better products.' One example would be Lot 22 Single Malt Whiskey, whose smoked element comes from German Bamberg malt smoked over alder and beech. 'Using peat would make it taste like someone else's whiskey,' he explains.

Blending lies at St. George's heart, with a portion of each batch being returned to cask, creating a library of super blends to draw from. One of the pair's favourite words is 'intentionality' – there is always a reason, an end goal. 'Whiskey,' says Dave, 'is bottled poetry, bottled sunshine put in barrel to have beautiful changes. You can't shortcut that. It's immortality, it's magic, it's a moment in time and it's to share.'

ST. GEORGE SPIRITS

★ ST. GEORGE SINGLE MALT LOT 22 43%

Light smoke. Thickly textured, super-ripe tropical fruits, almond extract, fleshy pear. Almost vinous but with balancing acidity to the layered fruits. Nicely oxidized.

FRUITY/SOFT

ST. GEORGE SINGLE MALT (TRAVEL RETAIL EXCLUSIVE) 46%

Mango and cream and that silky-textured distillery character alongside a note of decaying super-sweet fruits. Rich and elegant with sultana and toasty malt.

FRUITY/SOFT

★ ST. GEORGE 40TH ANNIVERSARY EDITION SINGLE MALT 48%

Filled with mellow fruitfulness. Apricot, mirabelle plum and greengage. Takes a darker turn on the palate with silky tannins, slivovitz, raisin, saffron and osmanthus. Sweetly concentrated.

FRUITY/SOFT

OREGON

The genesis of what would become **Westward Whiskey** started when in 2002 Lee Medoff (see also page 240) began bringing beer from the Renovation brewery in McMinnville, Oregon to Portland and distilling it, with the full knowledge of the Renovation team. In 2004, Lee, Christian Krogstad and Thomas Mooney founded House Spirits, best known for Aviation Gin, which it sold in 2016. Two years previously, Westward had appeared.

In 2015, ground was broken on a new site in Portland's Distiller's Row. 'It was a blank slate so we could sit down and design,' says master distiller/blender Miles Munroe. The key to his approach lies in the fact that Portland has 80 breweries.

'Making single malt has been inspired by the beer from the Northwest,' Miles says. 'I was a professional brewer who always wanted to distil. It was a calling. I wanted to make single malt and realized that craft beer will make an incredible one. Brewers make good distillers, but distillers don't make good brewers!' he laughs.

All the barley comes from the Pacific Northwest. 'Barley's less of a crop in Oregon, but that's changing because of growing demand,' he says. 'Any time Great Western maltings does an Oregon batch, we'll buy it. We can do single variety runs and are working with Oregon State University to develop new varieties like Oregon Promise, which give flavour and yield.' It's the same programme that WhistlePig's Liz Rhoades is working with (see page 192).

Even the terminology is brewer-speak – the process starts in a brewhouse not a mash house. 'We're using a high-coloured base malt to get more Maillard-related biscuity flavours,' Miles explains. 'We want to focus on the grain and the ferment. It's moving away from the cask-driven approach.'

He mashes in at 152° proof (76% ABV), then sparges at a higher temp than usual (170–175°F/77–79°C) to stop tannins in the husk being extracted. The wort's boiled in a brew kettle to pasteurize it, then put into a whirlpool to remove trub. Fermentation, using a pale ale

WESTWARD

WESTWARD ORIGINAL SINGLE MALT 45%

Sweet and thick with chocolate/hazelnut depth, light maple syrup (flapjack) and sticky apricot. Some creamed coconut. Nicely balanced.

RICH/DEEP

★ WESTWARD RIMROCK RYE SINGLE CASK (CASK SAMPLE)

Soft, velvety, lightly spiced. Freshly baked rye bread. Warming and slightly oily. Becomes more roasted with time. Subtle yet powerful.

SPICY/HERBAL

WESTWARD TWO MALTS 2023

(88% MALT, 12% MALTED RYE) 45%

Bold and layered with ripe black fruits, peach, some acacia, a little rye spice/chocolate and on the finish chunky roasted spices.

RICH/DEEP

Above *A brewer's mindset helps to create a new style of American single malt.*

Above *Sensory evaluation underway at Westward.*

Above *McCarthy's continues to push the envelope when it comes to single malt.*

yeast, is in closed, tall fermenters for five to six days with the temperature held at 70°F (21°C).

'We decided straight away that our approach to American single malt was not to imitate any other whiskey,' he says. 'We looked instead at Sierra Nevada Pale Ale and its stone fruit/tropical fruit character. Cold ferment allows those esters to form without souring. We want fruits.'

Originally one still did both distillations, but it has now been joined by a second. Both have steel bases with short copper necks. 'We *discourage* reflux,' he says with a smile. 'We want heavy, oily spirit, but we're getting less sulphur because of brewing at low temperatures. It's delightfully inefficient, but the key is making it our own and then finding that balance in maturation.'

AMERICAN SINGLE MALT

American single malt (ASM) is not (at the time of writing) legally defined. The existing definition of 'malt whiskey' follows the same template as straight whiskeys: a whiskey not exceeding 160° proof (80% ABV) from a fermented mash of not less than 51% malted barley, stored at not more than 125° proof (62.5% ABV) in charred new oak containers. Not single malt as the world recognizes it.

In 2016, the American Single Malt Whiskey Commission was founded and created a framework for new legislation. Under this, American single malt Whiskey should be:

– Made from 100% malted barley

– Distilled entirely at one distillery

– Be mashed, distilled and matured in the USA

– Matured in oak casks of a capacity not exceeding 700 litres [185 US gallons]

– Distilled to no more than 160° proof (80% ABV)

– Bottled at 80° proof (40% ABV) or more

Regulations establish a bond of trust between producer and consumer. They allow transparency, give reassurance and guide producers. They define but without hindering innovation.

When Lee Medoff left House Spirits in 2010, he started **Bull Run** to concentrate on single malt, though his brewing background is still strongly in evidence. Burnside Brewing makes the wash, which is then fermented at high temperatures using ale yeasts to drive up esters. It's a style that is unafraid of showing the cereal base, but with balancing fruits and weight.

Another distillery that can lay claim to being the first American single malt is **McCarthy's**, which was started in 1993 by the late Steve McCarthy at his Clear Creek distillery. It was sold in 2014 and is now part of Hood River Distillers, which has been producing fruit brandies since the 1930s. Now Caitlin Bartlemay is the head distiller.

Peated Scottish barley is used. A 6-year-old has the balance needed to bring out darker, ripe, sweet dark fruits alongside the smoke, while the use of oloroso casks amplifies both and adds lusciousness. The new-look McCarthy's bottlings nod towards Scotland, but have that Pacific Northwest balance of sweetness and depth.

MCCARTHY'S

MCCARTHY'S SINGLE MALT SINGLE BARREL #351 (COGNAC FINISH) 57.75%

Apricot and vine flower, big aromatic punch. Bouquet garni, then smoke and roasted malt. Becomes savoury/meaty with dried apple, violet and fruit juices.

RICH/DEEP

WASHINGTON STATE

Regionality is a logical way to look at whisky, but is trying to shoehorn whiskies into a preconceived idea of what they should all be like the correct approach? If regionality doesn't apply to Scotland, then why should it apply to America, given its greater geographical scale? If the narrow regional idea of whisky is hard to prove, then a bioregional one does have a degree of credence.

A bioregion is defined by watersheds, climate and geology but also culture. It is concerned with physical and environmental conditions but also with how people then react to those. Seattle's **Westland** is, for me, a distillery that takes this bioregional approach.

Established in 2011, its thinking has gone deep into Washington State's (or speaking bioregionally, Cascadia's) earth and forests to help create its range of whiskeys.

That's not to say that outside influences are not also grist to its mill – with single malt Scotch as inspiration, though not a template, Westland's complex partial triple distillation approach is unlike any other. Belgium has provided saison yeast for flavour, but the state's brewing culture has provided a matrix of different speciality malts to draw from.

It continues to evolve. 'In 2015 we began to use a "heavy mash" technique to create a spirit designed for longer term maturation in new oak,' says distiller Tyler Pederson. 'This uses increasingly hotter water for our second and third waters to extract more lipids and oils and then eschewing the cut we typically make in our wash runs to separate low wines from lower-proof feints. This then gives a low wine with more oils and fatty acids. It also works beautifully at capturing the phenolics, so it's now our de facto approach to peated malt production.'

The deeper thinking around place manifests itself in the Outpost Range, which explores the raw materials of whiskey making: oak, barley and peat.

The Garryana series uses local *Quercus garryana* to showcase the tree's intense, smoky/meaty mix of treacle, clove and smoke, while Colere is a forensic examination of barley in association with local cereal research station The Bread Lab. This explores

WESTLAND

★ WESTLAND SINGLE MALT 46%

Estery lift, marzipan and mirabelle plum, banana and coconut. Toasted marshmallow. Dense mid-palate, subtle wooded scent. Ginger and apricot, lightly peppery, then Earl Grey tea.

FRUITY/SOFT

WESTLAND COLERE 3RD EDITION (PILOT BARLEY) 50%

Lifted, scented. Mix of old paper, lemon curd, Calvados and chamomile. Soft mid-palate, a slippery feel, then nutmeg, peaches. Restrained.

FRAGRANT/FLORAL

★ WESTLAND GARRYANA 8TH EDITION 50%

Aromatic and dense. Some oils. Meaty, salami, briary fruits. Marmalade, bread pudding, pomegranate molasses. Backwoods stuff. Leather, bitter chocolate.

RICH/DEEP

Above *Westland's investigations into its bioregion include using local* garryana *oak.*

old varieties as well as new, sustainable crosses that work both for flavour and in terms of yield while reducing or eliminating inputs. The final series, Solum, investigates Washington State peat.

'There are more questions than answers still,' says head of marketing Erik Bennett. 'One thing is clear though, there is still so much to discover. American single malt's appeal lies in the ability to discover a wide array of flavours while staying within the same category. There is room to be curious, to really explore. Westland plans to continue leading the way beyond the expected.'

Although this is a feeling that could apply to distillers across America, there is something about the approaches taken by the distillers of the west where 'beyond the expected' is the norm.

A brewing heritage runs deep in Jason Parker's veins. This is the man after all who opened Seattle's Pike Place and Pyramid breweries, as well as Fish Tail Ales (Olympia) and Redhook (Woodinville) before making the jump to distilling in 2010. His **Copperworks** site started on the city's waterfront opposite the ferry terminal. A new, larger distillery was being constructed at the time of writing. Despite his Kentucky upbringing, there's no Bourbon made here. 'The rest of the world thinks whisky is malt, so...' he says with a smile, though with that brewing pedigree background would he make anything else?

'If I hadn't had brewing experience, I might not have had confidence to build craft spirits,' he continues. 'As a brewer, I felt qualified to make a single malt. I knew fermentation, but when we started, every distillery was playing with Scottish rapid ferment. We do it differently.' Fully lautered wort is boiled to get a sweet wort. A beer yeast is added and an initial week-long ferment is followed by two weeks of cold conditioning to mature the beer and reduce the amount of yeast going into the still. 'Basically, it's a three-week process with no lactobacillus.'

It's time-consuming, yet as he argues, 'the existing concept is all about chasing yield, but brewers don't do that. Brewers brew by flavour. Would you ever make tomato sauce because the tomatoes gave you the best yield?'

Single malt made sense for another reason. 'We grow grain here. We have the infrastructure in place to work with farms – there was our starting point.' It also reforged that broken link between grower and distiller. 'The distiller doesn't talk directly to the farmer and ask the questions about what's best to grow. It's hard to break that cycle, but we're trying.' Now he has moved to Vienna malt for flavour.

The wash, made to his specifications, comes from Elysian, Pike and Fremont breweries, but Copperworks' capacity was always constrained by that of the breweries. In 2023, he started a working partnership with Talking Cedar brewery (see box), which gives the distillery access to greater volume – there's a 60,000lb (27,215kg) grain silo just for the distillery.

Distillation is pot stills, with a middle cut between 147° and 121° proof (73.5 and 60.5% ABV), with the spirit going into high-toast, 18–24-month air-seasoned oak, but with five different variables in terms of char and use. 'We're putting it in at different entry points, so basically I have 20 variations to play with...and I can then blend for deliciousness. Every release is different, some single barrels, sometimes blends.

'It's a fun place to be. It's taken ten years to have the core flavour profile – it will be Copperworks five-malt, charred new American oak for three or four years... and then other things!' he laughs.

In a deviation from its American single malt colleagues, the **Woodinville** makes Bourbon. It even sells T-shirts with a map of Washington State with 'Bourbon' emblazoned on them. Started in 2010 by Orlin Sorensen and Brett Carlile, its rye is grown by Phyllis and Arnie Omlin in Quincy, who are now the biggest growers of rye in the state. What started with grocery bags of grain is now a distillery that takes in 30,000lb (13,608kg) a day. The expansion attracted the interest of LVMH, which bought Woodinville in 2017.

Another that was mentored by Dave Pickerell, the mash bill is 72% corn, 22% rye and 6% malted barley, and is a sweet mash. A secret yeast is added and the mash is given a four-day ferment at between 85 and 89°F (29 and 32°C) in order to capture glycol.

It's distilled in three 1,200-US-gallon (4,542-litre) stills, including a Kothe (schnapps-style) still with steam jacket and an eight-plate column with dephlegmator. The distillate can be run back into the pot. There's also a 1,320-US-gallon (4,997-litre) handmade pot still and a 16-plate bubble-cap 'Superaromator' rectification column for vodka. The white dog is pleasantly funky, like pancake

TALKING CEDAR

Washington State's $25 million **Talking Cedar** brewery/distillery 75 miles (120km) south of Seattle is owned by the Chehalis nation, making it the first tribally owned distillery in the USA. When fully operational, it will distil 1.8 million US gallons (6.8 million litres) a year. In order to open, the Chehalis had to overturn the 1834 Indian Trade and Intercourse Act that had banned distilleries on tribal land. Its repeal now gives all of America's 574 tribes the right to distil on their land. It joins Copper Crow, set up by Chippewa couple Curt and Linda Basina in Bayfield, Wisconsin, who got their licence in 2016 as the distillery is outside the Red Cliff reservation.

COPPERWORKS

★ COPPERWORKS SINGLE MALT CASK #047 50%

Leafy/estery. Fresh beech leaf, grassy light malt, then mandarin peel, celery and almond butter. Energetic. Some dandelion and burdock, charred pumpkin.

FRUITY/SOFT

★ COPPERWORKS SINGLE MALT RELEASE #034 (WA PEAT) 52%

Smoke is in there – reeds and grasses, black tea leaf, burning pine tree. Caramelized malty richness giving depth. Resinous, but controlled by the dark fruited weight.

SMOKY/PEATY

WOODINVILLE

WOODINVILLE BOURBON 45%

Minty, sweet, gentle. Red fruits, toffee and pecan pie, Swedish ginger cake, some rich char but with a silky feel. Finishes with red apple and crème brûlée.

FRUITY/SOFT

WOODINVILLE MOSCATEL FINISHED (MOSCATEL DE SETÚBAL CASKS) 50%

Light fruit syrup, thyme and scented tea. Palate is balanced with strawberry jam. By the end the floral side of the Moscatel comes through.

FRUITY/SOFT

mix with a grassy and quite cereal character, almost as if the rye is holding back in favour of the malt.

It's barrelled in at 110° (55% ABV) into 18-month, air-dried Missouri and Ozark oak with #4 char and toasted heads. As the climate is similar to Kentucky, with high summer temperatures and cold winters, strength rises. 'Five years old seems to be the sweet spot,' says sales manager Garrett Epling, 'We've a hunch that it would max out at ten, but it could go over that.' While Bourbon makes up 75% of production, there's also rye: 'That's a labour of love. It sticks to everything.'

If other West Coast distilleries are embarking on their examination of the possibilities of American single malt, here it's about a new style for Bourbon and rye. It's not Kentucky – the lower proofing, the yeast and temperature, the funk of the white dog and a grassiness to the rye that is reminiscent of those from Colorado. Maybe it's the stills, but maybe it's also the place.

ALASKA

Alaska might seem a step too far for distilling, but you'd be wrong. Four distilleries are currently in operation here, all using grain grown in the state. **Anchorage** distillery started in 2007 and makes both Bourbon and rye from grain grown in the valley between the Talkeetna and Chugach mountain ranges. **Port Chilkoot** blends sourced whiskey with its own make Bourbon, while **Uncharted Alaska** in Ketchikan lives up to its name by announcing itself with a sorghum whiskey.

Juneau is home to the **Amalga** distillery, which is also underway with its single malt exploration. Pot still, malt and chocolate malt are all used by Brandon Howard, who spent some time training at Kilchoman on Islay (see page 105) and, yes, there's some peat on occasion as well as alder smoke.

While Islay remains in his heart, he's also aware that this is an Alaskan whiskey. Inspiration is not copying or replicating. Distillation is in a 250-US-gallon (946-litre) Vendome pot and maturation in a wide mix of casks, starting in ex-Bourbon and then, most commonly, a period in ex-Cognac (though rum and Pineau de Charentes are also used). Barley trials have also included Golden Promise and the starting of a solera in ex-sherry casks. American single malt is all about exploration of new fields.

Above *Amalga is shining a light on Alaskan whiskey.*

CANADA

What is Canadian whisky? It's one of the whisky world's more peculiar phenomena that this country, a significant producer as well as being one of the oldest, has a nebulous identity. It has significant brands, including one of North America's largest distilleries, and yet it still isn't top of the mind when thinking or talking about the leading whisky countries around the globe. This is not to do with a lack of quality – Canada produces some world-class whiskies.

The country's approach to whisky making is different, which is in itself exciting. So why then is it the quiet one? It's, as ever, down to a combination of geography and the impacts of history and economics.

While there would have been farm distilling underway when the first settlers arrived, Canada's first commercial distillers of any size were millers – and most were English (Molson, Gooderham, Worts, Corby, Seagram) or German (Wiser). Wheat was the first grain used, to which they added some rye.

The first significant expansion in its whisky industry came as the result of the American Civil War (1861–5). With distilleries in the USA closing and distribution in disarray, Canada's distillers saw a business opportunity. It was the first country to introduce minimum ageing: one year (1887), increased to two in 1890 (the minimum is now three). By the late 19th century, wheat was displaced by mixed mash bills and then with the push west, corn began to be introduced. At the same time, blending had also started – led by Seagram and Hiram Walker. A Canadian style was developing.

Contrary to popular belief, Canadian whisky didn't flood into the USA during Prohibition. What did happen was that, post-Repeal, Canadian distillers took advantage of the opportunities created by the collapse of the American whiskey industry.

SIGNATURE STYLE

The way in which they achieved success was the final piece in the creation of a Canadian style. Seagram, by then owned by Sam Bronfman (see page 249), began making light and flavouring whiskies, often in the same distillery, ageing them separately, then blending them. This single-distillery blended whisky – a lighter whisky with just enough spice to extend the finish when mixed – became the Canadian norm, and it was ideal for a changing American palate.

One reason its qualities are almost unknown in the rest of the world is because the focus has always been supplying its thirsty, larger southern neighbour.

The great global whisky slump of the 1980s increased a process of consolidation. Distilleries closed, companies disappeared or were absorbed. The big players got bigger and, in a depressed market, prices

Right *Gentle autumnal scenes seem to reflect Canada's calmer style of whisky.*

fell. With them went image. As other industries pushed to premium, Canada found that its whisky had become codified as the supplier of low-priced, easy-drinking whisky. Also, as distillers say privately, the fact that none of the majors were Canadian-owned resulted in Canada always being the afterthought.

Then came craft distilling – in 1990 at Nova Scotia's Glenora making single malt, Forty Creek in 1999, Okanagan Spirits in 2004 and Still Waters in 2009. All looked for the alternatives to the big brands' approach, the wriggle room. At the same time, the majors – or the production teams in the majors – began to show what was actually in their warehouses. As they argued, Canadian whisky was different, and that should be celebrated, but it could also be even more different by embracing diversity, new thinking, identity and a willingness to ask questions.

The world will always want and enjoy easy-going, mixable whiskies, but the spirit's greatest asset is its ability to embrace the spread from that to the demanding, awkward and surprising.

The smaller players are out there challenging orthodoxies, but the large players have a role to play to build the category, not just their own brands. Education at generic level, opening distilleries, talking, glasses in hands. Talking up Canada, its grains, its approach.

PRESENT & FUTURE

What does 21st-century Canadian whisky look like? There are more distilleries (in excess of 250 at the last count) and greater volume and diversity. Logic suggests that there's a ready market for all of this south of the border, yet most US distillers are saying that their market is already

Above *Drifting into Collingwood.*

CANADA

THE CANADIAN APPROACH

Canada's legacy distillers approach their whisky making in a different way to the rest of the 'traditional' whisky countries. Each distillery will make a high-strength base whisky (usually corn, but rye and wheat are also found) in column stills. They will also make (or buy in) flavouring whiskies distilled to a lower strength, mostly in a beer still/doubler. These are usually a 'Bourbon-style' mixed mash bill, rye, corn or wheat. All of the whiskies are matured separately before being blended.

Confusingly, although Canadian whisky is often called 'rye', the rye flavouring content might make up only a small percentage of the blend.

saturated, new Irish brands are finding it increasingly hard to gain traction and smaller Scotch distilleries aren't even bothering.

If America's shelves are full, then what's Plan B? The USA accounts for 70% of Canadian whisky sales, but then you look at prices. The average price of the top ten Canadian brands there is under $20 a bottle. If your image is bottom-shelf whisky, then it gets progressively harder to convince people to trade up. The issue isn't quality, it's image, or rather, it's the image of Canadian whisky in the USA.

That suggests that if you are wanting to build a new idea of Canadian whisky, it might be wiser to look further afield where there is none of that baggage. Use the lack of knowledge to amaze people. That big decision needs to be made soon because it's not just America's shelves that are getting full – it's the world's.

Canada's distillers haven't missed the boat; rather in an act of politeness they held off setting sail and now there are many other passengers on board. It's one of the most exciting whisky-making countries in the world. It has heritage and quality. It just needs to get out and tell people.

THE 9.09% DISPENSATION

Canadian whiskies exported to the USA are permitted to have 9.09% of other liquids blended into them. In reality, only a few do. This dispensation was agreed post-Prohibition, allowing high-volume Canadian blends to benefit from a US tax break and compete price-wise with American blends, which are permitted to include up to 80% of unaged neutral spirit. These additions could be young rye whisky, American whiskey, sherry, port, Cognac, rum or wine.

THE EAST

It was in the east where Canadian whisky was born, not of Scottish and Irish émigrés but English and German mill owners finding an outlet for their excess grain. It was also where the whisky barons established their fiefdoms, often giving their names to entire towns, and creating the new 'light' style in the 1960s. And it's here where an alternative to the norm was first explored by 'craft' distillers, who in turn spurred the major players into making whiskies bolder in character.

NOVA SCOTIA

Nova Scotia is well named, the refuge of the first wave of Scottish settlers, and Cape Breton Island remains a place where the Gaelic language and music is still preserved. All that was missing was whisky. That omission was rectified in 1990 thanks to Bruce Jardine who founded **Glenora** distillery in Glenville, making it the first 'craft' distillery in Canada and its first single malt. Made to a classically Scottish template, it took a decade for the first single malt, bottled as Glen Breton, to arrive. It was also the first to finish whisky in Icewine casks. A 30-year-old Triginta appeared in 2020.

River John, on the north coast of Nova Scotia, is home to **Caldera Distilling**. A farm distillery, it grows much of its requirements on co-owner Jarret Stuart's family farm. The grain makeup will vary according to the harvest, but rye is significant. A house yeast is also used and distillation is in a tiny alembic. Hurricane #5 is a rye-forward blended whisky, while Champlain adds some Cognac.

QUEBEC

Canadian whisky's story is a tangled one, with companies merging, distilleries changing hands, brands that start in one place being made in another. Quebec's **Valleyfield** is a case in point. Martin Laberge, distillery manager, has been there for 27 years and has seen it all.

Located in the town of Salaberry-de-Valleyfield, 30 miles (48km) southwest of Montréal, it was built as an industrial alcohol plant in 1938. Post-war, as part of the Schenley group, it produced Golden Wedding and the revered Gibson's Finest. Now part of Diageo, it makes base whiskies for Crown Royal (see page 252) and other Diageo Canadian brands.

Two 100% corn-based whiskies are made. The lighter is run through a beer still and linked columns; a second, richer and more textured, is taken off the beer still and redistilled in a batch kettle (see box, page 254) with copper column. Only the middle cut is collected.

With Gimli (see page 252) at capacity, Valleyfield has expanded. 'We've added a copper beer still similar to Gimli's for rye and "Bourbon"

Above *The best whisky is made in winter...*

Above *Nova Scotia's Glenora was Canada's first 'craft' distillery.*

flavouring whiskies,' says Martin. There's also a new stillhouse for whisky innovation. 'It's kind of my baby,' he says. 'They asked me to be as flexible as possible and not to go crazy on costs.'

A column with doubler and a rectifying column allows him to make a range of distillates. 'I can use the beer still on its own, or with the doubler; use the doubler as a pot still, or link it to the column. We're now playing with innovations and mash bills and single grains.' There's also a wider range of casks. A distillery transformed.

'I'm delighted they decided to do it here. Scotland's been doing innovations in distillation for years – in North America, innovation was cask finishing.'

How challenging was it to match Gimli? 'Really challenging. And we can't quite do it exactly, in the rye especially. It's the same design as Gimli, but the grain's not the same, the water's not the same. Location and suppliers are important in making flavour.'

Not making an exact replica isn't a failure. In fact, it's the opposite. 'The blenders are happy,' he says. 'They don't want exactly the same profile as Gimli, as it opens up opportunities. Before the new stillhouse came in, I could ferment with my eyes closed. It was a steady single mash bill. Now it's really motivating. It's exciting.'

Might we be seeing Valleyfield again being home to its own brand? We can but hope.

Valleyfield is not the only reborn Montréal distillery. The Meagher distillery started in 1873, but by the late 1970s was used solely for bottling. In 2011, it was bought by Sazerac. The distillery was refurbished with a cooker, four washbacks, a copper column and a hybrid pot and a new name, **Old Montréal**. One reason for this volte-face was that Sazerac had by then established its Caribou Crossing brand and acquired Seagram's VO from Diageo.

It has completed a circle for master blender Drew Mayville, who is not only Canadian but started his career at Seagram. He's given VO extra weight and also created a new premium blend of Canadian and American whiskies named Mister Sam, in honour of Sam Bronfman (see box, below).

SAM BRONFMAN & SEAGRAM

Englishman Joseph Seagram joined the Hespeler and Randall distillery in Waterloo, Ontario in 1864. By 1878 he was its owner. In 1928, in the middle of Prohibition, the Seagram business merged with Montréal's Distillers Corporation Limited, owned by Sam Bronfman, one of whisky's most remarkable players.

The Bronfmans were from what is now Moldova and, fleeing pogroms, emigrated to Canada in 1890, eventually settling in Manitoba. The three sons, Sam, Abe and Harry, started as hoteliers, but when some of Canada's provinces began to experiment with Prohibition in 1916, they moved into liquor – selling by mail order, opening warehouses and 'boozeoriums' on the borders between dry and wet provinces. When America went dry, more 'boozeoriums' opened on the international border. They also became the Canadian agent for Scotland's Distillers Company Limited (DCL). Bootleggers? They weren't doing anything illegal, officer.

Sam was interested in making and bottling the whisky, not just selling it, and in 1923 shipped Louisville's Green Brier distillery to LaSalle in Montréal and started Canada's Distillers Corporation Limited.

By the 1930s, he had initiated research into yeast and the science behind whisky, and post-war Seagram embarked on an expansion programme into Scotland, the USA, West Indies and Venezuela, as well as establishing seven Canadian whisky distilleries. At its peak, the firm owned 39 distilleries and by 1965 was turning over C$1 billion.

Sam died in 1971. His grandson, Edgar Jr, made a fateful foray into the entertainment business and in 1999 the firm was broken up, with the distilling arm divided between Pernod Ricard and Diageo.

ONTARIO

Ontario's **Last Straw** began distilling in 2015. Don DiMonte's idea was to build an on-farm distillery, but realizing that a more central location might increase footfall, he acquired a site in a strip mall in Vaughan, to the north of Toronto.

'I was in my 40s and leaving a career in construction,' he says. 'My wife was returning to college, my eldest son was at high school. My wife said, "A distillery? That's the last straw."'

As a new player, it made sense to explore the open field that was Canadian single malt, which he makes from 95% 2-row barley with 5% chocolate malt added. It's distilled in a stainless-steel wash still, then a pot with four plates.

'The biggest learning is that mashing and fermenting is the most important element,' he says. 'If it's bad off the still, it's your fault. We tend to distil to a higher strength so it's cleaner going into the [full-sized] barrel. We're not using the barrel to fix the spirit but to give flavour.'

There's also a malted corn whisky and a rye, whose grains are sourced from family farms in Schomberg, and Beeton. 'A lot of the flavour is specific to that place,' says Don. 'We've started a single farm programme to help with regenerative farming practices, and saving heritage grains. We're moving out of the commodity market.'

In 2023, Last Straw entered into a joint venture with **Heretic Spirits** in Collingwood, Ontario, whose 7,000-litre (1,849-US-gallon) pots will supply contract whisky as well as spirit for both firms.

It's all part of the redefining of Canadian whisky. 'The old brands were mixing whiskies,' he says. 'Canadian whisky is going through a metamorphosis. It's to do with how you do it and treat it. It's the love. We put love into what we do.'

Forty Creek is seen as the distillery that showed there was an alternative to light, mixable, corn-based/flavouring blends. Located close to Niagara Falls, it began as a fruit spirit distillery in 1970. It didn't make money and, in 1992, a man called John K Hall came to wrap the business up. Instead, he bought it, turned it into a winery (Kittling Ridge) and made some whisky on the side.

In 1999, that whisky, Forty Creek, appeared. It was a momentous development. Hall used three mixed mash bills (corn, rye and barley), which he distilled either in column or pot stills. A mix of yeasts (brewer's and distiller's) were used, with wine yeast added for one stream. Ageing was in ex-Bourbon casks. In time, Canadian oak was brought into the mix, its thicker staves and denser structure adding extra richness and depth, seen in Forty Creek's Confederation Oak release. The Barrel Select is re-racked into ex-sherry casks.

Bought by Campari in 2014, it has doubled in size, but as Bill Ashburn, who arrived in 1987 as distiller, says, 'Nothing's changed. The whiskies continue to be produced using the methods developed by John.' The future? 'Keep pushing boundaries with artful blending, extensive use of wood, secondary ageing and wood treatments.'

Above *Last Straw is exploring place-related whisky.*

Some 115 miles (185km) north of Forty Creek, on the shores of Lake Huron, is the old shipbuilding town of **Collingwood**. It was here in 1967 that Barton Brands built its Canadian Mist distillery. It joined Sazerac in 2020, which renamed the distillery after the town.

It's a classic set-up, with 100% corn-based whisky, and flavourings from an unmalted rye, malted barley and corn mash bill, as well as a rye. The flavourings use the distillery's own yeast and beer still and doubler.

While still best known for Canadian Mist, in 2011 the Collingwood brand was launched. It includes a significant dollop of rye flavouring whisky before the blend is married in vats that have staves of toasted maple wood in them.

Three years later, a 21-year-old rye appeared, which apparently had been 'lost' in the warehouse. It was a game-changer, showing that Collingwood could play in the premium area.

In many ways, the distillery's story is Canadian whisky in a nutshell. Well-made, mass-market blends with gems hidden away; untapped potential that's now beginning to emerge. Sazerac has seen this. New developments are promised.

Ontario is also home to **Still Waters**, which in 2009 kicked off the modern 'craft' movement. Barry Stein and Barry Bernstein's first single malt, Stalk & Barrel, launched in 2012. A 100% rye soon followed and

LAST STRAW

LAST STRAW STOUT WHISKY 45%

Balanced. Liquorice, brown sherry, walnut. A chewy, rewarding texture. Sweet on the tongue. Malt extract on the finish.

RICH/DEEP

FORTY CREEK

FORTY CREEK BARREL SELECT 40%

Bready/yeasty. Lovage. A mix of dry elements and light barley sugar sweets. A slight oiliness, then corn nuts dusted with sweet spices. Nutty oak finish.

FRUITY/SOFT

FORTY CREEK DOUBLE BARREL 40%

Also that bready opening, but sweet and thick. Red liquorice. The palate has dark fruits, apple syrup, chocolate orange, toffee and finally green spices. Chewy and pleasing.

FRUITY/SOFT

now accounts for half of the production. The pair sold the brand and have switched to contract distilling.

Heading 260 miles (418km) to the south of Collingwood is the city of Windsor, which sits across the river from Detroit. Here you will find the **Hiram Walker** distillery, Canada's largest, now owned by Pernod Ricard. Hiram Walker's brand Canadian Club is made here for owner Suntory Global Spirits, but Pernod Ricard has its own brands headed by Wiser's, Pike Creek and Lot 40.

The volumes are hard to comprehend. There are rooms filled with vast lakes of yellow fermenting corn, multiple sets of stills. 'We're now at 51 million litres [13.5 million US gallons] per annum and 150 million litres [40 million US gallons] per annum for bulk,' says master blender Dr Don Livermore. 'We're almost at capacity.' Almost. The question here is how do you control this behemoth. Do you stay still or use your heft for change?

A 100% corn base is made in linked columns, the flavourings either direct from the beer still, or via it and the doubler. For Lot 40 rye, the first distillation is in a beer column, then its redistilled in a copper pot. Single malt has also been made (though not yet released). A new pilot plant allows Don to look at rare grains and the archive of yeasts.

Step back and you can see how variety is being built in, opening up possibilities for blending. 'I've been working with Don for nine years,' says global brand ambassador Dave Mitton. 'There have been over 50 new whiskies in that time.' Things are changing.

'We didn't promote ourselves through the 1990s/2010s like Scotch and Bourbon,' says Don. 'The larger corporations treated Canada as a cash cow.' Now there's an education campaign, a whisky school and a blending 101 course, and Dave is globetrotting. Lot 40, in particular, has opened up the conversation.

'Ten years ago, Canadian whisky was something to mix with ginger ale,' says Don. 'We need everyone to be innovative to help the category grow.'

And those innovations? 'I need to listen to my consumers and understand what they're looking for. Currently, they understand barrels, so we're working around finishes and exotic casks. Longer term, I'd love to talk about yeast and what it contributes to flavour. I don't think I'll see that in my time as a master blender, but I'll set that up for the next person.'

'Palates have changed,' says Dave. 'Younger drinkers like more complex blends and flavours. That's where you can have fun.'

HIRAM WALKER

PIKE CREEK 10 YEAR OLD 42%

Sweet and rummy with a charred element, some coconut, stewing orange, nutmeg and biscuit. Palate shows molasses. Light and easy-going.

SWEET/VANILLA

J.P. WISER'S 10 YEAR OLD TRIPLE BARREL 40%

Lean. Dry oak, dried beans. Crisp, oaky, a suggestion of rye. Some caramel and orange juice on the palate.

SWEET/VANILLA

J.P. WISER'S DISSERTATION 46.1%

Rye dust, grass, coriander seed and pepper, then toasted wood. Darjeeling tea, then clove, burned peels and chocolate with a buttery back note.

OAKY/DRY

LOT 40 RYE 43%

Perfumed, pungent. Light oak. There's dried mint, oregano and light rose. Palate has gently rolling spices, then cedar and lemon on the end.

SPICY/HERBAL

★ LOT 40 DARK OAK RYE 48%

Dense and more covered, almost phenolic. Roasted oolong tea, then black pepper, toasted cumin, treacle before a lift of verbena. Bold and rounded.

SPICY/HERBAL

Above *The mammoth Hiram Walker distillery in Windsor is Canada's largest.*

THE PRAIRIES

You would think that the vast wheat plains of the Prairies would make it a prime location for distilleries, but its biggest producer and Canada's largest-selling brand by far is corn-based. Even Alberta's status as the home of rye has seen that grain decline in acreage. It does, however, have a 100% rye distillery as well as one that's exclusively corn and another only wheat. Yet it's these and further variations on the grain theme being explored by newer players that's needed to enliven any category.

MANITOBA

Around 1,200 miles (1,930km) northwest of Windsor, almost in the middle of Canada, is **Gimli**. The town was given to Icelandic immigrants in 1875, who named it after the place where survivors of Ragnarok settled. *Gimli* means 'protected from fire', which is true enough. It isn't, however, protected from cold. It can be -40°C (-40°F) in winter. No wonder they make whisky.

It's home to Crown Royal, one of Sam Bronfman's (see page 249) attempts to curry favour with royalty, which was blended in 1939 to commemorate the visit of George VI and Queen Elizabeth to Canada. It's now the biggest-selling Canadian whisky by some distance, and became part of Diageo in the great carve-up of Seagram's liquor division.

The distillery, which was built in 1969, might be isolated on the edge of the Prairies, but it made sense to build it here. It's surrounded by ingredients. The corn comes from south Manitoba, the rye and barley from the Prairie provinces. It's also a distillery that, when you get into the detail of how it makes its whiskies, gives you an insight into blenders' minds.

Three main lines are made: 100% corn, 'Bourbon-style' (60% corn, 36% rye, 4% malt) and rye (95% rye, 5% malt). The last two are batch-cooked and fermented for 72 hours. The team draws from the 300-strong Seagram repository of yeasts for the Bourbon and rye to add further twists.

Above *The Icelandic settlement of Gimli is home to Crown Royal.*

Above *Last Mountain has come a long way from its start in a garage.*

Things then get more complicated in distillation. The corn is distilled in two beer stills and then either rectified in a four-column set-up to give a light clean base (94%), or from beer still to batch kettle (see box, page 254) to give a softer character.

'Bourbon-style' is distilled in either two beer stills (64%), or in the Coffey still, giving the confusingly named, heavy Coffey rye (94%). The rye is a clove-like fruity distillate from the two beer stills. The five distillates are then aged in a range of casks – new, charred or refill. These are filled and then refilled in a complex matrix of ten different barrel codes. It's a blender's dream, allowing Crown Royal's lush, honeyed character to be directed in many ways.

There are flavoured expressions, but also explorations of age (up to 29 years), and grains, seen in the Barley Edition (85% unmalted, 15% malted barley). 'It was a response to millennials wanting to try new things,' says Valleyfield's distillery manager Martin Laberge. 'They're happy to pay C$100, share it and move to the next thing.'

Blending isn't about closing down options but increasing them. In Crown Royal's case, that's important for the brand but also for the Canadian whisky category.

In 2025, Gimli and Valleyfield will be joined by a third distillery. The 20-million-litre (5.3-million-US-gallon) per annum Crown Royal St. Clair Distillery in Ontario will produce mainly base whiskies. It's a C$245 million investment and, like Valleyfield, will be carbon neutral.

GIMLI

★ **CROWN ROYAL LIMITED EDITION** 40%

Soft and honeyed with some baked vanilla, cream, corn nut, baking spices. The palate is rich and thick with melting toffee, light tangerine and ginger syrup on the finish.

SWEET/VANILLA

SASKATCHEWAN

It's 420 miles (676km) from Gimli to Lumsden, Saskatchewan and the **Last Mountain** distillery, founded by husband and wife team Colin and Meredith Schmidt in 2010. What started in their garage is now housed in an 892sq-metre (9,600sq-foot) building and the team has grown to 20. 'We started out of a genuine interest in and ignorance of the process,' says Colin. 'Our generation doesn't know as much as we should about how things are made; this is especially true of our food and drink.'

It remains deliberately small-batch in terms of production mashes, which include backset, are 2,400 litres (634 US gallons) and the hybrid still holds half that. Distillation is on-grain. 'With each batch being so small gives us the ability to do a wide variety of mash bills, barrel types, blends and finishes. Each release is slightly different.'

Place plays its part. 'We're surrounded by some of the best grain in the world and the 70-degree Celsius [158°F] swing accounts for a large portion of the terroir found in our whisky.' No surprise that 80% of this Prairie distillery's production is wheat, 'Canadian whisky's unsung hero,' as Colin puts it. 'On its own, wheat can be gentle, mellow and sweet. When aged in new oak barrels, we get vanilla and honeycomb characteristics and are finding that it hits its peak at around seven to ten years.' There's also a straight rye and a blend of their four distillates: wheat, malted barley, corn and rye, aged separately and blended.

ALBERTA

The craft distilling trend has also borne fruit in Alberta, mostly notably at the **Troubled Monk** in Red Deer. Brewer Charlie Bredo started distilling in 2018 as a way 'to explore the mystery and boldness of all that Canadian-made whisky can be. We aren't looking to fit the mould, but rather help shape the future of craft whisky.' Unsurprisingly, his brewing background was the inspiration. 'We play around with different types of malts, yeasts, fermentation schedules and mash temperatures,' he explains. All is up for discussion and experimentation.'

His Meeting Creek single malt, under the brand name GrainHenge, is based on its Open Road American Brown Ale with the two-row, amber, crystal, brown and chocolate malts increased to boost caramel, chocolate and coffee. The beer base is followed in Elevator Row, based on its Pesky Pig Pale Ale but again with the speciality angle increased – this time it's dark Munich malt – to build in dark fruit, mulled spices and nuts. There's also a mixed mash bill Alliance (72% corn, 17% malted rye, 11% malted barley) and a rye (Arrowwood). None are shy, neither do they conform to the amenable light sweet template that most people think of when 'Canadian whisky' is mentioned.

'We may be relatively new,' Charlie says, 'but we want to play a role in redefining Canadian whisky, and showcasing how it can be flavourful.'

Somewhat different in scale but not in terms of boldness of flavour is Calgary-based **Alberta Distillers Limited.** Founded in 1946 as an outlet for Alberta's farmers' grain, that meant rye. It's now the world's largest specialist rye producer. Its story is one of quiet service to the industry – so typical of many of Canada's largest plants.

It made world-class rye that would then be shipped and blended into other brands like Windsor Supreme or Canadian Club. This is also where Dave Pickerell sourced the first rye for WhistlePig (see page 192).

Any rye distiller will regale you with stories of stuck mashes, foaming fermenters, stickiness, endless cleaning. It's the price you pay for flavour. Canadian distillers can get round this problem by using enzymes that reduce the foam and viscosity. Alberta remains the only one that still produces its enzymes on site.

BATCH KETTLE

This type of still was invented by Seagram. In it, the spirit from the beer still is condensed and held in a 'kettle'. When filled, it's heated and the vapour passes into a short column. Only the middle cut is collected. There is one at the Gotemba distillery in Japan and Diplomático Rum distillery in Venezuela.

Above *Alberta Distillers...that's a lot of rye.*

Three main flavouring whiskies are made, all from 100% rye. One is taken directly from the beer still, another goes through the beer still and a batch kettle (see box, left) and a third is entirely from the batch kettle. There is also a rye base made in a three-column configuration.

This then allows all of rye's flavours to be revealed – spiced, grassy, earthy, fruits, floral. Again, a distillery made by blenders. No wonder it has been used around the world.

Other streams from corn and triticale are made and used in the Alberta Springs brand. The distillery is steadily emerging from the shadows. Owner Suntory Global Spirits has drastically cut the contract business in order to concentrate on brands headed by Alberta Premium and Reifel Rye, the latter showing the spirit at its richest, a fusillade of flavour, sweet yet spicy, vanilla and pepper. 'This,' it says, 'is rye.' More importantly, 'This is our rye.'

In Lethbridge you can find the **Black Velvet** distillery. In contrast to Alberta Distillers, it uses corn. Black Velvet was an established brand in the 1970s, produced in Gilbey's Toronto distillery. With sales growing, the decision was made to also produce it in Alberta to supply the US West Coast with Black Velvet and, for a time, Golden Wedding. Though the Toronto plant closed, Lethbridge stayed open.

'We have a production technique that isn't used by other Canadian distilleries,' says distillery manager Jonathan Goldberg, 'with different cooking, fermenting and distilling parameters for each of our three whiskies, a 100% corn light base whisky distilled in a four-column and two flavouring: "rye hi-wines" flavouring whisky [90% rye, 10% malted barley], and "corn hi-wines" [100% corn]. All have a different yeast and the flavouring whiskies are distilled to a low strength [57% for rye, 67% for corn].' What you end up with is a clean, sweet base; intense, citric, rye and a punchy lean menthol-accented corn.

Blending starts not at the end of maturation but throughout – Canadian élevage. 'The rye and corn hi-wines are barrelled off in once-used Bourbon barrels and matured for a minimum of two years. After this, the two flavouring whiskies are dumped and blended, along with a small amount of sherry, with freshly distilled base whisky.

'The mixture, referred to as "Blended fill", is then put back into barrels and further aged for a minimum of another three years for our Black Velvet Original and another ten years minimum for our Black Velvet Reserve.'

In 2019 it was bought by Heaven Hill and Jonathan promises great things. 'Whisky in Canada is undergoing a period of change,' he says. 'The increased popularity of US whiskey along with the expansion in craft distilleries is challenging some of the older approaches, by experimenting with more assertive profiles.

'Canada has not yet fully embraced Canadian whisky as a symbol of national pride and connoisseurship in the way that Scotch or Bourbon has, but the signs are there. I'm hopeful.'

The grain focus shifts once more at **Highwood Distillers** in the small town of High River. Here, the focus is wheat and it's been doing it this way since 1974. Production is idiosyncratic and somewhat violent. The wheat is pressure-cooked to burst the grains open. The slurry is then jetted under pressure at a metal plate that shatters any remaining solids. This is then fermented for 90 hours and subsequently distilled in a beer column and copper pot still.

For its blends, Highwood buys in rye and corn flavouring whiskies that are matured on site. The warehouses also hold the last of the casks from British Columbia's Potter's distillery, although its at the other end of the age spectrum where Highwood has received its greatest success – the rye- and corn-based, filtered white whisky White Owl. It also makes the slightly confusingly named Ninety 5-year-old.

Below *Black Velvet. Corn in a sea of wheat and rye.*

Above *Corn is key to Black Velvet's style.*

LAST MOUNTAIN

★ LAST MOUNTAIN UNMALTED WHITE WHEAT RELEASE NO. 0005 45%

Aniseed/fennel opening, then citrus peels, angelica. Perfumed but with a light bite. Breezy and gently fruited with a sweet honeyed core.

FRAGRANT/FLORAL

BLACK VELVET

BLACK VELVET RESERVE EXPORT 40%

Sweet. Light, bright and simple. Canned corn, thinned agave syrup, lemon butter cream. Palate is like dribbling light honey on milk bread.

SWEET/VANILLA

THE WEST

It's in Canada's west where the idea of what used to be called 'craft' or 'micro' distilling has taken its firmest hold largely thanks to liberalized laws in British Columbia encouraging the use of local grains, along with a strong brewing culture. Here you'll find single malt, 'Bourbon' styles, single grain varieties, ancient strains, peat and seaweed. This is where this thing called Canadian whisky is being examined most closely.

BRITISH COLUMBIA

With its access to grain and the climate, it's no wonder British Columbia has become the heartland of craft distilling in Canada, but it's also because of a change in law in 2013 defining 'craft' as being a distillery that produces less than 100,000 litres (26,417 US gallons) a year. In addition, the legislation stated that if you use 100% BC fermentables as your base ingredient, you can wholesale your products without paying provincial tax. It also permitted craft distilleries to sell on site rather than solely through the Liquor Board stores. The BC-grown legislation has consequently created a new relationship between distiller and farmer and the planting of varieties that can prosper in the province's conditions – which ultimately translates to flavour. It ended a decade of lobbying by the province's first craft distiller, Tyler Dyck, who with his family started **Okanagan Spirits** in 2004. He now heads BC's Craft Distillers Guild and the Canadian Craft Distillers Alliance.

The Okanagan Valley lies in the south of the province. Though a semi-arid desert, there's sufficient water from a chain of lakes to support agriculture, initially grain and fruit growing and in time a wine industry.

'We looked at this evolution of terroir-specific wines,' says Tyler, 'and looked at friends who had orchards or arable farms who were making no money. Distillation seemed a logical option to use fruit that supermarkets rejected.' It's both a modern solution and a return to the age-old practice of distilling what grows around you. 'We're creating a legacy project to show that there's a better way of doing things, and against homogenization. That change in law in 2013 was the first chink in the armour of the multinationals.

Above *Okanagan Spirits kickstarted British Columbia's 'craft' explosion.*

OKANAGAN

OKANAGAN SPIRITS BRBN 40%

This their 'Bourbon-style' corn whisky shows Canadian restraint. Hot buttered popcorn, toffee, overripe banana. A crisp back note and maple syrup! Easy-going with light rye spice.

SWEET/VANILLA

OKANAGAN SPIRITS LAIRD OF FINTRY RUM BARREL FINISH 42%

Dried cranberry, hazelnut, then sweet malt and oak. Notes of ghee, celery and light minerality. Palate is warm cinnamon roll. Finishes with toffee.

FRUITY/SOFT

Above *Government legislation helped to make British Columbia the heartland of the new Canadian whisky industry.*

'Even the big guys are realizing that they won't be legitimate in consumers' eyes if they make homogenized whisky. If they don't shift, they'll have the rug pulled out from under them. The next 40–100 years will be the greatest renaissance in whisky making. The time of the homogenized industry is up.'

Three years after starting, the firm expanded into whisky, but that experience with fruit gave them an understanding of the importance of the quality of the base ingredient. Whisky starts in the field. 'We started by working out what varieties of barley, rye and corn to grow in this place. Even within the Okanagan, the same variety will have different flavours when grown in the north or the south, so we celebrated it. This is Coldstream Valley rye, this is Ellison barley. It's allowing us to do smaller batch releases. The link to land is the reason why we get out of bed in the morning.'

Now with two distilleries in the valley, the whiskies are made with tight cuts – 'one ton makes one barrel' – but staying small suits the philosophy. There's a 'Bourbon style', rye and single malt, all lautered bar the corn and distilled off-grain. A range of yeasts are used.

Tyler has switched from refill to new deep toasted oak in recent years to add depth. In time, some of these casks will be used as refill. 'We have really hot summers,' he says. It means that peak maturation is at five years for the 'Bourbon', and a year more for the rye and single malt. 'People ask when are you coming out with a 10-year-old. The answer's never. We're not trying to emulate someone else's environment. If you do that, whose story are you telling? Champion something that's local and unique to yourself. We do what we do because we believe in terroir. What is around us is the sole important thing. We grow our crops in these fields. We've been here for four generations. It's our story to tell.'

When James Lester opened **Sons of Vancouver** in 2015, he didn't take the usual route of making gin or vodka to bring in cash while his whisky was maturing. Instead, he made amaretto, believing – correctly as it turned out – that it was better being the only one doing that than one of hundreds of craft gins and vodkas.

Head distiller and now co-owner Jenna Diubaldo joined in 2021 just as whisky production was moving from a couple of barrels a year to 80. The full switch to whisky is planned by 2025.

The same principle that drove the amaretto is repeated in the ideas behind the whisky. 'To stand out meant finding something no one else was doing,' says James. 'Everyone's gone in the direction of single malt. No one in BC was making rye. Also this, the second-biggest country in the world, had the same rye profile everywhere. We wanted to turn that on its head.'

The problem was finding a local supplier, leading them to work with a grain co-op that was supplying farmers with seed. 'Now we're seeing people in British Columbia farming for whisky,' says Jenna.

SONS OF VANCOUVER

SONS OF VANCOUVER CARIBBEAN CASK RYE 53.5%

Banana chips, banoffee pie, canned pineapple and drying spice. The palate shows coconut, rummy accents and light allspice. Appealing.

FRUITY/SOFT

Above *Sons of Vancouver is exploring all possibilities.*

'And the seed they're using is specific to distilling, rather than us trying to make bread seed work.'

The flavour vision was also different – fruit not spice. 'You get fruit in every whisky category except Canadian,' say James. 'We've ended up using a French saison yeast that when fermented is like a glass of pineapple juice.'

'We weren't seeing people using yeast in a big way,' adds Jenna. 'Typically the differences come from grain or barrels, age and finishing, but when it comes to beer styles, yeast is definitive in creating profile.'

It's now made on-grain with two mash bills, the main one of 100% rye, the other using 66% wheated rye with the rest wheat 'for sweeter, mellow, accessible whisky'. Starting with small casks, they've now switched to 200-litre (53-US-gallon) barrels.

There's this questing nature to their approach to whisky making, new distillers finding their style, their voice and helping to establish the diversity that's so vital at this stage of Canadian whisky's development. 'There's no bigger opportunity in whisky than expanding the Canadian market,' says James. 'People in Canada want to be proud of their whisky.'

Jenna concludes, 'We're aware of the reputation that Canadian whisky has: it's smooth and goes well with a mixer. We want to be known for more than that. Scotch plays on both sides, blends and malt, Canadian whisky only on one. In ten years' time, I'd like to see Canadian whisky punching at the same level as Scotch and American whiskey and with a style that's different to what people think now.'

Vancouver Island has also become a hotbed of distilling expertise. The first of the new distillers appeared on its north coast overlooking the Salish Sea where, in 2011, Patrick Evans started **Shelter Point** on his farm.

That link to the land has remained strong. 'We grow most of the barley we use and think that the proximity to the ocean and sandy soil adds a unique character to the grain,' says Chris Read at the distillery. 'The proximity to the ocean also benefits the maturation of the whisky as it ages. Our warehouses are about 500 metres (1,640 feet) from the shore and the salty air flows through them on a daily basis.'

Given that mild climate, maturation is gentle and slow. Production is along a classic Scottish single malt style, with double distillation in Forsyth pots, though fermentation at seven days is slightly longer than the average in Scotland.

There's a significant difference at the end of maturation. 'Our flagship classic single malt is made from a continuous vatting in a solera system.' In other words, the marrying vat is never fully emptied. The result is a complex whisky with grassiness, fruit, sweetness and a tiny seashore note. Whether it's due to distillation or that damp climate, the feel is thick.

SHELTER POINT

★ **SHELTER POINT SINGLE MALT** 46%

Nutty malt, sweet straw, oat milk and a light touch of oak. Cool and poised. There's just-ripe apricot on the palate with a little celery seed. Generous and soft.

MALTY/NUTTY

Above *Shelter Point distils the grains it grows.*

'The future looks promising for Canadian whisky,' says Chris. 'The new generation of distilleries are moving the needle in rapidly increasing the diversity of whisky.' As is Shelter Point. Distiller Leon Webb has added pot still and a 100% rye to the range.

Patrick sold a controlling interest in the distillery in 2022 to Chris Nelson of Nelson Investment Inc., but remains president.

Graeme Macaloney is a self-confessed fermentation geek, a passion that started while studying at Glasgow's Strathclyde University, which in turn led to a PhD in Biotechnology. He wanted to work in whisky 'but when I applied for jobs I was over-qualified'. He ended up working for Pfizer and emigrated to Canada where, in 2016, he decided to finally scratch that fermentation itch.

'I wanted to use the classic Scottish single malt process, but make it from here,' he says. With scientific rigour, he cast around for advice, ending up with the legendary Mike Nicolson and Dr Jim Swan. The result is **Macaloney's Island Distillery.**

It uses, as he never tires of saying, the Jim Swan method: clear wort, three-day ferment with two yeasts, 'one to give creamy mouthfeel', to condition and for esterification, slow distillation and narrow cuts.

'We could do a six-hour run, but we do twelve for maximum reflux to get those esters. It's half the productivity compared to competitors, but I'll take the hit.'

Malt fixed, he then went to Ireland to pick the brains of Brian Nation, then at Midleton (see page 140), returning to Vancouver Island to make pot still. Next on his 'to make' list was peated, which meant a trip to Islay and a rigorous quizzing of John Campbell (then at Laphroaig – see page 100) and Anthony Wills at Kilchoman (see page 105). 'Each time I wanted to understand the process and then how to operate that in Canada using local ingredients.' Ageing is in a mix of casks including STRs made from wine, port and oloroso.

'If I'm making a new expression, I'll taste a range of competitors and identify a flavour envelope where I want to be. Not copy what they're doing.'

This is seen most clearly with the peated range. Armed with data from Islay, he designed a smoker that was made by his brother-in-law in Alberta, who hauled it over the Rockies to the distillery. The peat also ties it to place. At the moment, it's sourced from Wright Bros. in Olympic Peninsula in Washington State (the same supplier as Westland – see page 241). He then decided to mix in seaweed gathered on Barkley Sound, which added a sweet maritime note to the Washington State signature bonfire/burning heather dryness.

MACALONEY'S

MACALONEY'S SINGLE MALT AN LOY 46%

Minty with some malt coming through balanced with rich PX. A crisp, starchy back note. Palate has barley, brown butter and red fruits.

RICH/DEEP

MACALONEY'S SINGLE MALT WASHINGTON PEAT 46%

A vegetal start, then sweet leather, chypre. The palate is sweet with a crisp back note. Moves to white mushroom, leaf litter and resin.

SMOKY/PEATY

Above *Large casks are now being used at Devine.*

Islay peat has also been sourced, and he's found peat on the island, 'which is organoleptically different again. We had a peat digging party and had to remove 3 feet [1 metre] of loam before we even hit the peat. It ended up like a World War I trench!'

'Call it romanticism, but I'm on an island. The Spanish Peninsula is almost an island on an island. You smell the ocean breeze and the finish of the whiskies is dry with a hint of salt. Something's going on there.'

His aim for an ever-expanding range is to be in 35 countries by 2025. You wouldn't bet against him.

Nine miles (14.5km) south in Saanichton on the outskirts of Victoria is **Devine** distillery, founded by Kevin Titcomb in 2014. 'We're located on a farm property in the middle of an active agricultural community, so we feel we are expressing our region through the spirits we produce, especially so with our whisky.'

A partnership with a local farmer who has a malthouse means everything is grown and malted within minutes of the distillery. As well as barley, they are using spelt, emmer, einkorn and khorasan that, with malted barley, makes up their Ancient Grains whisky – the only one in the world with this mash bill. All are either grown on the island or in central BC.

'Our first release was a traditional single malt whisky – Glen Saanich – but we wanted to explore something entirely different. We have an incredible variety of grain grown in BC, and we chose to look specifically at grains that date back further in history, or that remained largely unmodified over time.

'The essence of Ancient Grains is the balance between the character of the grain and the influence of the casks. There is a harmony there that really appeals to me.'

It's distilled in a Kothe hybrid copper pot still, with a four-plate column and dephlegmator, 'so new make character is generally on the lighter, more elegant side, but the combination of grains we use for Ancient Grains brings a roasted, nutty character to the new make'.

Maturation is in #3 char new American oak, 200-litre (53-US-gallon) casks for a minimum of three years. These have replaced the 57-litre (15-US-gallon) casks that he first used.

'It's an exciting time for whisky in Canada right now,' he says. 'We've seen a resurgence of consumer interest. There are so many unique and intriguing expressions available now, whether it be from a small, farm-based craft distillery like ourselves, or one of the larger-scale producers that have been producing whisky here for decades.'

DEVINE

★ DEVINE ANCIENT GRAINS 45%

Oils galore, mulberry, shoe polish, chicory, then jellied and candied fruits, beeswax and crema Catalana. Textured palate with tamarind, pomegranate and macadamia. Complex.

OILY/SPICY

TWO BREWERS

★ TWO BREWERS RELEASE NO. 36 46%

Broad and substantial with sherried accents and light oxidative elements. Nougat and macaron on the palate. Some caramelized malt, cocoa nibs, black banana. Complex.

RICH/DEEP

YUKON

Journey's end is Whitehorse in the Yukon and with Bob Baxter from **Two Brewers.** The name gives its origins away. He and Alan Hansen started Yukon Brewing in 1997. 'We love single malt whisky, and wanted to stretch our brains beyond beer,' says Bob, though the two brewers had to wait until 2009 for distillation to be legalized in the Yukon. There's little surprise as to the approach. 'We were used to using a variety of malted grains to make a number of different beers, and understood how much the grain contributed to the beer,' he explains.

'Our reasoning was that using a variety of malted grains to make whisky had to have a flavour impact on the new make – and that this would ultimately be discernible in the whisky. There's been mashes that were intentionally soured, hopped whisky and all-roasted malt whisky – think how a stout would taste if it became whisky instead of beer.' Malted rye and malted wheat have also been used.

Needless to say, different mashing techniques feature, and a range of yeasts, while a hybrid pot offers possibilities in terms of reflux, strength... and therefore flavour.

The diversity is increased when you consider the range of barrels used. 'We've done finishes in PX, Moscatel, brandy, even maple syrup and coffee barrels,' Bob says. 'We look at that pile of maturing barrels as a chef looks at spices. They let the final outcome dictate the spices needed. As such, every release we have done is small (usually 1,000 bottles or less) and uniquely numbered because we want every release to be different from anything previous.'

The casks are also aged in the brewery, meaning that they're 'subjected to varying humidity levels 52 weeks a year, depending on what we are doing. We don't know how or how much that affects the whisky while it ages, but our guess is quite a bit.' The impression is that they are having a huge amount of fun – and in a few short years have established themselves as one of the leading craft whisky producers.

'Canadian whisky is a specific thing, defined in our regulations,' he concludes. 'As such, we have always stressed that we do NOT make Canadian whisky, we make whisky made in Canada. There is a strong following here for Scotch, and our product we feel stands up well beside anything we have tried from Scotland.

'We just have to convince customers that there is no magic to the word Scotch...it can be every bit as good (or better) if it comes from elsewhere.' It's the perfect comment to end this brief tour of a whisky country that is, belatedly, making its mark by believing in itself.

ODD SOCIETY

ODD SOCIETY + WILD THING MUSHROOM WHISKY 46%

Petrichor, dried cep and the funk of age. Light fruits. Weird but good.

RICH/DEEP

ODD SOCIETY SMOKE & OAK RAUCH MALT 46%

Smoked with beechwood. Fresh start, some birch sap, green bean, then creamy oak. Delicate smoke.

SMOKY/PEATY

STILLHEAD

STILLHEAD DOUBLE OAK RYE 46%

Light malt-wine notes: juniper and lavender, then ripe red fruits, strawberry jam, Haribo-like chewiness, rosewater kulfi. Appetizing.

FRUITY/SOFT

OTHER BRITISH COLUMBIA DISTILLERIES

The spirit of innovation is carried through at **Stillhead** in Duncan, where distiller Brennan Colebank is making B-Word from 63% corn, 27% malted barley and 10% rye. An 80% rye/20% malted barley is also made, while in addition he blends unaged rye with blackberries. A selection of casks is used, including blackberry wine.

Triticale, a wheat x rye hybrid variety, is used by a number of distilleries in Canada, but it's not often seen on its own. Josh McLafferty at **Monashee**, however, is an exception. The distillery, in Revelstoke ski resort, started in 2017.

Odd Society in East Vancouver is conceivably the distillery pushing the idea of whisky the furthest out. Distiller Gordon Glanz is working with breweries to distil beer, as well as utilizing 'beer mashes'. Non-oak wooden inserts from Black Swan Cooperage are also used, as is peated Scottish barley suspended in a botanical basket to add smoke, plus reducing a whisky with mushroom water. Someone had to.

Above *Odd Society is pushing whisky ever further.*

MEXICO & SOUTH AMERICA

Whisky from the land of agave? It's not exactly new. During the time of US Prohibition there was whiskey made along the border and central valleys, but today's Mexican whisky is very different and, at its best, a celebration of the country. Mexico is also the spiritual home of corn with, at the last count, 64 heirloom varieties. Making whisky from such a rich natural and cultural resource isn't therefore strange – it's logical.

Each of Mexico's growing number of distillers is examining the potential of the country's native grain. They are not trying to replicate Bourbon, but making whiskies that have a distinctly local fingerprint. Additionally, distilling from corn is seen as a battle for heritage and preserving communities by resisting the monoculture of US-bred GMO corn.

I asked friend and distiller Alejandro Aispuro for his take. Mexican by birth, he has worked with agave spirits and is now blender and head of maturation at Feddie in Norway (see page 181).

'Like every other human endeavour, I believe there are those doing it for (arguably) noble reasons, others simply because they see a chance to make a profit. Several producers saw an opportunity to keep producing spirits and surviving, others to get into the spirits industry without flooding the local market with more agave spirit. A few saw the opportunity to try to preserve their heritage and some to take advantage of a gap in the market, the consumer's search for a connection to the land and the desire to consume local spirits.

'The preservation of corn varieties is a serious concern for some producers, either their local or family corn variety, or varieties from all over the country. I've tasted products (whisky, tortillas, atole, tamales) made from many varieties across Mexico and can tell you that each has a particular flavour. Heritage corn varieties' flavours are distinctive, from sweet vanilla, butterscotch and popcorn, to slightly cheesy, nutty or musty.'

He sees distilling as a logical outlet for those trying to preserve this patrimony. 'The market for unique varieties of corn in food is mostly restricted to high-end restaurants,' he says. 'Whisky has a better reach in consumer opportunities and in markets both local and international.'

How would he like to see it? Organically and with a strong sense of identity, of communion with the land, with responsibility to the farmer and the consumer, and with long-term thinking and actions that reflect it.

Right *Famed for its mezcal, could Oaxaca soon be as well known for its whisky?*

Achievements in preservation would be applauded, but beyond preserving the different corn varieties, it should be a search to preserve the flavours and aromas of a product and its terroir, as well as the biodiversity and sustainability of the ecosystem.

'More long-term planning is needed. A true long-term commitment, distanced from short-term profits and needs. We might see some bright stars.'

Travelling south to the adjoining continent, we enter a different whisky culture, since South America has long been a fertile market for Scotch blends and it's only in recent times that local whisky distilling has started to take root. Brazil has the oldest distilleries, but it's in Argentina – and particularly Patagonia – that single malt is beginning to establish itself. Its distilleries have recently been joined by those in Chile and, in an echo of the saviours of corn in Mexico, Peru.

MEXICO

The first Mexican whisky to appear on the wider international market was **Abasolo**, created by Dr Iván Saldaña (a world authority on mezcal) as part of the Casa Lumbre project, which he founded with Moisés Guindi and Daniel Schneeweiss in 2017 as a platform for Mexico's cultural, biological and sensory heritage. Situated at 2,377 metres (7,800 feet) in Jilotepec de Abasolo, 60 miles (97km) northeast of Mexico City, the distillery claims to be the country's first dedicated whisky distillery. After extensive trials, Abasolo uses the Cacahuazintle variety of corn, grown on three farms in Calimaya.

To help with conversion, a percentage is malted, while the remainder is nixtamalized, a technique used in Mexico for thousands of years to make the corn more nutritious and easier to grind (*nextli* meaning ashes, *tamale* corn dough). Dried kernels are boiled, then steeped in water and lime (calcium oxide/hydroxide). In a similar way to germination, the highly alkaline solution breaks down the corn's cell walls, releasing pectin and increasing the corn's taste and aroma.

It's subsequently roasted, mashed and fermented. Abasolo is then double-distilled in copper pots before a short ageing in first-fill ex-Bourbon casks resting in open-air warehouses.

Opposite *The ageing warehouse at Abasolo.*

ABASOLO

ABASOLO EL WHISKY DE MEXICO 43%

Virgin oak punchiness. Barbecued corn and popcorn, then some herbal elements before apricot, some vanilla, light tea, muffin and stewed apple. There's a light mineral element on the finish.

SWEET/VANILLA

Above *Abasolo is a leader in Mexican whisky.*

QUERÉTARO

Vinos y Licores Sabrosos's Juan del Campo brand is produced in El Marqués, Querétaro. Founded by Xaime Niembro, it's another releasing a range from different varieties: some single, some blended. He ferments using wild yeasts and double distils. Each bottle carries details of the farmer who grew the corn. Using a blend of blue and pink corn from San Juan del Río, Querétaro gives a mineral character; yellow Tuxqueño from La Esperanza, Chiapas is a mix of earth and pineapple, Pozolero Negro (black corn) from Tarimoro, Guanajuato has a more yeasty/doughy profile, while the same region's Huiochol purple corn is more estery and floral. Xaime is using a wide variety of casks, including mezcal and is also beginning to experiment with malt whisky.

Much of the thinking here parallels the pioneering work done by Ron Cooper for different agaves in his Del Maguey mezcals – the naming of variety and farmer/distiller is now common practice for the best mezcal brands. The potential here is huge. Watch this space.

TLAXCALA

Cuatro Volcanes is in San Juan Totolac, Tlaxcala at the foot of the (extinct) La Malinche volcano. Its whiskies are made from 80% locally

Above *Sierra Norte added whisky to its expertise in mezcal.*

sourced black, blue and Sangre de Cristo corn. The remaining 20% is malted barley.

'It's worth highlighting that our whiskey relies on corn varieties indigenous to our region, a commitment we actively promote,' says distiller Ernesto Vargas who co-founded the firm in 2017. 'We collaborate with local farmers to conserve those grains facing extinction or experiencing low production.'

Distillation is in hybrid stills and the new make is filtered through coconut charcoal. It's then barrelled at 55% into medium toast white oak with a short finish in walnut (apple and cherry woods are also being looked at).

'As the consumption of whiskey and other spirits grows in our country, it's the perfect time to create our expressions,' Ernesto adds. 'It's time to distill the spirit of Mexico into each bottle, sharing the passion and tradition that make our beverages truly unique.'

OAXACA

The state of Oaxaca has been identified as where corn was first domesticated 6,000 years ago. **Sierra Norte** in San Agustín de las Juntas is utilizing this in its exploration of corn's potential.

The distillery has been making mezcal since the late 1990s, but an agave shortage in 2013 saw distiller Douglas French looking for another option. Now he is distilling a six-strong selection of local varieties sourced from 25 growers across the states' eight regions. All are single barrels and each is made from a mash of 85% of one type of corn, with the addition of 15% malted barley. It's aged in mostly French oak wine casks.

Although all have corn's easily identifiable sweetness, the differences are stark: white corn has a sweet yet nutty note, black is creamier and yellow is the sweetest, lightly citric and floral, while purple flirts with rye-like spices and red fruits, with those spices amplified in red.

The motivation is preserving corn's heritage and Sierre Norte has established a seed bank to supply its increasing number of growers.

Pierde Almas is also in Oaxaca and was started by Jonathan Barbieri and his wife Yira Vallejo, who also have previous experience with mezcals. In 2014 he began using heirloom varieties as a reaction against the increasing monoculture of GMO plantings. Distilling is an effective way of preserving these landraces. Using surplus from farmers, the distillery sources yellow corn from Chinantla, black from Sierra Norte and white Bolita and red Sangre de Cristo from Oaxaca's central valleys. A small amount of malted barley is added and distillation is in copper pots. It was bought by Diageo Mexico in 2018.

Jonathan is also behind **Maíz Nation**, which again sources from Oaxaca's valleys, highlands and coast. Here, 6% of malted barley is added for conversion and ferments are kicked off with a 'sourdough' starter. Double distillation in pots follows and ageing is in new oak. Chalqueño, Bolita, Tepecintle and Olotillo are the corn varieties used, each specific to its own subregion. Saving the corn also means saving a community.

YUCATÁN

In 2000 in Merida, Yucatán, **Gran Maizal** has been using the 4,000-year-old Nal T'eel landrace (known as rooster corn to the Mayan people) alongside two others. Some of the corn is toasted, the rest nixtamalized and fermented with the distillery's own yeast. It's rested in terracotta, with vanilla pods and cacao nibs added to the spirit.

Above *Corn fermenting at Maíz Nation.*

BRAZIL

Originally a wine producer in Veranópolis, which sits in Brazil's Vale dos Vinhedos (valley of the vines), the **Union** distillery began in 1972 when the company was bought by the Borsato and Ziero families. They turned to the then Morrison Bowmore to assist in upgrading and expanding its facilities. Originally its business was providing bulk for secondary brands, but began to release its own range in 2019. A four-strong list includes wine cask finish, virgin oak finish and two peated variants. Its oldest expression dates from 2005.

The **Lamas** family opened its eponymous distillery in tropical Matozinhos (Minas Gerais) in 2019. Double pot still single malt is just one part of an extensive portfolio. The whisky range is made up of five variants, including two peated, and the Canem blend. Taking a lead from artisanal cachaca producers, they are ageing some whiskies in native Brazilian wood such as Bálsamo (*Myroxylon balsamum*) and Putumuju (*Centrolobium paraense*), the former adding a herbal/spicy quality, the latter giving intense colour and an earthier spice.

PERU

Michael Kuryla (aka **Don Michael**) hails originally from Miami but has long family connections to Peru, where he ran a construction company – a job that included fabricating stills. In 2015 he had the idea of using local corn to make the country's first whisky. After a number of trials, he decided on Peru's iconic black variety Maíz Negro de Junín. Just one of the country's 62 varieties, it only grows above 3,000 metres (10,000 feet) and is less sweet than

SOUTH AMERICA

DON MICHAEL

BLACK WHISKEY 45%

Immediate sweetness, roasted corn, creamed corn, light vanilla, nectarine, citrus and copper coin. Thick, with sweet drive and a balancing bitter finish.

SWEET/VANILLA

BLACK WHISKEY SINGLE BARREL #593

58.9%

Deep and polished. Some singed elements (cask char/corn), then nougat and salty liquorice. Highly concentrated. Robust with some burned toast on the finish.

RICH/DEEP

Above *Don Michael are Peruvian whisky pioneers.*

standard yellow corn. 'I didn't just want to make a good drink, but one with the identity of Peru, a whisky that could grab the world's attention.'

He started operations in Lurín (about an hour's drive south of Lima) with a 40-litre (10.5-US-gallon) still and zero experience. It was, in his own words, a tough learning curve. 'The first runs were so bad! It took me six or seven months to make it drinkable and work out the mash bill [60% corn, 30% malted wheat, 10% malted barley].'

Black Whisky was launched in 2019 just in time for when Covid hit. 'We had to get a permit to keep producing, and all we could do was distil, fill barrels and sell either online or by motorbike.'

Wary about an over-reliance on local sales, export has always been a focus and in 2021 the whisky won some awards and a subsequent TikTok clip went viral. Now Don Michael's brand is exported to seven countries and has replaced his original still with one that holds 10,000 litres (2,642 US gallons), housed in a purpose-built distillery. And with new investors on board, the plans are for further expansion. There are also trials with other corn varieties – he highlighted a white corn from Cuzco as being promising, and ageing in native hardwoods such as amburana. Maybe the era of six-day weeks and eighteen-hour days is over.

He has also created a definition for Andean whisky (being made from a mash bill of a minimum of 51% Andean corn), which he hopes will add weight to his export drive and encourage other producers to start up, as he says, 'There's plenty room.'

Above *La Alazana is bringing whisky to Patagonia.*

ARGENTINA

Argentina's venture into whisky started with the founding in 2011 of the **La Alazana** distillery in the Golondrinas valley, Patagonia by Nestor Serenelli, his wife Lila and her father Pablo Tognetti (see below). 'Nestor grew up on a farm in the country where drinking whisky was very common and he always dreamed of having his own productive farmland,' Lila explains. 'His grandfather made grappa and Nestor followed that tradition as well as trying corn whisky, but then decided to just make single malt.'

They grow and then malt their own barley and use Patagonian peat for the Haidd Merlys smoked versions. Inside the distillery there's a lauter tun as well as four stainless-steel washbacks where the ferments tick over for 100 hours. The stills have tall necks and long lyne arms, and the cut is high in order to get an estery new make.

There is an old Welsh (and Welsh-speaking) community in Patagonia, meaning it was inevitable that they collaborated with Penderyn (see page 150), while in 2022 two of their casks were transported to Argentina's Marambio Base in Antarctica to mature for another two years – just to see what might happen, or not.

La Alazana is a distillery deeply embedded in its locale. 'Terroir is everything to us,' says Lila. 'The exceptionally pure spring water that goes into our process and the mild mountain climate under which the spirit is aged are expressed in our whisky. At the same time, the valley climate is excellent for growing malting barley for whisky.

'Argentina and South America are following the direction that whisky consumption is having worldwide, which is inclining toward a younger age group, people who travel a lot and are very demanding of the quality of what they consume...and who love detail.' The family is opening a second distillery, Farm Alazana, in Montana, which will be run by Nestor and Lila's son and his partner.

Pablo Tognetti left La Alazana in 2016 to start his **Madoc** distillery next to the Manush brewery in Dina Huapi, Río Negro. He had started experimenting with whisky distilling at home in 1995 using his wife's pressure cooker as a still. Madoc, named after a legendary Welsh prince, has two pots with tall necks, each run filling a barrel.

A pressure cooker also features in the backstory of **Demian**, started by horse vet Max Demian Pelegrí in 2001. A self-confessed ferment geek, his head was turned by tasting craft beer. 'I thought, "Wow! This is real

DEMIAN

DEMIAN SINGLE MALT 43%

Powerful rich malt with some hard toffee, black cherry. Dry on the palate with some ink, then bittersweet berries, and some roasted malt on the finish.

RICH/DEEP

BUENOS AIRES DISTILLERY

Emilio Mignone in Luján, Buenos Aires province was started by Carlos and Santiago Mignone in 2015. The distillery has an open-topped mash tun, one washback and two pots.

Their range includes EM&C Pampa single malt, the Classic expression in ex-Bourbon, the Peated aged in PX. As an experiment, they are also ageing the peated casks in a warehouse on the Atlantic coast.

Above *Madoc: Welsh in name, Patagonian in approach.*

beer? Why is it so different from what the industry told me beer was?" Then I started to wonder: what would a craft whisky be like? Will that also be something different from the norm?'

There was only one way to find out. Without any reference point, he began to experiment with a pressure-cooker still and 'started testing and testing. Trial and error, trial and error. I still remember the first aroma that came out of that pot after distillation!' In 2018 he was ready to start ageing his new make.

The pressure cooker has gone and now he distils in two copper stills (50 and 150 litres/13.2 and 40 US gallons), but the handmade approach continues with everything from malting onwards taking place on site.

His first release, Classic, is 100% malted barley, with a malted rye ageing. In the spirit of experimentation, his corn whisky has been smoked with peat and espinillo wood (*Prosopis affinis*). Ageing is a minimum of three years in American oak, European oak and many others including ex-Malbec, mead, and apple or grape brandy casks.

URUGUAY

In 2023, Enrique **Larnaudie** launched Uruguay's first single malt. His background as a laboratory refrigeration technician brought him into contact with the scientific side of brewing, which he then started to research. Distilling took over from brewing in 2017 when the government-controlled monopoly was lifted.

He insists that this retirement project remains a hobby rather than a major commercial enterprise, and production is still small-scale, with only 30 bottles a month being sold by ballot. Though he has looked to what has happened in Argentina, his quest remains finding something distinctly Uruguayan.

CHILE

Like many countries, Chile has caught the gin-distilling bug, but two of its more remote producers have also turned their hands to whisky. As of 2024 they were still in cask. **Tepaluma** in La Junta is owned by husband and wife Mark Abernethy and Andrea Zavala Peña. Living in Belgium, they had a dream of making whisky, but the idea snapped into focus when, in 2008, Andrea was struck by lightning. During her recovery, they planned, researched and then returned to Andrea's native Chile to look for a site. In La Junta in Chilean Patagonia they are using local botanicals to make gin and raise money while their whisky – made from barley that they malt themselves – matures.

Aussies Kiera Shiels and Matt Oberg were backpacking in Chilean Patagonia in 2015. Needing a drink after completing an epic trek, they began to wonder why there wasn't anything akin to the distilling scene in Tasmania in Chile. Out of that came Puerto Natales' **Last Hope** distillery and bar. While gin was the first spirit released, whisky was always part of their plans. The thriving craft brewing scene means Patagonian malted barley is available, as is peat, while pisco stills are easy to come by. At the time of writing, the whisky is sleeping in cask. And so, at 51.7275° S, 72.4980° W in the world's most southerly distillery, we end this part of the tour.

JAPAN

An industry revived? Japan might have been slow to pick up on the global whisky explosion, but it's sure making up for that now. Its whisky map has changed completely in the last decade. 'There have never been more distilleries in Japan than now,' says Japanese whisky's leading writer Stefan Van Eycken. 'At the high point of consumption in 1983–4 there were 30-odd distilleries. We've gone from 7 in 2007 to over 70, 15 years later. And in the last two years [2022–3] it's been an exponential jump.'

Japan's whisky story is now well known: whisky arriving with the 19th-century's gunboat diplomats, the creation of the country's own whiskies from 1923 on, the building of an industry in the 1960s, the boom era of the 1970s and early 1980s and the subsequent 25 years of decline, before the export and Highball-fuelled resurgence in the 2000s.

Many of today's new producers are shochu or sake makers. With both drinks in decline, whisky seems a logical switch, but there are significant differences in the production of whisky that need to be understood. Already the most switched on of these new distillers are finding ways to adapt existing techniques to whisky production.

To some extent we have come full circle. Japanese whisky's two founders Shinjiro Torii (Suntory) and Masataka Taketsuru (Nikka) may have disagreed over style – the former wanting to make a whisky closely aligned with place and palate, the latter applying Scotch techniques – but both responses were reactions to the question 'What is possible here?' Today's distillers are asking the same question in a changed market and world.

Whisky is now being made from an island off the northwest coast of Hokkaido to the subtropics. New climates are being discovered, local barley varieties are being rediscovered, sake and shochu yeasts are used, shochu stills are making a new type of 'grain' whisky and there's rice whisky, mizunara oak and other woods.

What remains constant is the Japanese fundamental of clarity of character. Pure yet intense; transparent rather than bludgeoning; subtle but with penetration. Its whiskies are aligned inextricably with a cultural aesthetic, seen in Japan's art, food and poetry – simplicity without being simple.

There is also a resistance to staying still. 'I want to make better whisky tomorrow and then better again the day after,' as Suntory's chief blender Shinji Fukuyo told me. It's this belief in kaizen (continual improvement) and monozukuri (value of craftsmanship) that has kept Japanese whisky moving forward, even in tough times.

However, there's an elephant in the room – legislation or to be precise the lack of it, which has allowed 'laundering' in Stefan Van Eycken's phrase to take place (importing Scotch, blending it with

Right *Is the sun rising on a new era for Japanese whisky?*

Above *An ever-widening range of flavours is key.*

Japanese whisky and calling the result Japanese). Although there is a trade body agreement over labelling (adhered to by all the distilleries here), there is still no legally binding definition of Japanese whisky.

Pricing also remains an issue. During the period of stock shortages, prices were raised to dampen down demand but had the opposite effect, as the whiskies' premium price made them more alluring. Now new distillers are following suit. 'We now have distilleries selling 500ml of new make for ¥20,000 and 3-year-old single cask bottlings for ¥30,000,' says Stefan. 'This is unsustainable, because there is much better value for money to be had with whiskies made elsewhere in the world.'

Exports remain strong. 'More people have become fans,' says Liam McNulty of website nomunication.jp. 'The value of whisky exported in 2022 was 21 times what it was in 2012. That kind of growth won't continue forever, but at least many more people are taking the category seriously now.' Japanese whisky might still be hot, but you do wonder if it might self-combust.

Are we in a bubble? You could ask the same of any whisky-producing country. It is inevitable that there will be a rationalization and the current heavy reliance on the super-premium category is unsustainable. A shakeout will happen, as it will globally, but it will be the exciting ones with clear vision and long-term planning that will survive. They are here.

Above *Casks are once again being laid down.*

KEY
Malt Distillery
Hokkaido
Honshu
Kyushu
CHINA
RUSSIAN FEDERATION
Vladivostok
NORTH KOREA
SOUTH KOREA
SEA OF JAPAN
La Pérouse Strait
Sea of Okhotsk
Kamui Whisky
HOKKAIDO
Ishikari
Otaru
Yoichi
Sapporo
Akkeshi
Hakodate
Aomori
Akita
Ou Mountains
Sado
Miyagikyo
Sendai
Niigata Kameda
Niigata
Fukushima
Shinano
Iwaki
HONSHU
Saburomaru
Nagano
Komoro
Mars Komagatake
Komoro
Tone
Kiuchi Nukada Brewery
Hikari
Yasato Distillery
Chichibu
TOKYO
Mount Fuji
Hakushu
Yokohama
Ikawa
Mount Fuji
Shizuoka
Shizuoka
Nagahama
Biwa Lake
Kyoto
Nagoya
Chita
Eigashima
Yamazaki
Kobe
Osaka
Kii Mountains
Okayama
Okayama
Hiroshima
Sakurao
Oki Islands
Korea Strait
Tsushima
Nikka
Kitakyushu
Fukuoka
KYUSHU
SHIKOKU
Nagasaki
Kumamoto
Kyushu Highlands
Osuzuyama
Miyazaki
Nikka
Kanosuke
Kagoshima
Mars Tsunuki
Ontake
East China Sea
Nampo Shoto
PACIFIC OCEAN
N
0
miles
200
0
km
200

HOKKAIDO

The northern island of Hokkaido, the country's second largest after Honshu, is a place of thick forest, pastureland, snowfields, cold ocean and peaty mire. Think of that combination of characteristics in terms of conditions and you can see why, in 1934, Masataka Taketsuru headed here to make his vision of Japanese whisky a reality. It was, he wrote, the closest place in Japan he could find to Scotland.

Above *Yoichi's coal-fired stills add weight to its spirit.*

His choice of site was a small fishing port on Hokkaido's west coast called **Yoichi**. It was, however, a strange place to start the new venture. Japan's second whisky distillery was built as far away from the main markets as possible. Today, it's an essential visit on a Japanese whisky tour.

Three main styles are made – unpeated, medium peated and heavily peated (it used local Ishikari peat until the 1970s) – but with further variations within these broad groups thanks to various ferment times, combinations of house, brewer's and distiller's yeasts, and cut points, higher for the lighter style and lower to capture heavier phenols.

In the warehouses there are new oak, ex-Bourbon, puncheons, re-charred, ex-sherry and refill casks. Yoichi isn't one thing but many, though there is always a thickness to the spirit, an oiliness, a hint of cold air. It's clearly not Scotch.

YOICHI/NIKKA

YOICHI SINGLE MALT 45%

Light smoke with a briny accent. Oily and slightly leesy ripe banana-accent fruits. Leather oils, coal dust and light barley. Turnip/radish. Peppery with water.

SMOKY/PEATY

YOICHI 10 YEAR OLD 45%

Woodsmoke, fabric sticking plaster and a little brine, liquorice, dried apple, banana flakes and coal smoke. Medium-full-bodied with roasted cereal. Chocolate milk with water, chicory and smoke.

SMOKY/PEATY

NIKKA COFFEY GRAIN 45%

Soft. Clotted cream ice cream, bread dough/cold pastry backed with grapefruit and yuzu yogurt. Leaner than in the past. Mandarin and light coconut.

FRUITY/SOFT

RISHIRI ISLAND DISTILLERY

A candidate for Japan's most remote distillery, **Kamui Whisky**, on Rishiri Island off Hokkaido's northwest coast, started production in 2022, the result of the Islay bug biting founders American Casey Wahl and Miku Hirano. There's an intriguing mix of American thinking – Vendome hybrid still and the filtering of the new make (through volcanic rock) after distillation – and a bow towards Scotland in the use of peat and the worm tubs. There are plans to use local peat and supplement the imported Scottish barley with some from Hokkaido.

'Innovation starts from tradition,' says Emiko Kaji, Nikka's general manager for global sales. 'Our founder studied whisky making in Scotland, therefore our fundamental production methods are very similar. Due to the Japanese tradition of producers not exchanging casks for blending, we have developed our own ways of producing various types of whiskies. We're continuously developing new types. This has become our own tradition.' Nikka's single malts therefore, like Suntory's, are single distillery blends of those different malts.

Adherence to Taketsuru's ethos is shown by the rake and plough mash tun, and the coal fires that blaze in the stillhouse – the world's last coal-fired site.

Only one still ran until the 1960s. Now there are six, four wash and two spirit. All are the same dumpy shape, with downward-angled lyne arms and worms. No wonder the spirit is beefy.

Nikka was cautious when it came to returning to full production after the quarter-century slump in the domestic whisky market. 'The past ten years was a decade of reconstruction,' says Emiko. 'Our portfolio collapsed from the "Nikka shock" [shortage of stock resulted in the deletion of all age statements in 2015], which led to the redesigning of the future production plan. This long and painful period was a valuable time to look inside, reassess and sharpen our identity and values.' Finally, though, this famous old distillery is back.

Some 250 miles (400km) due east of Yoichi on Hokkaido's east coast is another pioneering distillery – **Akkeshi**. Though one of the first of Japan's new wave, in many ways the motivation behind it was similar to that which brought Taketsuru here.

In Akkeshi's case, it was founder Keiichi Toita's love of smoky Islay whiskies that induced him to build here in 2015. The distillery was in production a year later. Unpeated malt is also run, and both it and the smoky variant are given five-day ferments before distilling takes place in stills that bear a striking resemblance to Lagavulin's (see page 99).

He is examining the island not just for peat but seeking out different microclimates to see what effect those locations might have on maturation profiles.

The ultimate aim is to make a whisky in which everything is sourced from the immediate area: local barley and peat (there is a small drum malting on site), fermentation with a yeast found in the Hokkaido University's lab, with ageing in mizunara casks (Hokkaido has the largest stands of the tree). Some locally grown Ryofu barley has been sourced and, in a move that's replicated across Japan, there's work being done to build relationships with local farmers to grow more. While the 100% Akkeshi will be small-scale compared to the main release, the pull of place is there in the thinking and the liquid.

Above *Just testing. Assessing spirit quality at Akkeshi.*

HONSHU

Japan's largest island, Honshu is geographically varied and its whiskies are equally diverse. The island is home to the oldest distillery and many of the newest. It has the largest and some of the smallest as well. There are alpine distilleries and others on the coast. There is steam and direct fire, and there is heritage as well as new thinking. There is no Honshu style, but there is quality.

TOHUKU

Nikka's **Miyagikyo** distillery (Miyagi Prefecture) was founded by 1967 after Takeshi Taketsuru had been sent around Japan by his uncle and adoptive father Masataka in search of possible sites. With the home market growing, the firm needed a whisky to balance Yoichi's weight for its blends. This place, where the Nikkawa and Hirose rivers meet, was perfect.

Up and running within two years, it's effectively the reverse of Yoichi, being mostly made from unpeated barley, though some light peat is run. There is also an estery and (unusually for Japan) a malty new make produced, all coming from different yeast combinations. The fermentation times are shorter than Yoichi – between 48 and 60 hours.

The stills are modelled on one of Masataka Taketsuru's places of apprenticeship in Scotland – Longmorn, Elgin (see page 43). Big with boil bulb and upward lyne arms, they are all steam-driven. Four larger stills (of the same shape) were added to the original quartet in 1979. The streams are kept separate and aged on site in a mix of re-char, ex-Bourbon and ex-sherry casks.

The main malt for Nikka's blends is Miyagikyo's Malty & Soft, which has a gentle malted milk character. The single malt releases are based around Fruity & Rich: sweet persimmon, baked pineapple and quince. Miyagikyo is more lifted aromatically and, while not as oily as Yoichi, has a weighty core.

Nikka opened a grain distillery in Nishinomiya in 1964 to meet some of its requirements. On closing in 1998, its two Coffey stills were transferred here. Taketsuru spent a period at the Bo'ness distillery in Falkirk, Central Lowlands of Scotland, studying continuous distillation. The fact that he chose the

MIYAGIKYO

MIYAGIKYO SINGLE MALT 45%

Immediately fruity. Fresh wood, persimmon, papaya, brown banana and cashew. Palate has milk chocolate, tayberry and caramelized fruit. Water brings out oak and orange bitters.

FRUITY/SOFT

Above *Nikka's Miyagikyo distillery sits between two rivers.*

NATIVE BARLEY VARIETIES

If Japan is to start making its whisky in a sustainable manner, distillers need to be increasingly self-reliant on barley that necessitates a significant shift in agriculture – and more localized maltings – in a country where sustainability is not top of the political agenda.

Most new distilleries are exploring this area, with old native varieties being revived, such as Ryofu in Hokkaido, Harushizuku and Haruka in Kyushu and Koharu Nijo in the Alps.

Might Shinrai, Sai-no-Hoshi and the various Golden crosses – Sky Golden, Sachiho Golden, Kaneko Golden and the barley that was most widely used in the past, Golden Melon – become household names? Will they add another layer to Japanese whisky's flavour? We'll see. Is some of this window dressing? Perhaps. Should it stop? No.

Above *Kiuchi – from leading craft brewer to distiller.*

same design over the others available in 1969 reinforces his belief in tradition. Coffey stills give a richer, umami-like feel to the spirit and here five (at least) styles are made: light, medium, heavy, 'heavier' and 'super heavy'. Two are from a 100% malted barley mash, the rest are corn/malt. Different weights are obtained by varying yeasts and collecting at different points in the rectifying column. Some rye has also been run. As ever, maximizing flexibility is key.

Nikka's release of Coffey Grain (itself a blend of styles) didn't just open up another area for Japanese whisky but helped to create a new awareness of the quality of this style of whisky globally.

KANTO

The **Kiuchi** brewery's Hitachino Nest craft beer is the world's best-known Japanese craft beer. No great surprise then that it's now also making whisky. While distillation started in 2015 at trial level at its **Nukada** brewery in Naka (Ibaraki Prefecture), three years later a purpose-built facility in nearby Ishioka called **Yasato** started up. Unpeated malt is used for a single malt, while there's also a 50:50 rice to malted barley whisky.

In addition, two 'grain' whiskies are run through a column still installed in 2020: one made from rice hulls left after polishing for sake,

MIZUNARA

Japan's native oak (see left) has become a highly desirable element in a whisky-maker's armoury thanks to its remarkable aromas akin to the aloeswood-based incense that is burned in the country's temples, though as the scent only appears after 20 years, patience is required (distillers who want a quick fix by finishing in mizunara for a few months take note). The oak also appears to intensify the acidity of the whisky, adding a brightness to it.

It is a remarkable turnaround for a species that was only used due to a lack of American oak in the post-war period.

Quercus crispula (and the related *Quercus mongolicoides*) is slow-growing, taking up to 150 years to mature. It has no tyloses (the substance that blocks the pores of the tree's heartwood, making a cask watertight), meaning it's prone to leaking. It's hard to cooper, needing hand-splitting rather than sawing. It's also massively expensive. But that aroma...

and another from the Hitachino Nest 50:50 mixed mash bill of wheat and malted barley. Unsurprisingly, the firm uses a mix of proprietary yeasts and the wort is fermented in a mix of Slavonian, French oak and acacia. The new makes are blended before ageing, which takes place in a mix of ex-Bourbon, some beer, sherry and rum casks, while trials have started with locally grown Kaneko Golden barley (see box, page 277). 'Kiuchi's beers are in 40 countries,' says owner Toshiyuki Kiuchi. 'So we will also sell the whisky on overseas markets.'

The plans are big. At the time of writing, the distillery is processing 100 tonnes (110 US tons) a year and yet another expansion is planned. 'In 2023 our malting facility was completed. This will enable us to handle 500 tonnes [551 US tons] of local barley. We're also thinking that rice whisky might become an original Japanese style, like Bourbon using corn, or Irish pot still.' With that thinking, this is one to keep an eye on.

Ichiro Akuto has become a touchstone for the new Japanese whisky industry, not just for being the first to start a new venture but in his approach to what the country's whisky could be. His family come from Chichibu (Saitama Prefecture) and had been brewing sake there since 1625. In the 1940s, production was switched to the town of Hanyu where spirits were added to the portfolio. Whisky began to be distilled in the 1980s and was released in the '90s, just as the slump was starting. Also, the robust Hanyu style didn't chime with an ever-lightening palate. The distillery closed in 2000. Undaunted, in 2007 Ichiro returned to **Chichibu** and started again.

His vision was holistic – a whisky that reflected where it came from, a distillery where crafts could be learned and tradition reinterpreted.

It's a small site – the mash tun holds 400kg (882lb) and is hand-stirred. The eight washbacks are made from mizunara. Ferments average 90–100 hours, but the wash's readiness is measured by its acidity rather than time. If it needs longer, it gets it. Fruit, light spice and elegance are key here. These are whiskies with grace.

The two Forsyth stills are small, chunky pots with slightly down-angled lyne arms adding weight. Cut points are determined by taste. In winter, with colder water coming into the condenser, a heavier spirit is made and earmarked for longer maturation.

There's a vast number of different casks at work, more than in any other Japanese distillery – the standard ex-Bourbon, ex-sherry and refill, but also

CHICHIBU

★ CHICHIBU 2014 #3508 59.5%

Gently fruity. Tatami scents, persimmon, light apricot. Precise. Has the distillery's central weight. A light meatiness, incense and smoke. Vinous in the mouth, sequoia, mature, complex.

FRUITY/SOFT

★ CHICHIBU 10 YEAR OLD #2124 63%

Has classic Japanese transparency and lift. Frankincense, osmanthus, light cedar, tatami, artichoke, Cognac, then bubblegum before an intense savoury burst of spice.

FRUITY/SOFT

CHICHIBU 13 YEAR OLD (2009) #4406
(FIRST-FILL MIZUNARA) 62%

Tipple barley. Airy, precise. Orange blossom, sudachi and light coconut. Tighter and more austere with water. Palate shows deep savoury elements developing.

FRUITY/SOFT

★ CHICHIBU GOLDEN MELON #9429 5 YEAR OLD 63.5%

Plump. White fruits, lychee and melon rind. A light bitter edge, the oak adding a layer of sweet spices. Oolong, pepper and clotted cream.

FRUITY/SOFT

Above *Just some of Chichibu's vast range of different casks.*

quarter (chibidaru), red oak, ex-wine, -port, -Madeira, -agave spirits, -Cognac and, of course, mizunara.

Chichibu sits in a prime soba- (buckwheat-) growing area and Ichiro is working with farmers to also plant barley. A range of varieties including Sai-no-Hoshi and Golden Melon have already been grown locally and malted on site, some smoked with local peat. Sitting behind all this is his belief in training – the team went to the UK to learn how to floor malt, they work in the fields with farmers in Saitama and Niigata to learn the best way to grow barley, and Chichibu also has its own cooperage where the chibidaru casks and mizunara hogsheads are made.

In 2019, with demand rising, Ichiro built a significantly larger second distillery a few hundred yards down the road. Here, the mash tun is 2 tonnes (2.2 US tons), the washbacks French oak. Chichibu I will now concentrate on using Japanese and local barley, as well as the peated production, while Chichibu II will handle unpeated, medium and lightly peated. Here the style is slightly heavier than the original, thanks to the introduction of direct fire. In 2023 he announced plans to build a grain distillery in Hokkaido.

It seems like a headlong rush, but everything about Chichibu is carefully considered. There is no great pressure to get whisky out quickly. It's released when it's ready – some mizunara casks are laid down for bottling in 30 years' time.

'Being able to taste short-matured whisky is a good opportunity,' says global brand ambassador Yumi Yoshikawa. 'However, it seems that only a few producers are able to create their own direction by understanding both the short-term taste quality and the long-term taste quality. Evolution is still an exploratory phase for the industry as a whole.'

What Chichibu has done in a short time has been remarkable. It has also shown others what is possible. Where Ichiro Akuto goes, others follow.

TRANSFORMATIVE CHUBU DISTILLERS

Many new distillers have moved from (or added to) existing shochu and sake production. Sake plays a role in the story behind **Hikari**, based in Konosu (Saitama Prefecture). Owned by whisky-loving Malaysian Eric Chhoa (*sic*), it's a place where everything appears to be tried at the same time. The barley is imported and (mostly) unpeated, but there's also manuka-smoked barley from Gladfield (see box, page 341), as well as speciality malts distilled individually, then blended as new makes. Rye and wheat whisky are also made, as is one from a mixed mash bill of 65% malt and 35% rice. Ageing can be in wood – mostly ex-rye casks but chestnut is also being trialled, along with ex-barley shochu casks. Some new make is first rested and mellowed in ceramic (à la sake).

Niigata Kameda (Niigata Prefecture, Chubu), which started production in 2021, made a different jump – from producing traditional *hanko* (seals). The distillery is an intriguing mix of different component parts. The brewhouse comes from Germany (a full lauter giving clear wort), the washbacks are white oak, French oak and acacia, while the stills are from Forsyths. A range of distiller's yeasts are used and ferments are over 100 hours. As well as using imported malt (unpeated and peated), there have been runs of local barley (malted on site), which is added to the mash. There are also plans for a rice whisky, which makes sense in a famous sake-producing area.

CHUBU

Built at grand scale and yet hidden, Suntory's **Hakushu** distillery (Yamanashi Prefecture) is situated in deep forest on the slopes of Mount Kaikomagatake. One of Japan's alpine distilleries, it's also one that clearly reflects its cool climate and sylvan setting. Hakushu's story helps you chart the story of Japanese whisky since the 1970s, one of enterprise, hubris and rebirth.

The distillery was built in 1973 as a response to the upturn in sales of domestic blends. With six very large stills (30,000 litres/8,000 US gallons for wash and 20,000/5,300 US gallons for spirit), it made a light, clean spirit that acted as a sound base for Suntory's blends. By 1977, further expansion was required and a new stillhouse with another set of massive stills was erected. The two combined made Hakushu West the largest malt distillery in the world.

In 1981 it expanded again, but in a very different direction. Hakushu East had smaller stills, 18 wooden washbacks, greater variation. Why? Prescience? Indications that, with Scotch in decline, it might be Japan's turn next? Hakushu West was shut down in 1995, but the stills remain, a mausoleum to Japan's whisky boom.

A willingness to learn and change is fundamental to Japanese thinking about whisky. Call it kaizen, call it monozukuri. Whisky was changing, so Hakushu changed as well. Radically so.

The water comes from the mountain, the barley is peated to different levels, the wort is crystal clear and brewer's and distiller's yeasts are used. The stillhouse is baffling in its variety. The eight wash stills are direct-fired, which, allied to the downward slant of the lyne arms, gives a subtle solidity to the spirit's base. One even has a worm tub.

The eight on the spirit side are steam-driven. Two have lyne arms pointing up, the rest go down. Four types of malt, eight pairs of stills that could give thirty-two different options. Stir in five different types of cask and you can see why this is a blender's dream.

Hakushu is subtle. Even when heavily peated, there's a coolness to its delivery: mint, basil, green apple, pine, but textured and slinky. Like all of Japan's whisky, it was virtually silent due to the 1990s slump, but since 2010 it has been in full production. Stocks are once again building.

In 2013, Suntory installed a small two-column still. Because Chita is so large (see page 284), it's impossible to run experimental batches there and, once again, variety was needed. Here, different strengths are run as well as batches of 100% malt and even rye, wheat and corn.

None of this is innovation for innovation's sake. Rather it is thought-out, long-term thinking. We need to change? How then do we do it?

HAKUSHU

HAKUSHU DISTILLER'S RESERVE 43%

Soft. Sweet melon and kiwi, sencha, basil leaf. A gentle yielding quality to the palate. Light cucumber and mint.

FRAGRANT/FLORAL

★ HAKUSHU 12 YEAR OLD 43%

More herbal, some pine, green bamboo leaf and a touch of fragrant smoke. Fresh acids on the finish. Though seemingly light, it's complex and layered.

FRAGRANT/FLORAL

★ THE ESSENCE OF SUNTORY HAKUSHU RYE TYPE 57%

Rye malt/crispbread, hint of cherry, then geranium, white pepper and clove. Balanced, ordered, complex. Rye revealed.

SPICY/HERBAL

Above *The greenness of Hakushu's surroundings is reflected in its spirit.*

Above *Mars Komagatake: remade and remodelled.*

By admiring the past but not being bound by it, by understanding that tradition is a continuous process.

On the western side of the Minami Alps, 20 miles (32km) west of Hakushu, sitting at 798 metres (2,618 feet) is **Mars Komagatake**, formerly Mars Shinshu (Nagano Prefecture), a distillery that rose from the grave. Owner Hombo Shuzo is a long-established shochu distiller from Kagoshima. It was granted a licence to distil whisky in 1949, but it remains unclear quite what this whisky might have been.

A more concerted effort started in the 1960s when the firm opened a distillery in Yamanashi where it had a winery. The whisky side was run by Kiichiro Iwai, who had been Masataka Taketsuru's boss at Settsu Shuzo, the firm that sent Taketsuru on a study mission to Scotland in its bid to become Japan's first whisky distillery. When the young Taketsuru returned to Japan, he gave his report to Iwai (but kept one himself). Yamanashi, therefore, was run along 1920s Scottish lines. It operated for a decade, but closed just as whisky was taking off. Production restarted (in tiny quantities) at the Kagoshima distillery.

MARS KOMAGATAKE

★ MARS KOMAGATAKE YAKUSHIMA AGING 2015 (BOTTLED 2022) 53%

An ex-sherry cask gives sandalwood, myrrh, some molasses and chocolate. Komagatake's sweetness is there along with a scented element. Armagnac-esque, rice cracker and walnut liqueur.

RICH/DEEP

MARS KOMAGATAKE 1923 EDITION 50%

Fruity, citric with a mix of pillowy green fig, banana, light barley and a sumac-like back palate. Palate is rose petal with Komagatake's sweetness and energy.

FRUITY/SOFT

By 1985, Hombo decided to try again and built Mars Shinshu. At this stage a lighter whisky style was required, but by the time the whisky had matured the market was in free fall. In 1995 the distillery was mothballed, the stock eked out and supplemented with imported malt.

Hombo is nothing if not persistent and in 2011 there was life in Mars again. The distillery has been refurbished, with new stills installed in 2014. Today it makes heavily peated and unpeated (other levels of smoke were trialled) as well as experiments with Koharu Nijo two-row alpine barley. The old cast-iron washbacks have been replaced by new wooden ones and temperature-controlled stainless steel, allowing ferments to be extended. Lighter style is still the key – one giveaway being that the spirit still is larger than the wash, upping reflux opportunities, but that's balanced by a worm tub on the spirit. Complexity.

Ageing is on site, but some casks are sent to Kagoshima and Yakushima, allowing Hombo to show the variations between alpine and subtropical maturation. Finally, Mars seems settled.

One of the newest arrivals on the burgeoning Nagano scene is **Komoro**, which began production in July 2023. Founder Koji Shimaoka had the foresight to hire the legendary Ian Chang, former master distiller at Kavalan in Taiwan, to head up production. Koji worked closely with the local government, who saw the tourism benefits of a distillery in Komoro city and land was acquired on the lower slopes of Mount Asama.

The water comes from this (active) volcano and is used in a semi-lauter tun to give wort that is fermented over 72 hours with a cocktail of different yeasts. Distillation takes place in two Forsyth stills. Unusually, the wash still is smaller than the spirit.

Above *The view from Mount Fuji distillery.*

Ageing takes place on site in large arched warehouses – the half-barrel design means that snow will slide off the roof – which have been dedicated to Ian's mentor the late Dr Jim Swan.

While no whisky has been released yet, the new make shows an already complex fruit-laden and rich spirit. With Ian at the helm and Koji and wife Yoshie creating a whisky destination, there's a strong sense that Komoro could become the Chichibu of the region. Build it and they will come, make great whisky and they will flock there.

Mount Fuji (Shizuoka Prefecture) was built (as Fuji Gotemba) in 1972 by two firms who saw the opportunities presented by whisky in Japan – domestic brewer Kirin and giant Canadian distiller Seagram.

Unsurprisingly, given its parentage, inside it's a strange hybrid of Canadian and Scottish approaches yet the utilization of the equipment is wholly Japanese, partly driven by the lack of stock trading between distillers. Japan's distillers had to find ways to be self-sufficient in malt and grain as well, supplementing their volumes with bulk whisky. That was the model applied here.

The grain side has three sets of stills producing the same number of new makes. A 'Bourbon-style' high rye mash is run through a beer column and doubler producing an aromatically powerful (rose, berry fruits), heavy distillate at 70% ABV. This is aged in virgin and first-fill ex-Bourbon.

A batch kettle (see box, page 254) makes a medium-weight corn and malted barley distillate. This is run through a beer still, collected in the 'kettle', reheated and driven through a 61-plate rectifier. Though high strength, it's packed with flavour with an unctuous texture. It's aged in first-fill casks.

A five-column set-up gives a light spirit that is fresh, citric and slightly hay-like. This is aged in refill. You could be in Gimli, Canada (see page 252). No surprise really, as that was also owned by Seagram.

With one heavy- and one medium-bodied grain, the malt side has always gone light – the reverse of the norm. Here, the stills were modelled on those at another Seagram site, Strathisla in Keith, Aberdeenshire (see page 63) – though Fuji's are larger. Different yeasts are used from their own repository and via Four Roses in Kentucky (see page 208), which Kirin snapped up in the Seagram fire sale.

Though a large producer, it's always been a quiet performer, but now there is intent and investment. 'After 50 years, the past decade has seen the most evolution in whisky making and quality,' says master blender Jota Tanaka. 'Over the past five years, we have invested over $60 million in distillery operations, including new equipment to diversify the malt flavour profile, and a state-of-the-art ageing warehouse to increase our capacity.'

MOUNT FUJI

FUJI SINGLE GRAIN 46%

Soft. Clean tangerine, red cherry, the top of crème brûlée, ghee, sumac, fat corn and cassia. Clings to the palate. Tangerine peel, corn ribs, butterscotch. Balanced.

SWEET/VANILLA

★ FUJI SINGLE BLENDED 43%

Medium-bodied. Initially, soft grain adds depth. Lily of the valley, persimmon, guava, toasted marshmallow, then spices (nutmeg and cumin). Toasted fruit/oak as malt comes in.

FRUITY/SOFT

FUJI SINGLE MALT 46%

More weight and structure than in the past. Estery top notes (apple, green banana, nail varnish). Firm and spicy with water. Rich coconut, then char and raspberry.

FRUITY/SOFT

Above *Shizuoka's wood-fired wash still is on the left.*

Flavour-wise, the drive is for richness. There are new wooden washbacks to change the lactobacillus influence, while the four new pot stills are smaller than the existing pair. One of the wash stills is small and short-necked, the other tall. 'We're now producing a new type of fruity malt whisky and another rich type coming from the new wooden washbacks and stills.'

On the grain side, there have been changes in the 'Bourbon type'. 'In collaboration with the Four Roses team, we've created a very fruity whisky that is even more distinct from US Bourbon,' he says. With Jota a leading voice in the push for legislation, there's no surprise that the new Mount Fuji approach revolves around 100% Japanese whisky. A distillery transformed and one that you feel now is, finally, in control of its own destiny.

Starting a new distillery is a major undertaking. Working out what it should make is conceivably an even bigger challenge. When Taiko Nakamura caught the whisky bug, his response was to travel. 'I visited 50 distilleries in Scotland and more in Japan. One of the first I went to in Scotland was Kilchoman [see page 105]. It was small, old-fashioned, with no automation. That's when I thought what my small distillery could be like.'

A site in a forestry plantation 15 miles (24km) west of this home city of **Shizuoka** was found. Cedar-clad, the building fits perfectly into the landscape, but it's what's happening inside that matters. His approach takes that 'old-fashioned, manual' approach to a different level.

There are ten washbacks, four Oregon pine and six cedar. 'I wanted to make a local whisky, so using local wood was one way,' Taiko says. He didn't, however, expect cedar to add its own layer to the flavour. 'The pH is lower in the cedar washbacks,' he explains. 'When we analysed the wash, we found they had different LAB [lactobacillus] to the others – these washbacks are influencing flavour.'

JAPAN'S WHISKY LEGISLATION

The Japanese whisky boom of the 1970s and 1980s could not have been fuelled wholly by domestically produced spirits. Imported bulk whisky (predominantly from Scotland and Canada) was used particularly, but not exclusively, for low-priced blends.

There was an assumption that some form of legislative framework was in place for the other whiskies. It transpired that there wasn't.

This became a concern when Japanese whisky's global following began to grow. There was nothing to prevent a bottler or distiller importing Scotch and relabelling it as Japanese, or blending Scotch with Japanese and calling it Japanese. Some shochu producers, exploiting a loophole in American law, started selling their aged products as whisky there.

Thanks to pressure from distillers seeing the damage that was being done to Japanese whisky's reputation, in 2021 the members of the Japan Spirits & Liqueurs Makers Association (JSLMA) reached an agreement on labelling requirements, as follows:

1. To qualify as 'Japanese Whisky', the spirit must be made from malted grains, other cereal grains (including rice) and water (the last being Japanese).
2. Only malt can be used for saccharification, and fermentation and distillation must take place in a distillery in Japan.
3. The spirit must be distilled to no higher than 95% ABV and aged in wooden casks (of no more than 700 litres [185 US gallons]) in Japan for a minimum of 3 years.
4. It must also be bottled in Japan at no less than 40% ABV.

This remains a voluntary agreement that applies to members of the JSLMA. It's a step in the right direction, but full-scale legal protection is still needed.

OTHER DISTILLERIES IN THE CHUBU REGION

Rivalling Kamui for Japan's most remote distillery (see Rishiri Island Distillery box, page 275), **Ikawa** lies deep in the forested mountains beyond Shizuoka distillery and, at 1,200 metres (3,937 feet), is the country's highest. Its location makes sense when you realize it's owned by Tokushu Tokai, a paper manufacturer that owns the 94sq-mile (244sq-km) forest where it's situated.

After two years of construction, spirit began to run in 2020. There's a full lauter tun, handling heavy-, medium- and unpeated imported barley that is fermented (separately) in temperature-controlled washbacks with the help of a mix of ale and distiller's yeast. Distillation takes place in straight-sided Miyake stills. Casks are mostly ex-Bourbon, but some chestnut casks are being looked at. It would be no surprise if more wood types would be used, given the location. It will be also be intriguing to see the maturation profile, considering that summer temperatures hit 30°C (86°F) while winter can drop to -15°C (5°F).

Saburomaru (Toyama Prefecture) in Tonami is owned by long-established sake and shochu producer Wakatsuru Shuzo. Its Sunshine whisky, reliant on imported bulk and a smoky spirit from a small column still, was made from the 1950s until the early 2000s.

A full commitment to whisky started in 2016 with the installation of the world's first wholly cast-copper still, made for the distillery by Oigo Works, which specializes in casting temple bells.

The wash still uses both direct injection of steam (a shochu technique) and indirect steam heating. Another indication of cross-fertilization comes in the approach to fermentation, which, like some shochu, starts in enamel before finishing in Oregon pine. The length has been extended as the search for a rich fruited style develops.

Yeasts include brewer's and distiller's as well as a 'local yeast' cultivated on site. Unpeated and peated are made, with some barley now being locally sourced and malted by the distiller at a local maltings. This is both being distilled separately and mixed with imported. Here's a distillery that encapsulates much of the new thinking underway in the Japanese industry.

Several yeasts are used, with more coming. 'Shizuoka Prefecture is the kingdom of ginjo sake, and there are also many craft brewers, so there is already a lot of experimenting with yeast. I'm always kaizening!'

Trials are also starting with Shizuoka-grown malt (Sachiho Golden), with the aim of it making up 50% of the volume needed. 'The flavour is different,' he says. 'Scottish malt has a strong taste. Japanese is more delicate and fruity.

'We're making over ten types of spirit,' he adds. 'We have imported distilling malt, unpeated and peated, imported brewing malt, Japanese barley and Shizuoka barley.'

The wort is clear, but how it's made is the only part of the process he won't divulge. 'Clear wort is very important, and mashing is the most mysterious part of the process, so we did kaizen every year,' is all he'll say. 'It's savoury. It's the dashi.'

With ferments running between 48 and 120 hours, there are a wide variety of flavours before it hits the stills – where there are more surprises. One wash still is made from the usable parts of three of the old Karuizawa stills, which were taken from that distillery when it closed in 2002, and is heated with steam. Fruity and estery, it is bottled as Prologue K.

The other has a wood-fuelled furnace. 'When I told Ichiro this is what I was planning, he said it was unbelievable. I think he meant impossible!' With temperatures reaching 800°C [1,472°F), each run uses 300kg (660lb) of wood per 6–7-hour run. The final spirit, Prologue W, is hefty yet sweet with a silky texture and roasted flavours. The genius is how well this richness is balanced with delicacy. Even sumo is as much about delicate shifts in weight and hold as brute force.

'I've learned a lot,' Taiko says. 'I've drunk many old whiskies and saw the difference with new. I want to make whisky exciting again.' Kaizening his way forward. Shizuoka is not just new, it's significant.

Built in 1972, Suntory's giant **Chita** grain distillery near Nagoya (Aichi Prefecture) was established as a joint venture with the agricultural group Zen-Noh. With whisky on the upswing, the distiller needed a significant increase in grain whisky supply, so building next to Zen-Noh's dock made sense. The corn used comes from Canada and the USA.

This is distilling at scale. Conversion and cooking takes place in a coiled 150-metre- (492-foot-) long pipe. The fermenters take 24 hours to fill. After three to four days, a 10%ABV wash can be distilled.

With no exchanging of stock, it's not just single malt that needs a palette of flavours but grain as well. Chita accordingly makes 'clean', 'tasty' and 'heavy'. The heavy type is taken from the first two of the four columns (the analyser and rectifier); 'tasty' uses the analyser, an extraction column, then the rectifier; the 'clean' type uses all four. Chita was expanded in 1977 and 1985, and in 2022 a new Coffey-style still was installed to add another flavour character. This is essentially a larger version of the one at Hakushu.

Grain, often the anonymous player, has steadily emerged into the world spotlight, mainly thanks to the decision by Japanese distillers to give it the focus it deserves. Chita finally got its turn in 2019. The bottled expression is, unsurprisingly, a blend of the three styles.

KANSAI

The tiny **Nagahama** distillery (Shiga Prefecture) started in 2016 as an extension of the town's craft brewery Roman Beer with the help of a crowdfunding scheme. Though diminutive, it's spinning many whisky plates. Three small, direct-fired Hoga stills (two wash, one spirit) give 200 litres (53 US gallons) of new make per run – a barrel's worth. Working back from that, you get to a 400kg (882lb) mash with different barley variants being run. Unpeated (sourced from Germany and

SHIZUOKA

★ SHIZUOKA POT STILL W
(IMPORTED BARLEY) 55.5%
Light oak, some weight. Vanilla, light suede, woodsmoke, boot polish and chocolate. Palate is mixed herb, ripe fruit, bright citrus, a burned/smoky edge and dark cherry. Leathery, roasted. Very good.
RICH/DEEP

SHIZUOKA POT STILL K
(JAPANESE BARLEY) 55.5%
Light fruit, green plum, moss, salsa verde, sencha, some butterscotch on the palate, juniper, soft malt, but still good central weight.
FRUITY/SOFT

★ SHIZUOKA CONTACT S
(BLEND OF K AND W WASH STILLS) 55.5%
Ripe, almost cedary. A broad feel with red fruit, banana, ripe pear and acetone, then the weight pressing. Dubbin and a hint of kumquat, cacao, stone fruits.
FRUITY/SOFT

Scotland) has been the mainstay, but heavy peat has also been run – both Scottish and Japanese. Fermentation is in three enamel washbacks.

The building in of variety doesn't stop at the distillery. Nagahama is exploring the effects of different locations on its maturing stock. Some is matured in a local (abandoned) school, another portion in a disused railway tunnel, while the rest is split between dunnage-style warehouses in three island locations: one near Nagahama, another at Chikubu on Lake Biwa and in the tropical climes of Okinawa. Even coopering is being brought in-house.

Breaking with tradition, master distiller Takashi Kiyoi has exchanged stock with fellow distillers Saburomaru and Eigashima for the collaborative Inazuma series. The distillery also runs a 'make your own' whisky school.

Though production is small, he has big plans. 'The market could also use some expansion of the market base,' he says. 'Otherwise makers are just stealing other makers' customers.' That means export. 'We have 20% overseas sales at the moment, but I'd like to increase that to 50%. Communicating the culture and tradition of whisky in Japan will be key for success overseas, but there's so much paperwork!' He remains optimistic. 'Next year (2024) we'll do 300 runs and by 2033 we should have a standard single malt expression.'

Japan's whisky story starts with Shinjiro Torii's building of the **Yamazaki** distillery (Osaka Prefecture) in 1923. With Masataka Taketsuru as distiller, it was the first purpose-built whisky distillery in the country, and a statement of Japan's new engagement with the modern world. What had been 'foreign spirits' were starting their journey to become local. It's this that drove Torii's vision – and conceivably caused the falling out with his distiller, who adhered to a Scottish model.

The siting of the distillery was no coincidence. Situating it between Osaka and Kyoto gave Torii access to the major markets. The flat land next to the Uji, Katsura and Kizu rivers created mist that he felt would help with maturation. The Rikyu no Mizu water source was regarded as one of the finest in Japan.

There was a deeper reason. This was where Sen no Rikyu, the monk who formalized the Japanese tea ceremony, had his first tea house. His approach to tea was based around modesty, humbleness and simplicity. It became the founding principle behind ceramics, art, poetry and, in Torii's mind, whisky.

After Suntory's first blend Shirofuda failed by being too heavy and smoky, Taketsuru left and Torii began to create one that suited the Japanese palate, occasion and climate. In 1937, Kakubin was released. It was the start of whisky in Japan becoming Japanese.

Yamazaki has grown significantly since those days, each expansion mirroring the changes in the industry's development. Like Hakushu, by the 1970s it was producing volume for Suntory's

Above *The narrow road to Yamazaki.*

blends. By the 1980s, however, blends were faltering and single malt seemed a better bet. Yamazaki's was launched in 1984. Five years later, in true Japanese style, the old ways were swept away and the distillery reconfigured. Now richer, fruitier spirit was the aim.

In came a second, smaller mash tun, wooden washbacks and a new stillhouse with smaller stills, all run in pairs and of different shapes and sizes. In 2006, more were added to give more grunt, with a final pair, the same shape as Torii's originals, arriving in 2013, making a total of 16.

Yamazaki is mostly unpeated, but different strains of barley are used (including Golden Promise). Clear wort helps to drive esteriness and transparency, but on occasion cloudy wort is run for a malty note. Ferments are a minimum of three days, and at least two types of yeast are employed.

Clues as to Yamazaki's styles can be gleaned by looking at the angle of the lyne arms. All face downwards (some steeply), apart from one wash.

CHITA

THE CHITA 43%

Cassia, crème caramel, cashew cream, toffee, delicate coconut and macadamia. The palate shows banana flambé, plantain chips and dry spices. Sweeter with water.

FRUITY/SOFT

NAGAHAMA

NAGAHAMA FIRST BATCH 50%

Slightly nutty, amber malt caramelization, red fruits, mushroom, a hint of smoke, cassia and baked banana. Hints of incense. Palate is deep. Bay leaf, light (old) spice.

FRUITY/SOFT

Above *The filling store at Yamazaki.*

Wash stills 3 and 6 have worms, as do 3 and 5 spirit. All of this is to do with manipulating reflux. Weight also comes from the direct fire on the wash side.

The variety of new makes are aged in puncheon (new American oak), ex-Bourbon, ex-sherry, refill and mizunara casks – Suntory being the firm that rekindled interest in this unfashionable oak. The coconut and incense character is a significant aromatic element in all of the Yamazaki expressions; there's Bordeaux red wine, as well as umeshu-seasoned, cedar and cherry. As with Hakushu, each expression is a different blend of styles and casks.

All of this is a reminder that kaizen or monozukuri can be dramatic. It also shows a willingness to learn from the past, not only a relentless push for efficiency. If anything, Yamazaki is less 'efficient' than it could be. The reason? Flavour.

There is constant experimentation going on. A new small floor malting was introduced in 2023, restarting on-site production for the first time since 1972. Much of the learning from here is now being applied to Suntory Global Spirits' Scotch estate.

'Japanese whisky supply has not yet caught up with demand,' says chief blender Shinji Fukuyo. 'In order to maintain and develop this popularity, I think it's all about the pursuit of quality. Neglecting quality in pursuit of volume is a mistake.

'Our culture is different and we make things our own. Also, as a culture we pursue everything to try and make it better and better.' Yamazaki continues to lead the way.

YAMAZAKI

YAMAZAKI DISTILLER'S RESERVE 43%

A heightened, scented aroma of stewing berries, fruit salad and a hint of smoke. The palate is soft with deep fruits in the centre.

FRUITY/SOFT

★ YAMAZAKI 12 YEAR OLD 43%

More fruit-driven in recent times with an added mizunara incense/coconut element. Plenty of pineapple, succulent ripe fruits, layered exotic woods. Clear and complex.

FRUITY/SOFT

In a parallel universe, **Eigashima** (Hyogo Prefecture) would be the ancestral home of Japanese whisky. After all, Eigashima Shuzo had a licence for whisky distillation in 1919. It wasn't until 1961, though, that some form of malt spirit was made (possibly along barley shochu lines). In 1984, the White Oak distillery started up on site and ran intermittently until the 21st-century malt boom, with some releases being 100% from the distillery, others world blends.

In 2018, however, new stills were installed and White Oak became Eigashima. Production remains shared with sake production – the latter running from late autumn to early spring. The majority is unpeated, but some heavy peat adds to the spread of flavours.

One production quirk (seen in some other Japanese sites) is not to use the third water for mashing but instead add it to the yeast preparation tank. With local Shinrai six-row barley now being used, maybe finally Eigashima's day has come.

CHUGOKU

Hatsukaichi's **Sakurao** (Hiroshima Prefecture) is another that reflects the evolution of Japan's whisky story. Chugoku Jozo is a sake and shochu producer that, faced with a decline in sales, saw single malt whisky as an opportunity. The firm already had a 'whisky' brand called Glory, and from 2003 was releasing Togouchi as a 'Japanese' malt – though it was actually made from imported spirit.

It wouldn't be until 2017 that single malt production started in earnest and Sakurao is now a prominent member of the Japan Spirits & Liqueurs Makers Association and a vocal promoter of the new agreement (see box, page 283). Since 2021, its releases have been 100 per cent Japanese.

In order to do this, it makes both malt and grain. The former is distilled in a pair of Holstein hybrid stills. Unusually, the heads and tails from the spirit run are added to the next run of the wash still.

The grain whisky is a blend of pearl barley and malted barley run through a stainless-steel vacuum still, then redistilled in one with a copper column (see Kanosuke, page 288). Triple-distilled malt is also being trialled.

Imported, unpeated and peated barley are used and, like Eigashima, a two-water mashing regime is used. Ferments last 72 hours, while ageing is mostly in ex-Bourbon and ex-sherry. It's in maturation where the biggest difference lies. The Sakurao brand is aged at the distillery, whose location next to the Inland Sea has a humid, warm climate helping to push maturation along.

The new-look Togouchi is aged in a 750-metre- (2,460-foot-) long disused railway tunnel in the mountains. Here the wind is a constant, as are damp humid conditions, so damp in fact that the strength drops but without any loss of liquid. The same new make aged in the two sites is completely different when mature. An encapsulation of the new thinking.

Okayama's owner (Okayama Prefecture) is another shochu/beer producer that moved into whisky in 2015. While most of its malt requirements continue to come from Germany, it's also using the Sky Golden variety, which now accounts for 50% of its total. The brewhouse handles mashing and fermentation (using beer and sake yeasts), which lasts for a week. Distillation is in a Holstein hybrid still and the new make is matured in a range of casks including some with cedar, sakura and chestnut heads.

SAKURAO

SAKURAO SINGLE MALT 43%

A distinct herbal lift (angelica, marjoram, laurel), the soft semi-tropical fruits pushed to the front. Brioche, canned pear juice, a hint of barberry.

SPICY/HERBAL

TOGOUCHI SINGLE MALT 43%

Hops, bright citrus, sweet raisin and panettone. A little hint of green melon. Light oak pressing in on the palate with restrained ripening fruits.

SPICY/HERBAL

Above *Sakurao makes both malt and grain.*

KYUSHU

Kyushu, the southernmost and third largest of Japan's main islands, has long been home to some of the country's finest shochu, a spirit made from sweet potato, rice and barley. With the market for their spirit struggling, shochu producers have begun to turn their hands to whisky production. Some, though, have simply relabelled aged shochu as whisky, adding to the general confusion over what Japanese whisky is.

Kanosuke, however, has very deliberately kept the two spirits separate – although some shochu techniques have been adapted into their whisky-making approach. The firm has always been innovative. In 1957, owner Yoshitsugu Komasa's grandfather was the first to age shochu, seeing it as a way to elevate what had been a working man's drink.

A warehouse was built with plans to move production of the aged portfolio there. With the downturn, the plans were revised and in 2017 a new whisky distillery was built there instead. Located on Kyushu's west coast, 35 miles (56km) southwest of Kagoshima, it overlooks a 50km (80km) beach.

'When I was a child, I saw his vision,' says Yoshitsugu, 'so I studied fermentation at university. Then a trip to Scotland to learn about other spirits planted a seed. I needed to apply my knowledge to whisky.'

With a cash injection from Diageo's Distil Ventures, there is now scale to his project. In a bid to increase the Japanese component, 10% of the barley used is local, the rest imported, and talks are underway with sweet potato farmers to plant barley when fields are lying fallow.

Aiming for clear wort, they also add the third water to the yeast tank in which is a mix of dried and beer

KANOSUKE

★ KANOSUKE 2001 SECOND EDITION 57%

The little touch of sherry adds currants and mulberry jam to nectarine and papaya. Summer pudding tart/sweet interplay. Drifts into roasted tea on the finish. Balanced.

RICH/DEEP

KANOSUKE SINGLE MALT DISTILLERY EXCLUSIVE 50%

The fruit begins to build in weight and succulence thanks to subtropical ageing: mango and papaya, but there's delicacy behind. Some crisp linen gives a contrast to the lush fruits.

FRUITY/SOFT

KANOSUKE ARTIST EDITION #002 50%

Aged in PX, so raisin but also date and a gamey/pomegranate molasses element. The palate shows the thickness of the wine and warmth of the distillate. Sweet power.

RICH/DEEP

Above *On the beach – Kanosuke.*

OTHER KYUSHU DISTILLERIES

Kyushu is now one of Japan's fastest-growing whisky regions. Kuroki Honten's **Osuzuyama** is another shochu maker to make the jump. Located in Miyazaki Prefecture on the island's southeast coast, whisky production began in 2019. Already growing much of its own requirements for shochu, it's ahead of the game when it comes to utilizing local barley (Harushizuku and Haruka two-row), which they malt in small boxes that look like a cross between a manual Saladin box and the table used for mixing rice with koji. With open-topped cedar washbacks in use, a stainless-steel wash still (both injected steam and indirect heat) and copper spirit still, it's tipped as one to watch.

Ontake (see left), near Kagoshima is another of the new wave. Owned by Nishi Shuzo, it's one of only two distilleries in the world situated on a golf course (the other is Kingsbarns, Fife – see page 109). Using Scottish, Australian and Belgian malt, it runs a five-day ferment with its own yeast strain. Lightness is key here, with the spirit still having both a boil ball and upward-slanting lyne arm. While ex-sherry casks are preferred, there are some ex-Bourbon, mizunara and Pinot Noir casks from the firm's New Zealand winery.

yeasts. The ferments are long – 96 hours – in cooling-jacketed stainless-steel washbacks.

Another shochu technique is used to up the ester levels. 'I know that wood is better for LAB [lactobacillus],' he explains. 'So we take a percentage of the wash that's ready to be distilled and put it into a younger washback to increase the acidity.'

Variety is hard-wired into the thinking behind distillation. The wash still works as normal, the middle (of the three) can run wash and spirit, while the third makes only spirit. This means Kanosuke can run double, partial triple and triple distil. All have worms 'because that's what we have for shochu', but the angle of the lyne arms is different. The wash's is straight, the middle still's is downward-facing giving heaviness and the third's goes up to increase reflux.

Ageing is in a mix of re-charred 450-litre (119-US-gallon) shochu casks, with ex-sherry, Bourbon, Cognac, beer, red and white wine, mizunara, Madeira and their own rum used for finishing. 'There's a lot to play with,' he says.

Play is the right word, for at the firm's Hioki distillery they are making 'grain' whisky from a 10:1 ratio of steamed pearl barley to malted barley, to which is added yeast and water. This is a traditional shochu and sake technique of simultaneously saccharifying and fermenting. A five- to seven-day ferment gives a mash of 15% ABV before distillation in stainless-steel vacuum stills, one of which is horizontal. 'We used our experience to create something new,' he says. 'We believe that Japanese grain will be important in time. Everyone is doing it in columns, but what about in pots?' Is it grain, or a Japanese take on Irish pot still? Whatever, it's delicious, sharing with the single malt a warm tropical-fruit character. A classic example of learning from the past and adapting an old tradition into something new.

MARS

MARS TSUNUKI 2023 EDITION 50%

Bold, with more concentrated fruits, some mango, but also plum jam, macadamia and a little smoke.

FRUITY/SOFT

MARS THE Y.A. #2 (BLEND OF KOMAGATAKE AND TSUNUKI AGED ON YAKUSHIMA) 49%

Starts with light smoke, some eucalyptus, togarashi spice (chilli, salt, tangerine) and also light florals. Moves into retsina-like flavours and texture.

SPICY/HERBAL

The pioneer of Kyushu whisky is Hombo's **Mars Tsunuki** distillery, which started in 2016 using small pots salvaged from the firm's Yamanashi distillery. These ran intermittently, but in 2015, with Mars Shinshu up and running, the firm saw a chance to not only produce more but in a different style.

While a subtropical climate might suggest the need for lightness, here a heavier style is the aim. As with Kanosuke, the souring of a fresh washback with fully fermented wash is used as part of a four-day, temperature-controlled regime with a mix of shochu and distiller's yeasts. The stills have a steep slope on the lyne arm running into worms.

One interesting factor is that the inside of the spirit still is never cleaned, giving a build-up of residue, eliminating any chance of copper conversation (as with Mortlach – see page 52). Add in the worm and meatiness begins to appear.

Speciality malts are run, as well as imported barley, although the wide range of peating levels has been slimmed down to unpeated, heavily peated and occasionally super-heavily peated. Sometimes the new makes are blended prior to ageing.

With 7% annual evaporation, the spirit moves quickly along its maturation path. Hombo, however, goes the extra mile, ageing in mostly ex-Bourbon casks on Yakushima Island even further south where summer temperatures can hit 41°C (106°F).

In 2023, Nikka unveiled a new blended grain with spirit from Miyagikyo and two of its shochu distilleries in Kyushu – **Moji** and **Satsuma Tuskasa**. Both use stainless-steel stills. Suntory bought the **Osumi** distillery to the east of Kagoshima in 2014. Much to everyone's surprise, in 2020 it launched a rice whisky made from what was always considered to be a sweet potato shochu specialist. It's made in stainless-steel vacuum stills, without koji (a strain of *Aspergillus oryzae* used to start germination in sake and shochu). Who knows what else may emerge?

THE REST OF ASIA

The two most populous countries in the world are both now making their names as whisky producers. One has a long-standing love of the spirit; the other is a newcomer. Both are also key export markets for whisky's old guard, which can see potentially vast rewards should protectionist measures be overcome in India and whisky education be increased in China. A growth in domestic production seems set to make them both battlegrounds for whisky in the decades to come.

In the third century, India was one of the earliest (if not the first) country to distil a social spirit (made from sugar cane). Its relationship with whisky is considerably more recent, dating back to the 19th century and the era of the British Raj when whisky and rum were imported and in the latter case made on the subcontinent. Indian's first 'modern' distillery, Kausali, was home to Old Monk rum, which was given as a daily ration to troops.

Whisky, though, had a certain cachet – the allure of the imported, perhaps. Domestic brands appeared, but due to a reluctance to use grain that could be used for food, by the 20th century distillers made their 'whisky' from molasses to which was added imported Scotch. These IMFL (Indian-made Foreign Liquors) brands were often sold with Scottish names and continue to sell in millions of cases. India consumes 1.5 billion litres (396 million US gallons) of whisky a year, mostly IMFL. The use of molasses spirit, however, contravenes the global definition of whisky being a cereal-based product.

China's scale is hard to comprehend, but let's try. In whisky terms, if Scotch was to somehow get 1% of the total Chinese spirits market, the volume required would drain every distillery of its entire production. With statistics like this, you can see why people get excited about its potential, but it remains a new market. Spirits-wise, it's still the land of baijiu and moutai, and imported spirits are tiny in terms of volume. For many years the imported sector was dominated by Cognac, whose successes (and failures) in China should provide salutary lessons to gung-ho whisky marketeers. 'Should' is the operative word.

What is the Chinese plan? Produce whisky to satisfy its enormous population, or look to export in order to gain a degree of credibility? It's too early to say, but this is a new whisky-producing country with enormous potential, though with a number of barriers to surmount. Lessons might be learned by China's southern distilleries, as well as those in Thailand and Vietnam, from the Taiwanese, whose Kavalan distillery was at the forefront of subtropical whisky production (see page 298).

Right *Indian single malt is building its reputation at home and abroad.*

CHINA

Scotch tried to break into China in the early 2000s, targeting nightclubs by buying listings for blends. When the trend spluttered out, a prestige strategy was tried – appealing to newly wealthy consumers with top-end single malts. Recently it too has started to flag as image-fixated drinkers look for the next big thing. Meanwhile, a new malt-literate consumer also interested in American and Japanese whiskies is emerging.

A GROWTH BUSINESS

Chester Zhi runs the WhiskyENJOY media outlet as well as the annual WHISKY+ consumer show. He reports that of the 10,000-plus attendees at the show, most are aged between 25–40, and 45% are women – a significant demographic shift.

The downside is that the super-premium single malt approach is not targeting these new, interested drinkers. A more tempered approach of building a whisky market at different price levels is clearly the best way. It will take longer-term investment but with, hopefully, longer-term rewards.

That whisky is beginning to establish a foothold in China is borne out by the fact that, at the time of writing, there are now 27 operational whisky distilleries with many more planned – often cash-rich baijiu distillers branching out. Major international firms are also investing significantly in plant.

DEFINING CHINESE WHISKY

There are many questions being asked. Legislation for one. How to define Chinese whisky? How can you convince a country where ageing of spirits is unknown that maturation is a defining character – one recent proposal by budding distillers was that only malt whisky needed to be aged and grain could be used as new make.

This sensitive work is currently being done by the whisky branch of CADA (China Alcoholic Drinks Association), but progress is slow

Above *What will Chinese whisky's sense of place be?*

Above *Pernod Ricard's The Chuan is one of a number of international ventures in China.*

partly because of the multi-tier nature of legislation in China, which has four levels of regulatory bodies: national, industry, group and company, with the last three able to set their own parameters within an overall national structure. There's a clear need to tell new consumers what whisky is. How to get to that point is not clear.

China's scale also raises further issues. It's not possible for its distillers to only produce from imported barley. Not only is it expensive, has long lead times and puts pressure on global resources, but shipping massive amounts of raw material isn't sustainable from an environmental standpoint. Can China supply itself?

If there's already a shortage of American oak casks, where will all the wood required for casks come from? Where are the Chinese cooperages? Are there sufficient maltings?

The biggest philosophical question is what could Chinese whisky be like? Which barley varieties or other grains could be used? What about yeast strains, Chinese oak (*Quercus mongolica* and others) or other woods, smoke from wood or from tea? Yellow wine or plum wine casks rather than sherry? How will whiskies mature in the wide range of climates? All of these are opportunities rather than problems.

Maybe there are lessons to be learned from tea culture. At its root, single malt whiskies are variations on a theme of barley. Teas are variations on a theme of *Camellia sinensis*. It too has different varieties, is fermented and roasted to different degrees; it can be aged and its character is impacted by climate; it has its own range of ritual servings and is a drink that spans the everyday to the exclusive. The parallels are clear – and nowhere is tea more profoundly understood than China.

CONSUMER CONSIDERATIONS

Then there is the question of how to engage the new consumer and tie Chinese whisky into a drinking culture where spirits are (virtually exclusively) consumed with meals. It isn't as simple as matching whisky with food because there is no such thing as 'Chinese food' – that's a lazy Western construct. Instead, there are highly defined and refined regional specialities. Can you, or should you, match the whisky to suit all of these?

It will be those distillers who learn from their location and those older, deeper cultures of food and drink that will make the most compelling Chinese whiskies.

Whisky is new to people's palates, something that allows distillers a degree of latitude when it comes to creating flavours that might then appeal. At the same time, it also means success might not be immediate – steady persuasion is the order, but this is a culture based around long-term planning. For China's new distillers, it's a matter of finding out what the possibilities are.

HUNAN PROVINCE

The first Chinese single malt to appear on the international market was **Goalong**, which is made in Liuyang City, close to Hunan's capital Changsha. The Goalong Group was established as a producer of 'nightclub spirits' and an exporter of bulk spirit, but began its investment in single malt in 2018. At the time of writing, it's the country's largest single malt distillery, and with export already underway and a Goalong Whiskey Club site in Changsha, it's clear that the aim is to be a mainstream brand.

Australian barley, malted in China, is used and after visiting distilleries around the world, including The Macallan in Speyside (see page 54), the decision was taken to use small, 5,000-litre (1,321-US-gallon) stills with external heating. The style is, however, not as heavy, perhaps due to the fact that distiller Luo Feng has come to whisky from a career as a brewer and has also spent time with barley shochu. He has developed a yeast with Angel Yeast that promotes fragrant fruits, and it's this character that comes across in Goalong's early bottlings, which have an almost rum-like banana character. He is also working with rice.

The key was finding out what consumers wanted and building the character around that. The question is, will the delicate Goalong style go with Hunan's insanely spicy cuisine?

SICHUAN PROVINCE

Laizhou in Qionglai has scale. Built by the largest player in China's ready-to-drink (RTD) market, Shanghai Bacchus, construction started in 2016 and was completed in 2021 after being delayed by Covid. It was carbon neutral from the outset.

Inside, the distillery is split into two: a grain/neutral side with seven column stills capable of producing 20 million litres (5.3 million US gallons) per annum, and the malt side with a capacity of 5 million litres (1.3 million US gallons) per annum. The latter takes inspiration from the Japanese approach of making a range of new makes on a single site. The mashing is controlled to produce different degrees of wort clarity, while ferments use specific yeasts to promote a range of flavours. Smoke and peat will also be included.

Of the eight pot stills, four are small(er) with stainless steel as well as copper shell and tube condensers allowing meaty and fruity to be made. The other quartet have long, slightly upward-facing lyne arms, allowing a more estery new make. All of the stills are externally heated. This building of flavour variations continues in maturation. With wineries nearby, Laizhou has also built a cooperage, complete with an STR area.

GOALONG

GOALONG SINGLE GRAIN (CASK SAMPLE) 62.7%

Sweet, with plenty of oak, unripe lemon and then light pineapple. Water shows some youthful edge, then mint and lychee. Precocious.

SWEET/VANILLA

GOALONG SINGLE MALT 5 YEAR OLD (CASK SAMPLE) 48%

Light cereal sponge cake, lemon, cantaloupe melon. Develops rummy elements, pineapple, yuzu and limeade. Light oak and a short finish.

FRUITY/SOFT

CHINA, TAIWAN, THAILAND, VIETNAM & SOUTH KOREA

MONGOLIA
CHINA
NORTH KOREA
SOUTH KOREA
TAIWAN
BURMA (MYANMAR)
LAOS
VIETNAM
THAILAND
CAMBODIA
PHILIPPINES
0 miles 500
0 km 500
BEIJING
PYONGYANG
SEOUL
TAIPEI
HANOI
BANGKOK
Chengdu
Shanghai
Guangzhou
Hong Kong
Kaohsiung
Ho Chi Minh City
Yellow Sea
East China Sea
South China Sea
Philippine Sea
Gulf of Thailand
Yellow River
Yangtze
Mekong
Salween
Irrawaddy
MengTai, Ordos
Tianyoude, Huzhu
Leaven Fu, Dalian
Long Whisky
Jisshangbao, Yantai
Yuzhijin
Three Societies, Namyangju
Tsingtao, Qingdao
Guqi, Bozhou
Defulai Whiskey, Nanyang
Wulliangchan, Anqing
Longwei, Hangzhou
Weigu
Laizhou, Qionglai
The Chuan, Emeishan
Ruins Whiskey, Meishan
Jiangji, Chongjing
World Roasted, Shaoshan
Liuyang Goalong Liquor Distillery Co., Changsha
Eryuan, Dali
Entellus
Panda Brew, Anshun
Nine Rivers
Dexi, Longyuan
Daiking
Paihuan, Zhangzhou
King Car/Kavalan, Yuanshan Township
Nantou
Vé Đế Đi
Red Bull, Samut Sakhon
N

These casks join ex-Bourbon and a range of different sherry types. In addition, the distillery is trialling maturation in casks that have been seasoned for 18 months with either Jiafan (semi-dry) or Xiangxue (sweet) yellow wine. There is also work with Chinese oak (*Quercus mongolica*), with forests being planted. The warehouse is planned to take 100,000 casks. A visitor centre opened in 2024.

Pernod Ricard broke ground on a site in Emeishan in 2019, with the first release using spirit from its ¥1.5 billion **The Chuan** distillery appearing in 2023.

In the middle of a noted wine-growing area in the southern foothills of Daqin Mountain, The Chuan has ten washbacks and two pot stills. The firm's decision to make its first release a world whisky blend of Chinese and Scotch (with little mention of the latter) was a strange one at a sensitive time in the negotiations over a definition of Chinese whisky, bringing to mind the problems that have bedevilled Japanese whisky in recent years.

Above *The Chuan – hints of Scotland in China.*

SHANDONG PROVINCE

The best-known Chinese alcoholic beverage globally is **Tsingtao** beer, but it's a lesser-known fact that the brewer was also the first to make whisky in China, in 1912. Qingdao had been seized by the Germans in 1897 and with them came brewing and whisky distilling. The distillery continued to produce under Japanese occupation from the 1930s until the end of World War II. By 1947 there were apparently another 15 breweries making whisky in China. This is less of a surprise than it may seem when you think about the considerable importance of the Chinese market for blended Scotch from the early 1900s.

The Tsingtao distillery closed in the 1980s, although old stocks were retained. It restarted in 2022 and now has three brands on the market.

FUJIAN PROVINCE

Daiking in Daxi, southerly Fujian was the idea of Taiwanese businessman Wu Songbai. Situated at 1,544 metres (5,066 feet) on the slopes of Daiking Mountain in an old tea garden, it began production in 2014. A large player, at 8 million litres (2.1 million US gallons) per annum, it has 5 mash tuns, 48 washbacks, ten 10,000-litre (2,642-US-gallon) wash and 8 spirit stills, half from Forsyths, the rest from China.

Hong Kong-listed Grace Wine, which owns the Grace Vineyards in Shanxi and Ningxia, bought the existing Longyan distillery in Fujian, now named **Dexi**. With Mike Nicolson (see page 259) on board as consultant, the first whisky from its pair of pots appeared in 2023.

YUNNAN PROVINCE

Major international players are seeing the opportunities within China. As well as the country being a significant market for Diageo's Scotch, in 2023 the firm opened its **Eryuan** distillery in a high fertile plateau in westerly Yunnan, close to Dali. The search for a site had started in 2015, with the Chinese government recommending Yunnan to

NEW VENTURES

Another ambitious project is the new distillery planned for Ordos in Inner Mongolia (the region's first) by the energy business **MengTai** Group, which has extensive interests in the region. 'Scotland is the homeland of whisky,' says director and whisky maker Richard Lin Lu. 'MengTai has a similar climate to Scotland, with suitable temperatures. We believe that MengTai can produce the best-quality whisky in China, and be recognized worldwide.'

The ¥200 million project (covering an area of over 40 hectares/100 acres) was signed off in 2019 and started production in 2023. All of the equipment has been sourced and shipped from Forsyths in Scotland. It will be capable of producing 2 million litres (528,340 US gallons) per annum.

'This project is in response to the Chinese government's "rural revitalization" policy,' explains Richard, 'investing in and developing modern agricultural industries to boost the income and prosperity of local farmers and herders.'

That in turn means grains not just for whisky but also the by-product. 'Barley and grains are the main raw materials for whisky production,' he adds, 'but the spent grains can also be processed into green feed, reducing the breeding costs for local farmers and herders.' Circular thinking.

It's being set up to produce a wide range of flavour styles: floral, fruity and what Richard describes as 'region-specific, peaty flavours based on the regional characteristics of Ordos, which is rich in coal'. He is also exploring what might come from the grasslands of Inner Mongolia and using Mongolian oak barrels.

Even more substantial is **Nine Rivers** in Longyan City (Fujian province). The idea to build a distillery came from a group of whisky lovers and retailer Oak and Barley. Steadily they built a board of industry players including Dr Harley Seyedin, president of the American Chamber of Commerce in South China, and Stephen Davies, CEO of Penderyn in Wales (see page 150). With a mix of institutional investment and crowdfunding from whisky lovers, in 2019 the firm was registered.

At 600 metres (1,968 feet), the site benefits from a relatively cool climate, potentially helping with longer maturation. When built, it will be the largest malt distillery in the country complete with a malting and kilning operation, a cooperage and a stillhouse with a dozen stills of different shapes and capacities, with the intention of producing three styles of new make. While the domestic market will be important, the scale will also allow decent volumes for export.

In late 2023, Scotch distiller Angus Dundee announced its intention to build a distillery on an 8-hectare (20-acre) site in Qiandao Lake (Zhejiang province) complete with a maturation area in a mountain cave nearby, while in a link-up with Gujinggong, China's fourth-largest baijiu producer, Camus Cognac invested $30 million building the new **Guqi** distillery in Bozhou (Anhui province).

Above *Laizhou is one of the country's massive new whisky distilleries.*

assist with economic investment in this region, which borders Tibet, Myanmar, Laos and Vietnam.

Carbon neutral from the outset, all the power comes from wind or hydro and all waste and water is recycled. Equipped with a 6.6-tonne (7.3-US-ton) full lauter tun, 8 washbacks and ferments between 80 and 120 hours, there are two pairs of stills giving a capacity of 2.5 million litres (660,430 US gallons) per annum.

The aim is to tie the distillery to its surroundings and, while the early distillations used imported barley, the goal is to source from China. Trials are already underway to find local yeasts, and as well as maturation in ex-Bourbon casks, they are examining the potential of Yunnanese and other Chinese oaks such as *Quercus mongolica*, and also wine casks from the Ao Yun winery in Shangri-La, northern Yunnan.

While some of this is driven by sustainability factors – the long lead times for shipping, and the squeeze on the availability of American oak – there is also the need to make a whisky that aligns with the Chinese palate. This means that speciality malts could become part of the mash bill, as could yellow wine casks, or smoking with different fuels rather than peat.

Master distiller Andrew Millsopp says, 'We want to put Chinese whisky to the centre of conversation. We are not just replicating Scotch in China. This might be the first whisky for many Chinese people, and the first Chinese whisky for many around the world, so it has to be good.'

Vivien Duan of the local liquid team adds, 'We want to get the flavours of provenance in the whisky, and make this a new type of liquid with its own unique style. Yunnan is the place that will give that.'

LAIZHOU

★ LAIZHOU CASK STRENGTH (SINGLE CASK) 60%

A fermented aroma, wax crayon, chocolate ganache, beeswax polish, black banana, cream toffee, cedar, fresh mushroom and black bean. Mixes sweet, savoury and sour.

RICH/DEEP

DAIKING

DAIKING 'FOREST GREEN' SINGLE CASK (CASK SAMPLE) 51%

Oak moss, fern and hot green bracken. Light pine sap with a little dry malt on the palate. Becomes more cocoa- and cereal-accented.

SPICY/HERBAL

DAIKING SUN ROSE RED SINGLE CASK (CASK SAMPLE) 50%

On the fruity side of sherry with some vermouth, sloe, bitter chocolate, pine resin and malt extract. Palate has fresh orange, black cherry and bitter roots.

FRUITY/SOFT

TAIWAN

Now that it's a familiar presence, it's hard to recall what a sensation Kavalan was when it first appeared in European markets in 2006. The wider world was just getting used to Japanese whisky, but a subtropically aged whisky from a distillery where you could almost see the alcohol evaporating? Moreover, one of such quality coming from conditions that, in what was still a Scotch-centric world, seemed impossible for producing single malt? Yet there it was.

The idea and the money behind this challenge to whisky norms was Lee Tien-tsai and his son Lee Yu-ting, whose King Car Group was already a major conglomerate in the island state.

They had found a site north of Taipei in the hills of Yilan, humid but with no extremes of heat. Their vision was an ambitious one – to make **Kavalan** a global brand rather than a local speciality that might appear in small quantities around the world. The key to their success was the hiring of two men: Dr Jim Swan (whose name by now will be familiar) and distiller Ian Chang (now at Komoro in Japan – see page 281). Jim set out trying to understand the peculiar conditions in order to make a whisky that was clean, fruit-driven, mildly floral and, just as importantly, where that freshness of character was balanced with the potentially aggressive attentions of oak. In achieving this, he and Ian opened up not just Kavalan but the concept of subtropical whisky. Over one million people now visit the distillery each year.

There have been continual expansions, and Kavalan is now in the top ten of single malt producers in the world in terms of volume. There are 5 mash tuns and 40 fermenters running 60–72-hour ferments using a yeast that Ian's team isolated from the surrounding area, which helps to boost the fruity elements in the new make.

Above *Kavalan was the pioneer of sub-tropical whiskies.*

Above *The Kavalan distillery receives more than 1 million visitors a year.*

Above *Nantou's Omar is deservedly building a global reputation.*

The stillhouse has 10 pairs of pot stills of the same shape and size (10,000 litres/2,642 US gallons for wash, 7,000 litres/1,849 US gallons for spirit), with slightly downward-sloping lyne arms. In line with Jim's philosophy, the wort is clear and the spirit cut is tight.

A wide range of cask types is used including, of course, STR. While ex-Bourbon makes up the largest percentage, a variety of sherry types, port and wine are also used, either for full-term maturation or finishing.

The casks are placed in large five-floor concrete warehouses. With temperatures peaking at 42°C (108°F) on average, 13% per year is lost. Although trials are underway to try and find a way to slow this rapid evaporation, it's part of the Kavalan character.

The mechanisms that dictate mature style are accelerated here – Kavalan's whiskies don't just lack aggression (removed by evaporation and charred casks) but the impact of the oak and previous fill (additive maturation) plays a major role. However, even in the short time the whisky is in cask there is interaction between all of the elements – oak compounds, new make character and air to create more esterification and complexity.

The **Nantou** distillery in the central east of the island is a state-owned enterprise – a division of Taiwan Tobacco & Liquor Corporation (TTL). Taiwan's alcohol production was under governmental control from 1947 to 2002. The distillery started production of a range of spirits in 1978, but the international success of Kavalan was undoubtedly a spur to it beginning to specialize in single malt.

Its Omar brand first appeared locally in 2008 using wash sourced from a TTL-owned brewery. In 2010, all production was brought into the distillery itself. After 60–72 hours in 8 washbacks, distillation is in two differently sized wash stills: one is 7,000 litres (1,849 US gallons) in capacity, the other 5,000 litres (1,321 US gallons). The spirit stills are also oddly sized – the largest at 5,000 litres (1,321 US gallons), the smaller 2,000 litres (528 US gallons). Slightly varying new make characters result, which are then blended and aged. Some peated malt is also imported.

Maturation takes place at the distillery as well, with the bulk of the new make going into ex-Bourbon or sherry, though as the distillery continues to make a range of fruit spirits and liqueurs, casks that have either matured or been seasoned with those are used for the Liqueur Barrel Finished range.

KAVALAN

KAVALAN SOLIST VINHO BARRIQUE 50–59.9%

Big oak extract. Chicory coffee, dried berries, fermented black beans, then tomato, sloe, red apple and rose. Fruits come forward on the palate with burned oak.

OAKY/DRY

NANTOU

OMAR BOURBON CASK 46%

Very up and sweet. Tropical fruits (papaya, mango) then ripe banana and cassia. Palate has coconut cream, white chocolate and a little honey.

FRUITY/SOFT

THAILAND, VIETNAM & SOUTH KOREA

Whisky's apparently inexorable spread has begun to reach further into Asia. Thailand, Vietnam and Laos have long had local 'whisky' brands that have either used bulk Scotch to boost a local distillate or appropriated the cachet of the name for that domestic spirit. Now recognizable whisky distilleries have emerged in the first two (Laotian whisky is yet to do so), while Korea, where Scotch has long had a presence, has become the latest member of the world whisky family.

THAILAND

The **Red Bull** distillery in Tambon Nadee, northern Thailand is owned by International Beverage (IB) and sits next to its massive brewery. It has been in production since 2013, producing a base spirit that's mixed with Scotch for the firm's Blend 285. It now plans to release a 100% Thai single malt.

The malt distillery is modelled on the firm's Speyburn site (see page 44) and, with Scottish barley (some peated), Forsyth stills and a wood mix of ex-Bourbon and ex-sherry, it might seem like a Scottish distillery in a different land – albeit with 8% annual losses rather than 2% – but the results, IB's master blender Stuart Harvey says, are happily different.

'Whisky isn't predictable in any way,' he says. 'The new make has this amazing tropical fruit, lots of pineapple and banana especially. It just goes to show that you can't reproduce Scotch. Every whisky is a reflection of its own conditions.

'Our parent company is investing in making single malt around the world, but have always wanted to make it at home.' In doing so, another new flavour has been revealed.

VIETNAM

American expat Michael Rosen started the **Về Để Đi** distillery in Hanoi in 2020 when the old Sơn Tinh distillery was repurposed into Vietnam's first whisky plant, with two pots making single malt and corn whiskies.

Asking what led him to this point results in a philosophical debate. 'I struggle at what is a proper way to spend a life,' he says. 'Whisky, it

Above *Về Để Đi – Vietnam's first whisky distillery.*

Above *Creating Korean whisky culture at Three Societies.*

seemed, was no longer for the elite; it no longer sat on its own, both shy and ever so slightly precious. Any old boundaries had disappeared.'

He wondered if he could have embarked on work around 'the world's more apparent immediate crises' rather than just distilling. 'Should I be spending so much effort, emotion, money and more at this and not doing things that make the world better?'

The solution to this existential crisis came with the realization that whisky was more than just a drink, an idea that other distillers around the world have articulated in many different ways.

'I have to believe that there is another way,' he says. 'That people don't just want to know what whisky – or any other great spirit – is and what it's like, but *why* it's the way it is. That there is a need to see how any spirit in all of its myriad forms is part of a bigger picture, that we have to continue to find out why it touches people's hearts and souls, and makes them speak and sing and cry and smile.

'These are magical liquids that manage, somehow, to span the world and act as a form of communication between people; which eliminate class and sexuality, religion and politics, and help make us, in some way, human. That sounds grand, but I believe it. It's what keeps me going. It's been a long road, but this felt like the start of something new.'

SOUTH KOREA

Bryan Do's epiphany in Korea, which resulted in the birth of the **Three Societies** distillery, was slightly different. An exec at Microsoft, he had begun to explore the world of craft brewing and, inspired by what he discovered, quit the world of tech to found Seoul's Hand & Malt Brewing Co. This led to him wondering whether the same 'craft' approach could be applied to whisky. Korea has had a vexed relationship with the spirit. It was a popular spirit in clubs and KTV outlets, but the image of hard-drinking and badly behaved businessmen didn't translate to a new generation. Attempts to revive the category with low-alcohol variants also failed.

By 2020, however, the market had swung back, with single malt leading the way. The time, therefore, was ripe for Bryan to see if his hunch was right and, after selling Hand & Malt to AB InBev, he started building his distillery in the hills around Namyangju, east of Seoul.

The name comes from the background of the three founders: Bryan (American-born to Korean parents), 'spiritual advisor' Andrew Shand, who has over 40 years' experience in distilling in Scotland (Speyside – see page 66), Japan (Nikka – see page 289) and the USA (Virginia and Copper Fox – see pages 218), and the Korean workforce personified by distiller Seo Hong-jun.

The key was what to make, which for Bryan meant a whisky that needed to work with a cuisine noted for its spiciness. How he gets that is a secret, though longer ferments and wood choice both play their part. This spiciness might also be helped by the fact that with summer temperatures hitting 30°C (86°F) and winter -20°C (-4°F) there is a rapid uptake of wood.

Maturation takes place in virgin oak, ex-Bourbon and ex-sherry, but there is also work underway with casks that have either held or been seasoned with domestically produced spirits. As well as the local market, export is already planned. Ki One (aged in virgin oak) was launched in 2023.

THREE SOCIETIES

★ KI ONE 3 YEAR OLD EX-BOURBON
(CASK SAMPLE) 56.5%

Fruity and full with candied nuts, sandalwood incense, raspberry and chocolate. Water adds mid-palate roundness. Impressive.

FRUITY/SOFT

INDIA

India has enormous whisky potential, but buying Scotch remains an expensive option thanks to high import taxes and tariffs imposed by each state. That some are also dry creates even more barriers. The hurdles to achieving a degree of parity are considerable: tax, politics, a wider trade agreement, plus the insistence by some Indian distillers that their 'broader' interpretation of whisky should be permitted, allowing their brands to be sold as such internationally.

Scotch firms are still betting heavily on liberalization happening. If it does, then theoretically, India could become one of the world's biggest markets for whisky.

Were this to happen, would Scotch and whiskies from other countries have it all their own way? Not necessarily. In 2023, Indian single malts accounted for over half the sales of single malt domestically, outselling Scotch, evidence of a new-found belief on the part of the local consumer in the quality of their own product.

It represents a shift in consumer thinking from 2004 when **Amrut**, the first Indian distillery to launch a single malt, decided to sell it in export only – it was only launched on its home market in 2010.

THE SOUTH

The Amrut distillery, founded in Bangalore (now Bengaluru) in 1948 by the Jagdale family, had long been a producer of local whisky and other spirits, and branched into dedicated single malt production in 2001.

Ashok Chokalingam, as head of sales, travelled continually around the world for many years spreading the word not just of Amrut but seeding the idea that the subcontinent could make, and was making, world-class whisky.

'We've strongly cemented our position by releasing a number of expressions that were enticing and entertaining,' he says 'I think we've

Above *Checking casks at Piccadily.*

TAJIKISTAN
CHINA
AFGHANISTAN
ISLAMABAD
Muree, Rawalpindi
Plateau of Tibet
Himalayas
Mekong
Salween
Lahore
Kasauli
PAKISTAN
Piccadily, Haryana
Mohan Meakin, Ghaziabad
NEW DELHI
Indus
NEPAL
Rampur/ Radico Khaitan
United Spirits/Alwar, Goondpur
Jaipur
THIMPHU
Gelephu
BHUTAN
Sikkim Distilleries Ltd, Rangpo
Kanpur
Thar Desert
Karachi
Ganges
BANGLADESH
Ahmedabad
Kolkata
INDIA
BURMA (MYANMAR)
South Seas Distilleries, Ganjad
Pernod Ricard/Dindori, Nashik
Bay of Bengal
Mumbai
Arabian Sea
Hyderabad
Paul John, Cuncolim
Chennai
Amrut, Bengaluru
N
0 miles 500
0 km 500
SRI LANKA
MALDIVES
INDIAN OCEAN
KEY
Malt Distillery

moved away from being a 'novel' single malt distillery into a serious player. This is supported by exponential sales growth over the last decade.'

One of the keys to Amrut's success has been its willingness to innovate and, in a world that can be overly serious, have fun. Two Continents had casks shipped from India to finish maturation in Scotland; Intermediate Sherry is an exercise in élevage: starting in virgin oak, then sherry, then ex-Bourbon; Spectrum is aged in casks that are made up of staves from four different cask types: new American Oak (#3 char), light new French Oak, ex-oloroso and PX staves, and rye; Naarangi aged in an ex-oloroso sherry cask seasoned with wine and orange peel for three years; while Greedy Angels is a fascinating project showing the impact of subtropical maturation on flavour – the oldest, remarkably given that those angels suck up 14% a year, a 12-year-old.

'Our new make character is fruity, heavy and malty,' says Ashok, who is now Amrut's master distiller. 'We have one of the longest-fermented wash and slowest-distilled new make spirits. The ferments are over 140 hours.'

A major expansion took place in 2018, with a new mash tun, more washbacks and four stills taking capacity to over 1 million litres (264,172 US gallons) per annum.

Sustainability looms large in the thinking. 'It's the need of the hour,' as he puts it. 'Our first target was to reduce the stress on our watershed by completely reusing our treated water from our effluent treatment plant. We have reduced 60,000 litres [15,850 US gallons] of water a day from our watershed. The methane generated in our effluent treatment plant is now used in our hot water generator to use in our boiler feed. This is going to cut down our carbon emission by reducing the quantum of briquettes that we fire in the boiler by 2 metric tonnes [2.2 US tons] a day. 'The whole Indian category is stronger than ever,' he concludes, 'and it looks to me that it is going to be one of the strongest categories going forward.'

AMRUT

AMRUT FUSION 50%

Clean, nutty with warming smoke, biscotti, then petrichor, new sandals. All elements are integrated on the palate along with shiso leaf, biscuit, melting butter and light smoke.

MALTY/NUTTY

★ AMRUT TRIPARVA TRIPLE DISTILLED 50%

Hothouse, orchid, cassia bark and hessian, then a burst of tropical fruits with a balancing green note and sweet biscuity spices.

FRUITY/SOFT

PAUL JOHN

PAUL JOHN CHRISTMAS EDITION 2023 46%

Sweet with heady spice, slightly off-dry. Honeycomb and baked banana, orange blossom and light malt. Palate has cumin and fig leaf.

FRUITY/SOFT

★ PAUL JOHN MITHUNA 58%

Firm nose. Roast chestnut, dark marmalade, light wood resin and coffee. Full-bodied with good grip, then dark fruits flood forward. Finishes with a refreshing acidic drive.

RICH/DEEP

Above *Hand bottling at Amrut.*

Above *Goa's Paul John embracing the local climate.*

Above *Paul John's pots.*

THE WEST

In 1992, Goan-based **Paul John** began making (molasses-based) spirit for domestic 'whisky' in its three distilleries. It now sells 21 million cases a year. Its move into single malt took place in 2001 when a pair of pot stills were installed in its Goa site. This was then doubled, then doubled again, now giving the distillery a capacity of 3 million litres (792,516 US gallons) per annum.

Six-row Indian malted barley is used alongside different levels of imported peated, with 24,000 litres (6,340 US gallons) of wort split between two fermenters for 70 hours. The pots have upward-facing lyne arms, helping to give a lighter, fruity, sweet new make.

The climate is the final factor. 'Goa is tropical,' says the firm's Asa Abraham. 'It allows the whisky to mature differently. The annual angel share for us is between 7 and 8%. Due to the unique weather conditions of the region, the interaction between the wood and the liquid is rapid, giving richer and bolder flavours. There's also salt in the air due to the distillery's proximity to the sea and we believe this adds a mineral characteristic to the single malt.'

She is optimistic about the future for single malt. 'Whisky in India has been the mainstay of the industry and will continue to be a dominant category, with single malts an explosive subcategory.

'We also foresee the rise and establishment of the Indian single malt category across the world with more players coming in from independent distillers and even multinationals.'

MULTINATIONAL ACTIVITY

The global majors are not blind to India's new-found love of its own single malts. The country's largest spirit producer United Spirits is now a subsidiary of Diageo. The USA's McDowell's is the largest selling 'whisky' in India and the firm has plants across the country. Its **Alwar** distillery in Rajasthan, produces the spirit for Godawan single malt, which is distilled from local six-row barley and aged in casks that have been seasoned with botanicals. Sales help to conserve the critically endangered Great Indian Bustard.

Pernod Ricard launched Longitude 77, a blend of 'locally sourced ingredients with a double-matured single malt' made at a distillery in **Dindori**, Nashik where the firm has started work on a greenfield site to build a larger facility.

Scottish distiller Ian Macleod (owner of Glengoyne, Tamdhu and Rosebank – see pages 80, 58 and 115) has had a long supply relationship with India and in 2023 started work on a single malt distillery in **Una**, near Amritsar, which will have a capacity of 1 million litres (264,172 US gallons) per annum.

THE NORTH

Radico Khaitan's **Rampur** distillery, in the Uttar Pradesh town of the same name in the Himalayan foothills, started operations in 1943 with the usual mix of Indian whisky and other spirits. Single malt became part of its operations in 1992, with some of the production initially going into the firm's blends. It was in 2016 that its first single malt brand was launched.

It's a considerable business with a total capacity of 72 million litres (19 million US gallons) per annum of molasses spirit, and 30 million

Above *Rampur is flying the whisky flag in India's north.*

litres (8 million US gallons) per annum of grain. The malt plant is housed in a separate facility with an increased capacity of 9 million litres (2.4 million US gallons) per annum.

The ferments are triggered by the firm's own recipe of three different yeasts and are relatively short at between 52 and 56 hours, followed by slow distillation in large stills (25,000 litres/6,600 US gallons for wash, 16,000 litres/4,227 US gallons for the spirit). The combination gives a malty undertow to the ripe fruits. Maturation is predominantly in ex-Bourbon casks, but wine casks are being assiduously researched to some effect.

'The location plays a very crucial role,' says Sanjeev Banga, the distiller's president of international business. 'We have access to the Himalayan groundwater, and sit in the middle of a tropical fruit-growing area. The extreme temperatures in North India make our malts mature at an aggressive pace.' The work in widening the cask recipe – for full maturation and finishing – gives the sense of a distillery leaning into the possibilities of flavour.

The trio were joined on the export market in 2021 by **Piccadily** (sic) Distilleries' Indri brand, named after the town where the firm (Piccadily Agro) started in 1952. It was a leap of faith to some extent. 'In 2008, we took the decision to exit the regular Indian alcohol segment and enter the super-premium segment,' says Madhu Kanna, Piccadily's head of international business. This meant building a malt plant and they started to distill malt spirit in 2011.

There was a sound commercial reason for this switch. Malt was rising, as was premium whisky, and, by 2018, Piccadily claimed to be the largest supplier of bulk malt spirit in India, supplying to more than 100 distilleries in India and overseas, as well as growing its Indri brand.

Still independent, it now makes 4.5 million litres (1.2 million US gallons) per annum from two mash tuns, eight washbacks and three 25,000-litre (6,604-US-gallon) wash and three 15,000-litre (3,963-US-gallon) spirit stills. Further expansion is planned.

The blender is Surrinder Kumar, who moved here from the same position at Amrut. 'We followed a simple path of accepting the natural ingredients available in India – unpeated six-row barley, which has been grown in India for thousands of years. This gives us a more fruity and aromatic character. Clear wort and long ferments give an easy-drinking whisky, with sweet and fruity notes.

RAMPUR

RAMPUR DOUBLE CASK 45%

Slow and gentle. White cherry, warm cereal mash, then butternut squash, biscuity malt. Palate has chocolate, pistachio, then apple, mango and peach to finish.

RICH/DEEP

★ RAMPUR ASĀVA 45%

A fruit bomb. Perfumed. Cherry pie, heavy rose, then toffee, Pop-Tarts, coconut and strawberry. The palate shows some grip and oak-driven spice.

FRUITY/SOFT

RAMPUR JUGALBANDI #3
(PORT CASK MATURED) 54.8%

Warming soft black fruits mixing with black banana and mango. Thick concentrated palate, some fig jam. Long and gently drying finish.

RICH/DEEP

'The location, terroir and weather are what creates our whisky,' he continues. 'As soon as you exit the Himalayan range and enter the northern plains, you find Indri on the flood plains of the River Yamuna. The soil is extremely fertile and the area is known as the food bowl of India.'

Summer temperatures can reach 50°C (122°F), which can dip to freezing in winter. Other than the two months of monsoon, the climate is dry. All of this means that maturation is speedy.

'If you stand in the warehouse quietly, you can hear the wood and whisky whimpering to each other,' he says. I suspect he means whispering, but it would be rude to correct him – anyway, 'whimpering' sounds about right given the extremes.

PICCADILY

★ INDRI DIWALI COLLECTOR'S EDITION 2023 63.5%

Sweet and floral. Light smoke with some raisin and perfume from the PX. Fragrant, controlled and exotic. Palate is savoury, full and layered. A slow release of light smoke.

RICH/DEEP

INDRI SINGLE MALT TRINI THREE WOOD 46%

A soft delivery. Apple and granola, orange, sultana and stewed fruits. Perfumed wood, prune and pineapple. Rhubarb, cooked red fruits with a gingery finish.

FRUITY/SOFT

OTHER INDIAN DISTILLERIES

Kasauli distillery (formerly Mohan Meakin) was established in Himachal Pradesh as a brewery in the 1820s. India's oldest distiller, its brands include Old Monk rum and Solan No. 1 whisky. In 2019, it entered the single malt category with Solan Gold.

The town of Rangpo is home to **Sikkim Distilleries Ltd**, which produces Sikkim Himalayan Premium Old Gold Single Malt Whisky as well as a corn whisky.

Then there is **South Seas Distilleries** based in Aswa, Dahanu. Capable of 10 million litres (2.6 million US gallons) per annum total production, it claims to have the country's largest pot stills and as well as malt spirit makes grain alcohol, ethyl alcohol and whisky and brandy concentrates. A coastal, tropical site, in 2023 it launched its Crazy Cock brand (a name that might not work in export markets), made from six-row barley and matured in a mix of ex-Bourbon and ex-sherry casks. A peated variant, Crazy Cock Dhua, is also on the market.

'Earlier, the distinction between malt whisky, blended whisky or flavoured whiskey was quite blurred,' he adds, 'but whisky is now being more clearly understood thanks to the internet and the awareness of the younger generation. There is also more disposable income. People want to drink and eat better and don't want to consume artificial flavours in their drinks, so quality whiskies are gaining popularity.'

Above *Charring a cask at Piccadily.*

SOUTH AFRICA & THE MIDDLE EAST

Though Israel and Lebanon are separated from South Africa by an entire continent, there are a number of approaches that link their whiskies and it comes down to location. In recent years, Israel's whisky-making community has grown considerably and has also started to examine the possibilities of the locality. South Africa's whisky story might seem to be more orthodox, but it too is taking cognisance of its surroundings.

Whisky in Africa might be seen as a signifier of success – witness how sales of Scotch have grown any time a country has seen an economic upturn. In recent years, sales of Scotch and Irish whiskey have grown significantly in Angola, Nigeria, Ghana, Zambia and Kenya. Production, however, has been limited to South Africa's main distillery. Finally, though, it has been joined by others. Will this spread to the rest of Africa? Time will tell.

South Africa's distilling heritage dates back to the 17th century when Dutch colonists began distilling brandy. Barley arrived later with the British, though brandy remained the country's main spirit, helped by an advantageous tax system that effectively outlawed grain distillation.

This might have been one reason for Scotsman John Duff's attempt to make whisky in the Transvaal in the 1890s having left his native land after establishing Glenlossie (see page 43). His South African project failed but, undaunted, Duff went to America to try his hand there. When that too foundered, he returned home and built Longmorn (see page 43) and Benriach (see page 42).

Barley and corn are now used to make the country's whiskies, but new distillers are beginning to explore the possibilities of sorghum, which has been a staple for African beers.

They are not the first to try to use this grain. Thanks to baijiu, sorghum is the world's most widely used grain for distillation. Sorghum

Right *What could African whisky become?*

whisky has been made in Australia and the USA, and it would make a degree of sense for it to now become part of an African whisky scene.

SOUTH AFRICA

The country's recent whisky-drinking culture began post-Apartheid when Scotch became a status symbol for the new Black middle class. Its domestic industry has been slower to establish itself, but its recent expansion is down to one man, Andy Watts, former master distiller at the **James Sedgwick Distillery** (JSD) in Wellington. A 19th-century brandy distillery, it switched to whisky in 1990 when the distilling equipment from Robertson & Buxton (R&B) in Stellenbosch was transferred there.

R&B's Three Ships brand started as a blend of Scotch and local whisky, though today, apart from the 5-year-old, it's all from JSD, but it's Andy's Master's Collection of single malts that has excited most. 'It was about changing perceptions,' says Andy, 'and slowly changing them by showing that whisky is a world product and not limited to one country.' The range shows his skill in working with different casks and finding new angles on the Three Ships character. He also created maize-based Bain's Cape Mountain Whisky. It too has a premium range – Founder's Collection – all of which demonstrate the versatility and complexity of grain. Both ranges are world class.

'Our whiskies have evolved from being not a high standard to ones that on the international stage are being awarded the highest accolades,' Andy says. 'The last ten years have seen a complete change in how South African whiskies are being perceived.'

With the South African market sluggish, the major international distillers have become seduced by easier pickings in oil-rich Angola, Nigeria and Ghana, but when a market is on the slide is also the time when education is at its most important. It also means importers, retailers and educators need to work together, not factionally.

There is also a role for Three Ships and Bain's internationally. Success on a worldwide scale might just be the boost that's needed to stimulate sales at home – and Andy Watt's work deserves it.

In the early 2000s, the number of South African distilleries doubled when brewer Moritz Kallmeyer added distilling to his **Drayman's** brewery in Silverton, Pretoria.

JAMES SEDGWICK DISTILLERY

★ BAIN'S 15 YEAR OLD SINGLE GRAIN 52.5%
Maple syrup, dried mango, banana bread, creamed corn. Palate has crème brûlée, sticky apricot and butterscotch. Toasted wood/roast corn finish. World class.
FRUITY/SOFT

★ THREE SHIPS 11 YEAR OLD SINGLE MALT SHIRAZ FINISH 51.4%
Red fruits, hint of sooty smoke and barbecued pork. Silky feel, black grape, tamarind, earthiness, then toasted pecan and gentle smoke. Complex.
RICH/DEEP

DRAYMAN'S

DRAYMAN'S RED OCHRE 56%
Marzipan/almond oil alongside scented red fruits, raspberry leaf, then maraschino, dried mint and lemon peel. More herbal on the palate with effusive fruit.
SPICY/HERBAL

DRAYMAN'S PROTEA SMOKED BLENDED WHISKY 43%
Not smoky, more floral perfume, raspberry yogurt. Becomes more botanical in time with a toasted malt base and a light drift of smoke.
FRAGRANT/FLORAL

Above & right *James Sedgwick – home to world-class whiskies in South Africa.*

SOUTH AFRICA

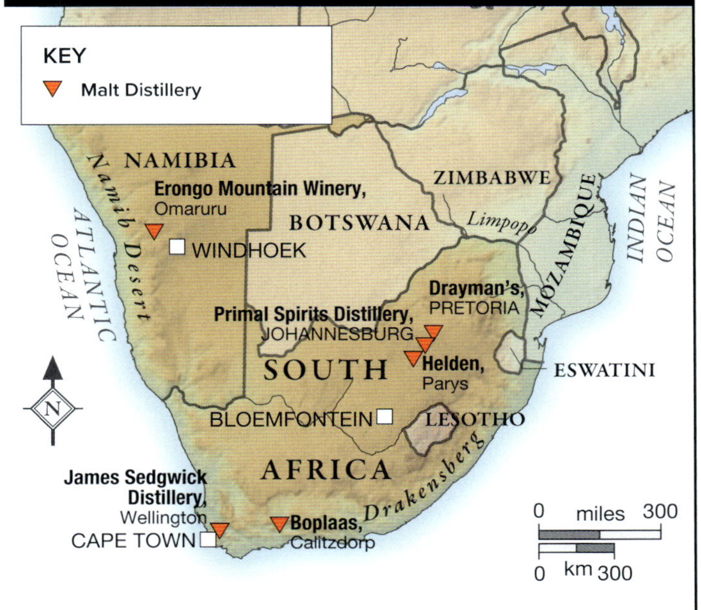

NAMIBIA

Namibia has also entered the world of whisky distilling. Ondjaba is made at the **Erongo Mountain Winery** from a mash bill of barley, maize and mahangu (Namibian pearl millet), which are smoked over elephant dung and column distilled. Maturation is in a mix of virgin oak and red wine casks from the estate.

Above *From brandy to whisky at Boplaas.*

While beer remains the main focus, his whiskies are made from two-row barley malt (some imported peat is used). Only two waters are used, giving a clear wort. A week-long ferment with ale yeast is followed by a two-day conditioning before double distillation. Batches are small, and prized.

The **Helden** distillery is located on the banks of the Vaal River, 75 miles (120km) south of Johannesburg in what was the Dog & Fig brewery. Owner/distiller Pieter van Helden, who has a PhD in chemistry, started distilling when he was at school and, prior to Helden, was distilling in a lab behind his house. Whisky, to him, 'has the greatest flex of any spirit, sitting in the middle between art and science. Distilling is the science. Art is maturation.'

He uses different roasts of local malt, fermented with a fruit-boosting yeast coming from a collaboration with a laboratory in Bloemfontein. It's distilled in a hybrid still made by Daeyoo with the head of the pot built to his design. 'I wanted to make the vapour cloud move more slowly to get maximum copper interaction,' he explains. 'The width of the cloud is what matters. It's the shape of flavour.'

Ageing is in 100-litre (26.4-US-gallon) casks either STR ex-red wine, or toasted and lightly charred new American oak. The combination of high altitude and low humidity gives rapid ageing. He has also started exploring location by distilling sorghum. His African Bonfire is a malt/sorghum mash aged in casks with camel thorn inserts. It smells of the bush after rain.

Primal Spirits Distillery is located in Victoria Yards in Johannesburg, where distiller Chris Muaku is adding whisky to the existing gin and vodka portfolio. Early trials have used imported peated barley run twice through Primal's hybrid still with plates disengaged. A rye, a field blend of wheat and rye and sorghum are also on the cards.

The Nel family began distilling brandy at their **Boplaas** estate in Calitzdorp in 1880 (though a ban on private distilling meant the stills were silent from the 1920s–70s). Whisky is more recent. Boplaas's single malt appeared in 2023, made from imported peated malt, aged in refill, then finished in their own port barrels for 18 months.

HELDEN

HELDEN SINGLE MALT 3 YEAR OLD CASK STRENGTH 71.3%

Mightily strong but filled with sticky caramelized fruit juices, hot earth, bacon and lime marmalade and grapefruit. Thick with a bitter kick. Promising.

FRUITY/SOFT

BOPLAAS

BOPLAAS 6 YEAR OLD SINGLE GRAIN (CASK SAMPLE) 58%

Surprisingly dark. Butterscotch, caramelized peaches, thick fruit and maple syrup. Water increases the stickiness.

SWEET/VANILLA

Above *Exploring different climate zones at Tel Aviv's Milk & Honey.*

ISRAEL & LEBANON

Distillers on both sides of the Israel/Lebanon border are looking at old grains, oak species and other woods. Israel's widely ranging climate is also being explored for its flavour potential by Israeli producers.

ISRAEL

A newcomer to the whisky scene, Israel now has a number of distilleries, the most visible of those being Tel Aviv's **Milk & Honey** (M&H), which was started in 2013 by Gal Kalkshtein. Seeing the rise of Israeli wineries and craft breweries, he felt that whisky would be the next logical step. Like so many distillers around the world, he contacted whisky consultant supremo Dr Jim Swan who worked with distiller Tomer Goren to create a whisky that was both internationally acceptable in terms of flavour, but also identifiably Israeli. In 2015 the distillery moved to its permanent location in Tel Aviv.

M&H uses imported barley. 'Israel has relatively low rainfall and a harsh climate,' Tomer explains. 'That doesn't allow vast barley cultivation, especially not strains that are suitable for whisky or brewing.' For one month a year, peated barley is run.

The wort is fermented with dried Belgian yeast to give a floral and fruity profile for a relatively short time (48–72 hours). Distillation uses two cultural approaches. The wash still is a Spanish brandy still, while the spirit distillation takes place in a Christian Carl still. A high cut is taken to make a powerfully aromatic new make. As maturation is relatively rapid (the angels remove 10–12% a year), not having components that need time to fully oxidize and esterify in cask makes sense.

Ageing is in a large range of casks: from ex-Bourbon and STR to ex-rum, Cognac, orange wine and pomegranate wine.

MILK & HONEY

★ M&H CLASSIC 46%

Slight heat, then honey, baked apple and herbs. Moves into cereal, sugared almond, then butterscotch, orange honey, cumin and pistachio. Complex.

FRUITY/SOFT

M&H APEX DEAD SEA 56.5%

The location brings out more resin and deeper honey. Slightly phenolic, toffee, epazote, wild oregano. Thick, unctuous with layers of spice and crystallized fruits.

RICH/DEEP

M&H APEX POMEGRANATE CASK 57.4%

Red fruits and also slightly oxidized. Baklava, lightly meaty, rose and blackberries. Molasses on the palate but with balancing acidity. Long.

RICH/DEEP

Above *What kinds of flavours could Lebanese oak produce?*

That Israeli approach is seen most clearly in M&H's ageing of its whisky in each of the five Israeli/Palestinian climate zones: the humid Galilee, the cooler climate of the Jerusalem mountains, Tel Aviv's benign Mediterranean conditions and the dryness of the desert and the extremes of temperature and air pressure of the Dead Sea.

Katzrin is home to the **Golani** distillery in the occupied Golan Heights, which was started by David and Alona Zibell in 2014. It uses unpeated barley and a pot-distilled mixed mash bill of 51% malted barley and 49% Israeli wheat. Trials with local barley have also taken place. Ageing is in a mix of local casks: wine, brandy, beer and cider.

The region is also home to Tal and Nir **Pelter**'s eponymous distillery in the Kibbutz Ein Zivan. What started as a winery in 2002 expanded into spirits in 2013. Tal uses a Cognacais still to make spirit from grapes, apples, Medjool dates and malt.

Edre'i distillery, also in the Golan Heights, was founded by Yechiel Luterman who malts local barley on site.

Yerushalmi in the Jerusalem Hills, set up in 2017, concentrates on peated single malt (sourced from Belgium and Scotland). It can make 100 hogsheads a year and is also exploring a range of casks: new French, its own ex-rum, local beer and Israeli sweet wine.

LEBANON

Lebanon's sole single malt whisky, **Athyr**, is produced by one of the country's oldest wineries, Raichi, which, as is commonplace in the country, also makes arak.

Situated in Khenchara, it uses locally grown and malted barley from the Beka'a Valley, the new make being aged in casks made from Lebanese oak (*Quercus libani*), with some of the mature spirit finished in Lebanese cedar.

ISRAEL & LEBANON

AUSTRALIA

To say that Australia has seen a boom in whisky distilleries in the past decade is something of an understatement. To try and grasp just how many there are now, I turned to two friends and experts, Luke McCarthy, who runs ozwhiskyreview.au, and industry consultant and whisky historian Chris Middleton. The conclusion was around 250 (including those in the pipeline). It's a very busy place whisky-wise.

What is often thought of as a new whisky country is, in fact, a revived one. Distillation in Australia started with the arrival of colonialists, though it was rum, not whisky, that was the country's first ardent spirit.

After a ban on distilling between 1796 and 1822, Tasmania's eight original distilleries had dwindled to two by the 1830s, Derwent in Hobart and Caledonian outside Launceston. A second period of Prohibition between 1839 and 1847 saw the end of whisky production on the island until the late 1980s.

Commercial production of whisky was more successful in Victoria, however, starting with Robert Dunn's Warrenheip distillery in Ballarat. In 1884, Thomas Aitken was making Irish-style pot still, while the Federal Distillery in Melbourne could make 4 million litres (1 million US gallons) per annum through its 5 stills. It also used seawater for condensing. In 1924, Victoria's distillers merged to create Federal Distilleries Pty Ltd. Between 1924 and 1933, Old Court (a vatted malt from three of them) was the country's top seller.

With Scotch sales restricted as a result of high import taxes, in 1929 Distillers Company Limited (DCL) opened the massive Corio plant in Geelong and, true to form, took a majority stake in Federal 16 months later. In 1937, British-based Gilbey's built an equally large distillery at Moorabbin. Corio's pots and columns would run until 1980, and Moorabbin's until 1985, but with tariffs dropped, the UK-based multinationals had no need for their Australian outposts.

THE NEW WAVE

The arrival of new distilleries in Tasmania from the late 1980s, therefore, wasn't the start of Australian whisky but a new beginning, which, thanks to lobbying by Bill Lark (see page 318), permitted small-scale distilling to take place. A template for much of the new Australian whisky was also set: wort or wash from brewery, small stills and casks, short-term maturation and single-cask releases, with most brands positioned at the super-premium end.

'The bulk of Australian whisky remains three to five times more expensive comparable to Scotch and Bourbon,' says Chris Middleton. 'While 94% of all whisky sales are under A$100, only six Australian brands can competitively price themselves there.' No matter the quality,

Right *Tasmania kick-started the new Aussie whisky boom.*

this is a precarious position to be in, not just on the domestic market but in export. Being truly competitive necessitates some scale in terms of production and brands.

RECENT KEY DEVELOPMENTS

The profile of the industry is, however, changing. Manildra in New South Wales distils 100 million litres (26.4 million US gallons) per annum of wheat-based ethanol. Tarac in the Barossa Valley, with a capacity of 55 million litres (14.5 million US gallons) per annum, which had previously made grape spirit has now begun processing grain. Ostra Distillers in Robinvale, Victoria will have capacity for 20 million litres (5.3 million US gallons) per annum, while in Tasmania, Greenbanks has started up as a contract producer.

Distilleries such as Archie Rose (see page 327) and Starward (see below and page 322) have also upped production and diversified their portfolios, introducing higher-volume brands that give leverage at home and in export. Other firms such as Tin Shed (see page 331), Morris (see page 325) and Manly Spirits Co. (see page 328) have also entered the mid-priced market.

Diversification is also happening in terms of style. Single malt might still dominate, but there are also rye, mixed mash bills, native grains and blends. Ex-fortified Australian wine casks are being joined by ex-Bourbon, ex-wine, refill, ex-sherry and many more. An adherence to small casks is giving way to larger sizes. Local peat is being used as well as other native smokes. All in all a better balance.

There's ever-greater focus on heritage grains and a willingness to take the hit on yield in return for flavour, part of what Luke McCarthy highlighted as a stylistic shift away from cask-driven styles to more grain/ferment-derived flavour. Refining the offering.

A WIDENING PERSPECTIVE

There has also been a generational shift. The pioneers of the new Australian industry – Bill Lark (see page 318), Casey Overeem (see page 320), Patrick Maguire (see page 318), David Baker (see page 323) *et al* – took a Scottish template and applied it to Australian conditions. The new wave is now taking that Australian interpretation of whisky and deepening that understanding. Scotland or Kentucky are not the inspiration; distilleries like Belgrove (see page 320) are.

This, in turn, has shown that if the future is to come from a deeper understanding of Australia, then farmers have to be involved. 'The more we're able to harness their exceptional work and skill and capture that in the whisky we produce, the closer we'll get to producing distinctive Australian whiskies,' says Luke. That in consequence means an increase in specialized maltings such as Voyager (see box, page 325).

'When I started Starward, my objective was to experiment with local ingredients: indigenous grain species and local hardwood containers. I was looking for a new and distinctive flavour vector that would prove appealing and potentially create an Australian whisky with distinctive flavour characteristics, not "whisky made in Australia",' adds Chris.

Above *Sullivan's Cove helped build Australia's reputation around the world.*

Though he's not convinced that early attempts have wholly worked, that doesn't mean dismissing the principle. 'Ultimately, I think the question of what exactly Australian whisky is still lingers,' says Luke. 'What styles, signatures and hallmarks do Australian whisky makers want to export to the world? It will be fascinating to see what shakes out in the next decade.'

It will come from this widening of perspective. There are parallels between Australia's whisky and its wines – moving from full-blooded, oak-driven concentration to refined, terroir-driven and diverse. It is this evolution that makes Australia such a hugely exciting whisky-making country. Those that follow are the pioneers and the key players in this next step. If I could include more, I would.

TASMANIA

Tasmania's role as the birthplace of the new Australian whisky industry is in some ways a very belated return to its whisky-making origins, which started in the late 18th century before the 1839 Prohibition Distillation Act banned distilling in the colony. Although it was lifted eight years later, legal distilling wouldn't recommence on the island for another 150 years. Now its burgeoning distilling scene offers diversity, range and quality.

HOBART REGION

It was in 1989 that Brian Poke's Darwin distillery (aka Franklin/The Small Concern/Cradle Mountain) started up, marking the beginning of an extraordinary period that has transformed the island. In 1992 it was followed by **Lark.**

Poke's licence covered experimental trials. Bill Lark, however, successfully lobbied to get a change in law permitting the use of small stills. By 2012, Tasmania had eight distilleries. Now there are 80.

The Lark brand's story since those early days has been a tangled one. In 2013, the family sold a controlling stake in the business. The same year Lark bought Overeem (see page 320), before Australian Whisky Holdings (AWH) came in as a major shareholder. Further acquisitions followed: Redlands (now Old Kempton), the Master Cask cooperage and, disastrously, Nant (now Bothwell), which was embroiled in a multimillion dollar fraud case.

AWH became Lark Distilling in 2020 and began to trade as an umbrella brand, producing or bottling/blending whiskies from its other distilleries, including Bill and Lyn Lark's original site now renamed Coal River. In 2021, it bought Shene Estate with plans to build a new facility (Pontville) as the main production centre.

Amidst these corporate shenanigans, Lark remained true to its original ethos. An open-topped, hand-stirred mash tun, small Knapp Lewer stills giving a fat, viscous spirit, given extra heft by ageing in small ex-fortified wine casks from Seppeltsfield.

There's peat as well, cut from the distillery's own bog whose smoke is added to the malt after kilning. Multiple new make styles are made, meaning that blending is also key. Different streams are identified during maturation before the final whisky is married.

LARK

LARK REBELLION (FINISHED IN CASKS SEASONED WITH CHINOTTO CIITRUS) 44%

Perfumed, citric and high-toned. Curaçao, Jaffa cakes and mixed peach. Liqueur-like. Confected but with Aussie warmth.

FRUITY/SOFT

LARK CLASSIC CASK 43%

Creamy. Orange skin/marmalade, then plump black fruits. Slightly oily. Robust yet textured with polished oak in check. Plump and soft.

FRUITY/SOFT

Above *Disgorging at Lark.*

'We have the freedom to operate beyond the confines of traditional whisky-making parameters,' says CEO Satya (Sash) Sharma. 'We see Lark doing to New World whisky what Yamazaki did to Japanese whisky. Australia fosters freedom, experimentation and innovation in whisky making. It's this dynamic approach that will propel Australian whiskies onto the global stage in future years, challenging old-fashioned stereotypes.'

Sullivans Cove took a different tack to Lark. Founded in 1994, it was brought to world attention by Patrick Maguire who went from distiller to owner in 2004. He left in 2016 and is now setting up Maguire & Co. as a distiller and independent bottler. Maguire's successor as distillery manager is Heather Tillott. With a move to a new site planned for 2024, there's no change in terms of production. 'We're upgrading facilities,' she says. 'Rather than just upping capacity, we're increasing capability.' The whole process she outlines makes it hard to simply make more. Low and slow is the watchword. Temperature-capped ferments are seven days, with a further two days of rest. A mix of different yeasts are used.

Above *Increasing capability: Sullivans Cove.*

Above *Do Sullivans Cove's stills add length and elegance to its spirit?*

'Longer ferments have benefits for the spirit. They give a breadth of character to the new make that's oily and billowy off the still.' The still itself, a Cognac-style alembic 'run low and slow', plays a big part in crafting this character. 'The bends and curves and the big onion head means immense reflux, so we get light, floral, honeyed character, but then the worm adds its robustness.' There's also a character she picks out as 'briny', which she didn't find in any other Australian or Scotch malts, 'but when I went to Cognac there it was. It's the still!'

Casks are full-sized, predominantly ex-Bourbon and ex-tawny port, allowing the whisky to mature for an average of 13 years – long by Aussie standards. Though she highlights the importance of the need for greater scale industry-wise, Sullivans Cove is sticking to its principles.

SULLIVANS COVE

★ SULLIVANS COVE TD0046 48.7%

Lightly phenolic but oily/fruity malt with a complex gentle palate: hot chocolate, poached pear. Silky feel. Then spiced apricot. Restrained. Classy.

FRUITY/SOFT

SULLIVANS COVE AMERICAN OAK SINGLE CASK TD0354 47.9%

Freshly planed wood. Lightly bready with some light tropical fruit notes. The feel is soft and slightly vinous. Refined, fruited and decidedly unflashy.

FRUITY/SOFT

'It's hard to scale up without losing what makes us different,' she says. 'We've an oily spirit with a high content of medium- and long-chain fatty acids, which means the chances of haze is high, so we drip dilute to bottling strength over six weeks, then sit it down for a couple of months before bottling and then rest the bottles for two months so that everything is retained. How do you scale that up? There's a limit to how big we can go.'

Heather is one of a number of winemakers who have become distillers. Is there a different mindset at work? 'Probably. When you've spent years nosing and tasting certain things, your framework for understanding and articulating what's on your palate is fixed in a certain way.

'Winemakers look for that peaceful space where everything is held in balance. Winemakers also see oak differently.' The fruits of her explorations into wine casks will appear over the next decade.

'As a winemaker, I'm also a believer in terroir, which is why I'm so anal about retaining those fatty acids and not chill-filtering. Their genesis is in the barley and how that barley is grown, so if you want to talk about terroir, you shouldn't be removing a large volume of compounds that express it.'

That, in turn, has led to partnering with local farmers to bring back older varietals, a long-term project that could offer up even more possibilities for flavour and blending. 'We're weird,' she says, 'and we've doubled down on our weirdness. The key is having a balanced product, but knowing what the extremes are and accentuating those. I think

Above *Overeem – a distillery reborn.*

we're getting weirder, but in the process getting more us. Happy people make wonderful products. The care comes through.'

It was in 2005 that Casey Overeem's dream of owning a distillery came to fruition when he built on his property overlooking the Derwent River in Hobart. Producing by 2007, it was sold to Lark in 2013 before his daughter Jane and her husband Mark Sawford bought it back in 2020.

'Long' is the **Overeem** philosophy in ferments, distillation and maturation. To get around Tasmania's climate fluctuations, the whisky's aged in purpose-built, cool-temperature barrel stores, allowing it to develop gently over time. The casks are heavily toasted and charred ex-sherry, ex-port and ex-Bourbon ranging from 100 to 500 litres (26.4–132 US gallons). The deliberate creation of cooler conditions rather than accepting Tasmania's temperature extremes is a distinct nod towards Scotland. Still, you can't keep Tasmania at bay. 'Tasmania's our essence,' says Jane. 'Rich, oily barley and soft, pure water from the slopes of Mount Wellington gives rich, complex flavours.' Again there is that signature Tasmanian viscosity to the whisky. 'We're incredibly positive about the future of whisky in Australia,' she says. 'There's some real momentum and excitement here at the moment. The rest of the world is starting to look to Australia now as a serious whisky-producing region.'

If Lark is flirting with luxury, you couldn't get more grounded than **Belgrove**, where fifth-generation farmer Peter Bignell has created one of the world's most remarkable distilleries and a model for sustainability. He grows the rye (and other cereals such as oats), the ferments are with wild yeast, the still is homemade and the burners for the flame are fuelled by used cooking oil from a roadhouse across the road – it also powers his vehicles. The spent mash goes for his sheep, rainwater is used for cooling and the cleaned waste water irrigates

OVEREEM

OVEREEM 5 YEAR OLD 50%

Thick, juicy, jammy. Dried fruits, plum skin with a little lanolin and wood spice. Palate mixes Pacharán, then pollen and sticky apricot. Light char with water.

RICH/DEEP

OVEREEM BOURBON CASK MATURED 43%

Warm, fruit-forward and mature (not woody) with an oily/waxy note (polished floor), then peach, citrus and lychee. Nicely balanced.

FRUITY/SOFT

BELGROVE

★ BELGROVE OAT WHISKY BATCH 6 54.4%

Creamy oat milk, strawberry milkshake with perky spices. Lovely interplay between softness and spice, then Starburst sweets, hoof glue and toasted porridge.

FRUITY/SOFT

★ BELGROVE RYE (CASK SAMPLE) 61%

Earthy rye, with light green touches. Warm and lightly nutty and savoury. Slight camphor, then fennel seed, tarragon and then raspberry. The rye adds a balancing burned note.

SPICY/HERBAL

his fields. Unusually, he malts his rye but doesn't then kiln it, believing that it reduces the complexity and fruitiness of the spirit. With trees planted and rye trash adding back carbon, Belgrove is potentially carbon negative. Small-scale but big thinking. Aged in refill and ex-Pinot Noir casks, the whisky is also, rightly, regarded as one of the leading ryes in the world. An outlier, or a leader?

THE NORTHWEST

While the 'small is beautiful' ethos ruled the thinking of the majority of the pioneering distilleries, **Hellyers Road** had an alternative view. It saw, looked and acted big from the outset. Based in Burnie, the distillery started operations in 1999 as an offshoot of dairy cooperative Betta Milk. It's now run independently.

All of the processes take place on site. The stills are huge – the wash can hold 60,000 litres (15,850 US gallons), the spirit a third of that. Originally they were all stainless steel, which had an impact on new make quality, but now have copper heads and lyne arms. They are also run extremely slowly – the spirit cut takes a day, allowing massive reflux.

While others were filling small casks and selling young to generate cash, Hellyers Road was laying down stock, giving it a deeper repository to draw from. On average, the whisky gets between seven and ten years before release and age statements have reached 15 and 21 years. Triple-distilled (The Journeyman) and peated (from Scottish barley) are also made. The long-term vision is now, seemingly, paying off.

OTHER TASMANIAN DISTILLERIES

Kristy Lark-Booth knew the whisky business from the inside out from her time at her family firm. All of that knowledge has since been applied at **Killara** distillery in Richmond, which she founded in 2016 – Australia's first solely female-owned and operated distillery.

Like many, mashing is contracted out – in Killara's case, it comes from Spring Bay's Cambridge site (see below). The wash still is a stocky beast with a short lyne arm, while the spirit still has a flat top and no lyne arm. No surprise that the new make is thick and meaty. Weight is key here, as it was for so many of the early players. The bulk goes into ex-Apera (Australian 'sherry') casks. Inside knowledge you see.

Cam and Suzy Brett started their **Spring Bay** distillery on the east coast in 2015, with whisky distilling beginning two years later. Mashing and fermenting is now in a second distillery in Cambridge, near Hobart. All of the new make is aged at Spring Bay. The whisky is reduced with rainwater, which, they claim, adds a little salinity. There's also a solera with mature whisky going first into ex-Chardonnay, then ex-tawny port and then ex-Muscat fixed casks, none of which are ever fully emptied.

Flinders Island is part of the Furneaux chain sitting off Tasmania's northeast coast. For **Furneaux** distillery's founder Damien Newton-Brown it's as close as he can get to Islay in the southern hemisphere. There are even peat bogs made up of compressed swamp marsh herbs, tussocks, coastal heath, tea tree and other myrtles. In Tasmanian fashion, smoke is added after kilning. It opened in 2019 using sourced whisky from Launceston, with 100% peated Flinders Island. For Smoky Wedding, a 50:50 blend of Scottish peated barley and Flinders peated is used. The aim is to use Flinders Island barley.

Bill **McHenry** started Australia's most southerly distillery in 2011. Here, in rainy humid conditions, strength drops and maturation slows. While this placed him at a short-term commercial disadvantage, longer ageing is reflected in a gentler, less oak-driven style and a balanced 10-year-old release in 2023.

Above *Hellyers Road started as an offshoot of a dairy co-op.*

VICTORIA

Victoria was Australia's major whisky producer thanks to the growth of domestically produced blended whisky from the 1930s on. The idea that Australian whisky is something new is simply untrue. The state's large distilleries were eventually owned by major international conglomerates, which closed them down during the downturn of the 1980s. What was big scale and blended is now balanced between larger and smaller producers with a huge diversity of approaches.

Above *Victoria – open spaces and open thinking.*

MELBOURNE

The first of Australia's new wave to have a coherent idea of the necessity of Australia to look at export – and therefore to have the requisite scale – was **Starward** in Melbourne, started in 2009 by Chris Middleton and Lark (see page 318) alumnus David Vitale. 'I'd discovered that single malt Scotch whiskies talked to the place they were made', says David. 'It inspired me to think what would a distinctly Australian whisky taste like if it were anchored in the place it was made? The key was to manufacture a good-quality whisky at a competitive price to imported Scottish malt. We wanted to make whatever we did have depth and complexity without intensity.' True to his word, in 2013 an A$80 whisky appeared.

STARWARD

★ STARWARD LEFT-FIELD 40%

Ripe with buzzing spices, citrus, caramelized fruits. Cherry-accented vinosity on the palate, some barley sugar sparkle, then banana and coconut finishing with star anise.

FRUITY/SOFT

STARWARD TWO-FOLD DOUBLE GRAIN 40%

Citric, waxy and medium-weight. Stewing fruits with light and slightly dry malt adding weight. Plate has red fruit freshness. Intended for mixing. A significant whisky.

FRUITY/SOFT

STARWARD NOVA 41%

Toffee, peach and fresh orange lead, then cacao, cherry, pomegranate. Thick and generous palate that ends with tropical fruits and banana jam.

RICH/DEEP

Like all distilleries there's been some tweaks since those early days, most significantly a move away from the Aussie whisky standard of using old Apera casks to ex-red wine casks. 'Almost every Scottish distillery was first-filling with ex-Bourbon, making this flavour too generic,' says co-founder Chris Middleton. 'Ageing in red wine barrels gives us a flavour and texture that is distinctive, and pretty delicious. From a process perspective, our climate and importantly the availability of wine barrels to age our whisky is pivotal to what makes Starward Starward. We age at a lower ABV. More of the bottle comes from the barrel compared to other 40% ABV whiskies.

'My home town of Melbourne with its vibrant and diverse food scene inspired me to be equally as bold with thinking about whisky as a drink.'

That boldness has been shown in the multiplicity of casks (from fortified wine to mesquite and ginger beer) and the introduction of a column-distilled wheat whisky and a single malt double distilled in pots fermented with Two-Fold, a blend of wheated whisky from Manildra (see page 316) and Starward's single malt aimed at the mainstream and mixing market. Mainstream might be anathema to some, but it's how most whisky is sold. 'My view was that if we wanted to be relevant to drinkers, we really had to find a way of challenging the status quo on what whisky was.'

While David takes cues from Melbourne, Chris is unconcerned with the notion of place. 'It's not a "selling point" for a modern whisky brand; it is a "telling point" for labels and tastings. Unfamiliar places of origin, such as Australia, raise the barrier of risk for purchase. Our name and brand must have appeal as modern, flavoursome whisky that transcends traditional labels and often parochial messaging. The hugely successful labels sell their psychological brand advantages.'

Following an injection of capital from Diageo's incubator arm Distill Ventures, Starward is now globally distributed, the first Australian whisky most people will try. There's a big responsibility to get things right.

'It's really exciting to see producers interpreting what Australian whisky means to them', says David. 'It's usually anchored in three key areas – local grains, local barrels and importantly the wide and varied climate that our continent experiences. The best years of Australian whisky are ahead of us.'

Is it wrong to call **Bakery Hill** a forgotten pioneer distillery – out there at the start, quietly working away, then overlooked when the second wave came with greater public attention? What's certainly true is that David Baker and his son Andrew have been making excellent whiskies since 1999 when their distillery became the first of the new wave on the Australian mainland.

'I set up Bakery Hill to prove that we could make world-class single malt whiskies to match anything coming out of Scotland,' says David. 'For generations, we were told the only place you could make whisky was Scotland because of the "special air" and the "special water". As a scientist I knew this was just rubbish. Here in Australia, we make excellent beers and wines, so why not excellent single malt whiskies?' Then he says something that echoes with all of the world's best distillers. 'It ain't the air, and it ain't the water – it's the process. I just needed to adapt to the local conditions and work with them rather than against them.'

That in turn meant getting to grip with the vagaries of the Melbourne climate, which can be freezing cold and then baking hot on the same day,

Above *Bakery Hill: the quiet pioneers.*

leading to the whisky maturing at twice the rate it does in Scotland (and with twice as much angel's share).

'The location of the distillery and location of barrel storage is also critical,' he adds. 'If the ambient temperature is too cold, the whisky doesn't mature or matures at a very slow rate. If the temperatures are too high, then curtain chemical reactions take over and the whisky is unbalanced.'

Though made to a 'Scottish' single malt template – 'why change a good thing if it works so well?' – he and Andrew use unpeated and peated malt, ageing for an average of eight years mainly in American oak. 'We want them to be sweet, three-dimensional, rich and have a beautiful aroma.'

BAKERY HILL

BAKERY HILL CASK STRENGTH 60%

Classically framed. Light coconut oak with some espresso, light smoke, toasted malt, green apple and golden syrup. Restrained.

FRUITY/SOFT

★ BAKERY HILL DEATH OR GLORY 15 YEAR OLD 50%

More fruit-driven with a hint of oxidation. Warm bread, Nutella and tropical fruits (mango and papaya) at its heart. Flirts with old Bowmore territory.

FRUITY/SOFT

Above *Chief's Son is bringing whisky to the Mornington Peninsula.*

Now in a new, larger site in central Melbourne things seem set fair. 'Because we grow our own grain and the climatic conditions are perfect for maturing our spirit, Australia is the perfect place to produce single malts,' he concludes.

'After two decades of consistent growth, the industry in Australia is really taking shape. Distilleries are finding their feet in all categories and price points, and consumers have a huge range of whiskies readily available to try and even visit.

'As distilleries produce more whisky, many distilleries including ours would like to enter or re-enter the export market so that Australian whisky can be more easily accessible to those off our shores.'

A perfect example of the widening of the Australian whisky flavour map comes in the shape of Melbourne's **The Gospel**. It's one thing to decide, as founders Andrew Fitzgerald and Ben Bowles did, to quit their jobs in engineering and start a distillery, but quite another to then decide that they're only going to make 100% rye whiskey.

'We started distilling with a process that wasn't right for rye,' Andrew admits. 'Now at the new site in Coburg, there's a hammer mill (less heat transfer to grain), mash vessel (direct steam injection, high speed agitator) and a 6-metre [20-foot] column still (no scorching). We learned the hard way.'

The unmalted rye comes from a single farm in the Murray Mallee region. 'It's a dry area,' he says, 'and

THE GOSPEL

★ THE GOSPEL PROJECTS LEGACY RYE 56%

Huge coffee notes. Dry and earthy with masses of allspice, tamarind and Jamaican five-spice powder. Fruity ripeness appears when water is added. Elegant.

SPICY/HERBAL

★ THE GOSPEL STRAIGHT RYE 45%

This has sweetness, blackcurrant mixed with rye accents, some dill in the back. Complex and layered with a cooling effect. Silky fruit and spice in harmony.

SPICY/HERBAL

CHIEF'S SON

CHIEF'S SON 900 STANDARD SINGLE CASK 49.5%

Big-boned and slightly oily. Overripe fruits, Apera softness, then boiled sweet. The palate shows heavy florals, vanilla and rosehip syrup. Silky texture.

RICH/DEEP

CHIEF'S SON PEATED MALT EXPERIENCE 66.32%

Dry and smoky with some eucalyptus and embrocation. Deeply darkly fruited. Water softens and smooths. Nicely textured.

SMOKY/PEATY

MORRIS

MORRIS SIGNATURE 40%

Ripe soft fruits with a hint of oxidation. Dessert-like. Maple syrup and popcorn, dark marmalade. The texture is clinging with bruised dark fruits, light tobacco, then apricot stickiness.

RICH/DEEP

★ MORRIS MUSCAT BARREL 46%

Deep. Manuka honey, sweet liquorice, treacle/molasses and Muscat(!) Medjool dates, leather and pipe tobacco on the palate. Almost smoky. Balances sweet and bitter well.

RICH/DEEP

VOYAGER MALT

Australia is the second-largest exporter of barley in the world, one reason why the commodity market is dominant, something that can work against the demands of distillers who are interested in flavour, different grains and heritage varieties – and in smaller batches.

In 2014, fourth-generation Riverina farmers Stu Whitecross and Brad Woolner spotted the gap in the market and started Voyager Craft Malt. Knowing that craft malting has to work for farmers first, they act as an intermediary, encouraging distillers to meet the people growing their grain. With all of the players understanding the needs of their colleagues came a deeper understanding of possibilities that can translate into flavour.

Flexibility in type of grain, batch size and specialized kilning has been built in from the start. Different/heritage varieties, speciality malts and smokes from local hardwoods and walnut shells.

In 2020, the firm opened Whitton Malt House, not only with increased overall capacity but as a tourist destination complete with accommodation, an adventure playground, bars focusing on beers and whiskies using its malts. A maltings has been transformed from an industrial unit into a hub for a region.

by default produces a distinctly small and dry grain. Sustainability was one of the contributing factors. In this country, rye makes a lot of sense. It's grown where other crops fail and can reintroduce nutrients into soils that have been over-farmed.'

Fermented on grain, it then runs through the (self-made) column and is finished in a pot still, coming off at 78% ABV before being barrelled at 58% ABV into heavily toasted and lightly charred new American oak casks. Melbourne sits between polar weather and desert heat, and these extreme temperature swings mean control of oak extract is vital. It's ready to bottle at three years.

It's clearly rye, but there's a lusciousness to The Gospel's take, the oak controlled, a syrupy sweetness yet with rye's green twang. A revelation, and another variation on the grain's personality.

'My hope is for a more defined "Australian" character,' says Andrew. 'Too often people seek to imitate what is already on the world stage. I'm hoping that, over time, confidence will push more distillers to do what is truly irreplaceable. We choose our grain source and curate the style of whiskey for that reason. I want it to be a reflection of the place we come from. Yes, we're a rye whiskey, but I argue we taste like no others out there.'

Whisky distilleries are steadily moving into what were considered wine regions. One such is the Mornington Peninsula, home since 2016 to Stuart and Naomi McIntosh's **Chief's Son**. The Gaelic *Mhic an Tòisich* means 'Son of the Chief', so there's no surprise that their approach is Scottish – single malt, double distilled in pots. 'However,' adds Stuart, 'we modernize and push the boundaries.' In simple terms, what Chief's Son does is look at Scotch's efficiency-driven approach and take the opposite path (to be fair, there are plenty of newer Scottish distillers doing the same). Malt, made to the distillery's spec, comes from Voyager Malts (see box, left), with Scottish peated for some expressions, and is fermented at the Mornington Peninsula Brewery.

'We're different to many distilleries in that we create four different single malt recipes. It's logistically hard, but allows us to appeal to a much broader range of whisky drinkers.' In practical terms, that means using speciality malts and varying percentages of peated malt, as well as different approaches to distillation. The Pure Malt, for example, uses dark roasted malt instead of a peated percentage and doesn't include foreshots and feints from the previous run in the second distillation.

Maturation is mostly in French oak Apera barrels, but its Cask Expression series adds in casks sourced from local breweries, cideries and wineries. 'Getting the whisky into the barrels is a very specific science, but a science without constants,' says Stuart. 'The whisky coming out of the barrels, on the other hand, emerges with its own personality and character.' This is where craft lies…the midpoint between art and science.

RUTHERGLEN AREA

Was it a surprise when, in 2016, the **Morris** winery revealed that it was making whisky? Revered as a world leader in sweet fortified wines – Muscat to be precise – this Rutherglen winery has been operational since 1859. Making fortified wine necessitates the addition of spirit, in this

OTHER VICTORIA DISTILLERIES

Owned by Top Shelf International, **NED Whisky** started in 2015 as a ready-to-drink (RTD) whisky and cola brand. Bottled product appeared four years later complete with a new distillery in Campbellfield, north Melbourne. The whisky is a blend of sourced product from Robinvale's Ostra Distillers and the Campbellfield facility, made to a mash bill of 71% corn, 17% wheat and 12% barley run through a stripping column and doubler. It's aged in American oak with a high percentage of new wood.

Deep in the bush to the northeast of Melbourne is **Kinglake Distillery** (see left), owned and run by Sam Lowe and Chantal Daniels since 2018. Located in an old gold mining and lumber area, it's off-grid, using a spring-fed dam for all the water, and is powered by solar. One-third of the property has been given to a reforestation charity to regenerate with indigenous shrubs and plants.

A complex mash bill of 46% Atlas, 25% Vienna and 4% chocolate malt (from Voyager – see box) is supplemented with peated malt from Simpsons in the UK.

All production is in-house with wild yeasts encouraged before double distillation in a 2,500-litre (660.4-US-gallon) Knapp Lewer pot still. A wide cut is taken to capture a heavy, oily spirit, which is allowed to shine in refill and small casks. There are also some red gum casks being experimented with. The angels drink up 6–10% a year and strength drops over time.

Above *Backwoods – Australian thinking for Australian whisky.*

case to stop fermentation before all the sugar has been converted. The likelihood is that Morris had always distilled. Certainly in the 1940s, the spirit was produced in-house from a hybrid copper pot still, which was then abandoned when spirit was bought in.

It was Morris's new owner Casella Family Brands (Yellow Tail wine, Redlands/Old Kempton whisky) who refurbished it and pressed it into service for whisky. A second, restored hybrid has since been added. With consulting help from John McDougall (ex-Laphroaig – see page 100) and Dr Jim Swan, distilling is in the hands of industry veteran Darren Peck.

Rutherglen is a rich cereal-growing region and the wash is sourced from the (Casella-owned) Yenda brewery. Jim Swan's hand is visible in the use of STR-ed Shiraz and Cabernet Sauvignon casks, with finishing taking place in ex-Muscat. The result is a more distillate-driven spirit than an oak/wine bomb, a case of old knowledge making a new whisky.

Leigh Attwood is one of the second wave of distillers who are looking within the country for their inspiration. For his **Backwoods Distilling** he turned to Belgrove. It's a hugely significant step change in thinking. The distillery opened in Yackandandah, a historic gold rush town about 30 miles (50km) from Rutherglen, northeast Victoria that coincidentally was once known as Whisky Flat. There's grain from nearby Riverina region and wineries for casks.

'Our goal is to create whiskies that capture the essence of our surroundings,' Leigh says. 'We source all of our ingredients within a three-hour radius of our distillery and primarily use casks that have previously held Australian products. This process results in a uniquely Australian taste that is deeply rooted in our locale.'

This also means using heritage grain varieties. Three barleys are used for the single malt, as well as different roasts. The rye was a forgotten low-yielding strain that was rediscovered growing wild in an unused paddock. 'It's half the size of a New World variety, but its distinct flavour has been intensified. This grain has evolved in local conditions and has been shaped by years of neglect. It's unique to our region.'

Smoking is over red gum hardwood – again tying it to location. The mash is soured in open-topped fermenters for a week; the stills are run slow and cut deep. Oils and fullness are key here. Casks used are solely from Australia, including Muscat and Shiraz casks, with the rye in refill Australian casks.

'We aim to honour tradition while also introducing fresh elements to the mix,' Leigh says. 'We hold the belief that the grain plays a crucial role in determining the final flavour. We put a lot of thought into designing our mash bills, selecting some malts that have been kilned for longer and some that have been roasted. Our goal is to find the perfect balance of flavours right from the start, rather than relying solely on the barrel to do the work.'

BACKWOODS

★ BACKWOODS RYE 46%/55%

Dense, wild berry fruits, a little suet and blackberry, dark plum and cherry stone/hawthorn jelly. More rye bread on the palate. All-enveloping. Bittersweet, hinting at meatiness.

SPICY/HERBAL

★ BACKWOODS SINGLE MALT WHITE OAK 46%

Crystallized fruits. Peachy with verbena and lemon balm in the background. Palate has cooked apple, mint, gummy bears. Broad, long and balanced with a mentholic finish.

FRUITY/SOFT

NEW SOUTH WALES

While whiskies were made (and consumed in large quantities) in New South Wales from the late 18th to the mid-19th century, it too had to wait until the last whisky boom in the 1990s for the practice to restart. Here, distillers are utilizing close links with the wine industry as well as embracing the climate. There is no regional style: a distillery's place on the map is one thing, its flavours are quite another.

SYDNEY

Having scale gives you the ability to reach more people, have whiskies at a mid-priced bracket and help build the category in export. The downside is that larger operations are, wrongly, not considered 'authentic'. The reality is that you need both. And anyway who says that big can't also be creative?

Archie Rose's distillery in the Sydney suburb of Banksmeadow is both built on a grand scale (it can make 2 million litres/528,340 US gallons per annum) and home to some of the country's most creative and locally focused thinking. 'First witnessing harvest here was a transformative experience,' says master distiller Dave Withers. 'I was standing in a field in 40°C [104°F] heat with red ochre earth at my feet, the big deep blue sky above, gum trees lining the paddock. How could grain grown under these conditions taste the same as that from overseas?'

What's followed was an exploration, in tandem with Voyager Malt (see box, page 325), into the possibilities of heirloom varieties, 'lost' grains, old ryes, malting techniques and a pivot away from wood-led whisky to one where the flavours of grain and ferment were key.

'It's easy to merely look at grain as a sugar source, a vehicle to alcohol. I've always seen it as the most fundamentally important ingredient,' he says. 'What you get from the grain determines how it should be mashed, the flavours of fermentation and the cut points in the still. By allowing the grain to call the shots, we've unlocked the nature of the grain.'

This has led to a distillation process where the six malted barleys and three malted ryes are mashed, fermented, distilled and aged separately, then blended. Five yeasts are used, and each spirit is barrelled at different strengths into the cask type best suited. It gives the blenders 15 whiskies to play with.

Above *Going boldly where none have gone before – Archie Rose.*

Above *Organic grains are key to Corowa's approach.*

Above *Vaporous thoughts.*

'That question "why?" is our constant companion,' Dave continues. 'If the answer is merely tradition, then the very next question should be "but is there a better way?" Australian whisky has a discontinuous history. This gives us a freedom. Often we've found ourselves reintroducing traditions long lost to find a better way of making spirit.'

So, a distillery with volume, but also flex. That seems very 21st century, as does the love of blending. Scale allows small-batch releases, but also the Double Malt blend of pale and amber malt and rye malt whiskies aimed at the mainstream. Scale but with quality.

The **Manly Spirits Co.** in the beachside suburb of Sydney fired up its stills in 2017. Stills plural, as from the outset founders David Whittaker and Vanessa Wilton wanted to create a diverse portfolio: gin, vodka, whisky and liqueurs.

ARCHIE ROSE

★ ARCHIE ROSE STRINGYBARK SMOKED BATCH 1 54%

Aromatic smoke: eucalyptus, balsam, leather, bush fire. Feral, yet strangely calm. Palate is oily/meaty with sour fruit, and sesame at the back.

SMOKY/PEATY

★ ARCHIE ROSE RYE MALT BATCH 12 46%

Big cherry and bold oak. Sleek with roasted rye, garam masala. Sophisticated, balanced and deep. Ever so slightly haughty.

SPICY/HERBAL

All production is in-house and each spirit has its own dedicated still – there are two tall pots for the whisky, with a boil bulb on the spirit adding reflux. The wash, made from an 'ale malt', has a five-day ferment. Everything is pointing towards lighter and fruity, which is now carried forward to maturation with a move away from 100-litre (26-US-gallon) casks to 200-litre (53-US-gallon) ex-Bourbon and 300-litre (79-US-gallon) Apera and tawny port casks. There are even trials with seaweed and wild mushroom-smoked casks. It's another distillery looking at export and also that important mid-priced market with its Coastal Stone Xplore brand, a 60:40 blend of wheat whisky from Manildra (see page 316) and its own single malt.

VICTORIA BORDER

Corowa is demonstrative of the point that, while dividing Australian whisky by state makes some sense, it has to be understood that not all whiskies within any given state taste the same. That's down to the thinking of the distiller and the distillery's particular location. So Corowa is closer in flavour to Morris and Backwoods in Victoria than it is to Archie Rose 370 miles (595km) away.

It was started by Dean Druce 'because we're about using what we have on our farm for a better use. The fact that we love whisky helps.' There is an element of taking back control as well. From the 1960s to the 1980s, a lot of Australian barley ended up in Scotch or Japanese whisky. Housed in an old flour mill, it started operation in 2016 with stills to his design made at Burns Welding, in Griffith.

'We've set up a Scotch-style Australian distillery that inboxes the Kilchoman way of making whisky [he did part of his training at the Islay distillery – see page 105], while giving the spirit a fruitier, sweeter new make character.' The key is the grain. His father Alan is a pioneer of organic production. Schooner barley is central to the sweet fruit, but there is also (organic) oats, spelt and wheat.

'We're only just scratching the surface of what we can do and will continue to see even more growth when the public realize that they can drink a more flavoursome whisky that is made in their backyard. We don't hide away from the fact that we are wanting big bold characters in our whisky. Just like the people who make it.'

CENTRAL NEW SOUTH WALES

Black Gate Distillery in Mendooran (population 559), 230 miles (370km) northwest of Sydney, was the brainchild of Brian and Genise Hollingworth. Though minuscule – they only make 1,000–3,000 litres (264–792.5 US gallons) a year – it was a name regularly dropped by other distillers. This is a dry part of the state, with highs of 40°C (104°F) plus during the summer months and -5°C (23°F) in winter. 'Not your typically ideal whisky-producing conditions,' says Brian. 'They do dictate aggressive maturation with rather heavy oak extraction that's responsible for the "big" flavour our spirits have become known for.' With 15% a year flying heavenwards, there's no surprise that he's using larger casks and now feeding in second and third fills. A direct-fired still helps to give a weighty base. He imports peated malt because he wanted grunt and the 'Lark method' (see page 318) was too light.

'Our whisky production has always been based on the Scottish model,' he says. 'I love the history and tradition of Scottish whisky, and am not sure I could come up with any innovation that would genuinely improve it. The old "If it ain't broke, don't fix it" adage is relevant in this case.' And yet, for all that (genuine) modesty, Black Gate's whiskies couldn't be anything other than Australian.

Small-scale, local, quality-focused, there are links here to the earliest distillers anywhere in the world, making for their community.

COROWA

COROWA 4 YEAR OLD BATCH 2 46.8%

A balance between richness and warmth. Some gentle mint, milk chocolate. Palate has roasted malt, cinnamon, then wellies. Nicely poised.

FRUITY/SOFT

BLACK GATE

★ BLACK GATE 4 YEAR OLD 2ND-FILL APERA CASK 61.64%

A lick of paint, bread and butter with softness and weight from the cask. There's some orchard fruits, light malt and saison-like lift and energy. Long and elegant.

RICH/DEEP

BLACK GATE BOURBON CASK VATTING 3 YEAR OLD 47.8%

Juicy, canned fruit. Overripe melon, dying embers, sweet malt. Some ripe pear and soft fruits, smoked cream and a clinging quality. Graceful.

FRUITY/SOFT

Above *Black Gate – tiny but with a mighty reputation.*

Above & below *The rainforest distillery of Cape Byron benefited from advice from Jim McEwan.*

NORTHEAST COAST

It was in 1988 that Pamela and Martin Brook bought a run-down dairy farm close to Byron Bay to start cultivating macadamia nuts. Since then they have planted 35,000 trees, reforesting what was once the largest subtropical rainforest in Australia. It was their son, Eddie who saw an opportunity to diversify the business. He contacted legendary distiller Jim McEwan (Bowmore and Bruichladdich, see page 103) to see if he might care to consult. The result was Brookie's gin, using botanicals from the forest. In 2019, whisky distilling commenced at the **Cape Byron Distillery** using wort from the neighbouring Stone & Wood Brewery. Knapp Lewer stills are used for distilling and the bulk of the new make goes into ex-Bourbon casks. Some (Simpsons, UK) peated malt casks have been laid down, but experiments are underway with Voyager Malt (see box, page 325) to create macadamia wood-smoked malt.

CAPE BYRON

★ CAPE BYRON CHARDONNAY CASK 48%

Bold and slightly leesy. Macadamia nuts, butter churn, overripe yellow fruits. Dries to zingy pepper, sake lees and grilled fruits.

FRUITY/SOFT

CAPE BYRON THE ORIGINAL BATCH 3 47%

Lean. Dry. Green malt and light oak, then fresh washed herbs. Has typical Aussie sweetness alongside roasted spices and a fresh crisp apple finish.

FRUITY/SOFT

SOUTH AUSTRALIA

Adelaide had a distillery in the 1840s, and even though most distilled spirit was either brandy or used for fortifying local wines, whisky was made in South Australia up until the 1960s. Even famous wine producers like Penfolds and Yalumba had whisky brands. Now Apera (aka 'Australian sherry') casks are widely used across the country's new whisky industry – a nice piece of circularity.

Angela and Gareth Andrews spent seven years at their Steam Exchange brewery in Goolwa before making the move into whisky. 'It felt like a natural progression,' says Gareth. 'The knowledge from brewing traditional-style ales provided a solid foundation.'

The creation of their **Fleurieu** single malt has been the result of a rigorous, date-driven approach. 'Everything we do is recorded for future reference,' says Angela, 'There are no mass changes or ad hoc decisions. A new yeast trial, for example, takes years before results can truly be assessed.'

Fleurieu is made from South Australian malted barley (with some imported peated malt) entirely at the distillery. The couple's brewing background rings strongly with the use of different mash bills and yeast strains that are then matched to specific casks, or perhaps it's the other way around.

'If we have a first-fill fortified barrel, we'll use our base malt bill and a combination of brewer's ale yeast and distiller's yeast,' Gareth explains. 'Refills are less active, so adjustments are made to our brewhouse process to allow different aspects of the spirit to shine through. Blending is important to create rich, balanced whisky. All of our single malts are a blend of different casks, sizes and ages depending on what it is we are trying to achieve.' They have also branched into collaborative bottlings with Black Gate, Backwoods and Timboon.

'Whisky is simultaneously very old and very new,' he adds. 'Brewing and distilling are rich in history and tradition, and to be able to be part of that in a modern age is special.'

Tin Shed's Ian Schmidt comes from a line of distillers, not always legal ones. 'Both my father and grandfather had played with a small still making tiny batches of Calvados and brandy, so distilling

FLEURIEU

FLEURIEU NEVER A DULL MOMENT 47.6%

Fruit skin and Muscat-like perfume. Hint of toffee with balancing mineral firmness. Palate has scented cherries and becoming floral with water.

FRUITY/SOFT

FLEURIEU I AM THE WALRUS 47.6%

Yellow matter custard? Mineralic, hard pear with low smoke. A fresh and crisp attack with a thick birchwater-like palate. Smoke oak and spirit all well integrated.

SMOKY/PEATY

Above *Whisky and water – Fleurieu.*

Above *Tin Shed – a well-named distillery.*

OTHER SOUTH AUSTRALIAN DISTILLERIES

David Pearse and Steven Griguol set up their **5Nines Distilling** in the former's garage in 2016. The Adelaide Hills distillery uses a single pot that they designed and built. Malt is coming from Yorke Premium and sourced from a single farm that's working with the University of Adelaide's seed bank to rediscover heritage grains. It first appeared in 2020.

Founded in 1911, **St Agnes** in Renmark is Australia's oldest family-owned, continuously operating distillery. Famous for brandy, since 2015 it has also made whisky. The wort comes from the **Coopers Brewery** in Adelaide, which is building a A$50 million whisky distillery, visitor centre and microbrewery at its Regency Park. The aim is to produce 200,000 bottles per year.

Renmark is also home to **23rd Street Distillery**. A distillery on the site of the old Château Tanunda brandy distillery has been built by Vok Beverages/Bickford's (which also owns the country's oldest operational distillery Beenleigh rum) using 1950s pot stills and brandy stills.

The wine connection continues with **Smith's Angaston**, owned by Hill-Smith Family Estates, which made Smith's Imperial Vat Whisky in a 1930s brandy still from the late 1950s until the 1970s. It was briefly revived under production director the late Peter Wall in 1998 and 2000. More runs took place in 2011, 2014, 2017 and 2021. Might it be a permanent addition? We can but hope.

may have come to me through osmosis,' he says. In 2013, he and business partner Vic Orlow began distilling on that old still on the mezzanine floor of Ian's flagpole factory, but the distillery has now relocated to Nairne in the Adelaide Hills.

All is done on site, distilled in double pot stills and smoked, when required, with local peat or hardwoods. The beefiness of the spirit is accentuated by the temperature extremes. 'It can be wet and cold, then as dry as a dead dingo's donga,' as he puts it. 'These extremes make for rapid extraction of colour and flavour as well as oxidation, but it takes effort to ensure the whisky doesn't get cooked.

'Most of our whiskies have been the in-your-face types that Australia is famous for, but we've included some more tempered expressions. They're still bold in comparison to many of the basic beige Scottish drams, but pretty tame when compared to what is often on offer here in Australia.'

Sustainability is also key. 'We've a moral obligation not to deplete our resources. Also, sustainability is an economic imperative. Not only is it cheaper to run a sustainable distillery, but the world is starting to demand it. Ten years from now some sustainability credentials will be required to sell all products, not just whisky.' The new site is solar powered, the still electric.

'We set out to make an Australian version of Scottish single malt, just better, with an Australian twist,' he says. 'To be honest, it's the most fun you can have with your pants on and your feet dry!'

The search for the 'Australian' in Australian whisky has been pushed the furthest by **78° Distillery** in Hay Valley. The main line is a single malt made from a pot-still-esque mash bill of 80:20 unmalted to malted barley, with the added twist of speciality malts (toffee and biscuit to up sweetness; dark malts for their chocolate notes). While that might not be orthodox, it was another avenue of exploration that made the whisky world take notice.

'We've always tried to create something that speaks of where we are,' says ambassador Stuart 'Chewy' Morrow. 'Which is where the Native Grains project started, using cereals that were endemic to Australia such as wattleseed and weeping grass.' Every craft needs its avant-garde. Boundaries have to be tested, orthodoxies challenged. The fringes are where deeper ideas are born. The question 78° asked was what would Australian whisky be if barley had never been planted?

Roasted wattleseed (technically a pseudo-cereal and under Australian rules not a whisky) was in a mixed mash with malted barley. Weeping grass was then tried (also with malted barley), giving soft fruits and a dusty spiciness. Could in time there be a 100% native grain whisky? That's up to new owner Mighty Craft, the 'accelerator' firm that owns small breweries and distilleries across the country, and its head of production, Islay-born distiller George Campbell. Fingers are crossed.

TIN SHED

INIQUITY FLUSTERCLUCK 49.1%

Some mossiness/green silage, then comes treacle toffee, wild fire, gum, scented grasses and wood ash. The sweetness balances the smoke. Has some grunt.

SMOKY/PEATY

INIQUITY GOLD BATCH NO. 007 56%

Pepper and mulberry. Fruits in syrup, bruised peach and pie filling (apple crumble and demerara sugar) before brightening into fruit tea and Amalfi lemon.

FRUITY/SOFT

NATIVE GRAIN

★ 78° NATIVE GRAIN 46.2%

Roasted cacao bean. Rich with rye-like spices, alongside an enveloping wintergreen warmth. Spices flare in the palate along with green lentil, tonka bean. An important whisky.

SPICY/HERBAL

78° MUSCAT FINISH 44%

Some nutty dryness along with dried apple, peach kernel, bletted fruits, butternut squash and dry grass. Palate mixes coffee with condensed milk and a herbal zippiness.

RICH/DEEP

WESTERN AUSTRALIA

Western Australia was a significant consumer of whisky at the start of the 20th century, much to the fervent disapproval of the state government. Attempts to control what was claimed to be rampant over-consumption simply saw illicit production prosper – people will always find a way. Today, however, legal distilling is back underway with whisky making playing a central role.

Owner of **Limeburners**, Cameron Syme was destined to be a distiller. He was brought up on a farm where his dad grew malting barley and wheat. There was also, apparently, an ancestor who moonshined in Scotland. Limeburners started in Albany in 2007 making single malt, and is now one of three distilleries in Cameron's Great Southern Distilling Company portfolio: one making gin in Margaret River, the flagship distillery in Albany and the home of Tiger Snake and Dugite in Porongurup, 37 miles (60km) north of Albany. 'Location is key,' Cameron says. 'A distillery represents the terroir of the place that goes into the bottle. This extends from the sources of the raw materials, the grain and water and the barrels, to the location where fermentation, distilling and maturation occurs.'

At Limeburners, that involves using local grains, distiller's yeast, week-long ferments and distilling in Knapp Lewer wash and a locally made all-copper spirit still. The heads are diverted as a renewable heating source. In an industry that's new and a climate favouring short maturation, he's planning 12- and 18-year-old releases. Tiger Snake has a pot and a column and produces single malt, mixed grain corn whisky and rum.

'There are some big players and bigger volumes are coming onto the Australian and world markets,' he says. 'Many more people will be exposed to Australian whisky, which increases the knowledge and willingness of people to hunt out small-batch whiskies like mine.'

Above *Limeburners is helping to revive whisky-making in Western Australia.*

Born in Western Australia's Wheatbelt, Jimmy McKeown opened **Whipper Snapper** with brother-in-law Alasdair Malloch in 2014. 'No one else was making whiskey [they use the "e"] in Perth,' he says. 'In fact, there wasn't a whiskey distillery in Western Australia using solely Western Australian grains. Using local grains was something I was passionate about.'

Based on a recipe featuring locally sourced grains created by an American airman after the war, then refined by ex-Springbank (see page 83) distiller Frank McHardy, core brand Upshot uses 80% corn, 10% malted barley and 10% wheat, but the investigations into local grains has resulted in a fascinating range: a wheat whiskey, a rye (85% rye grain), single malt and a Jimmy Red corn. All the farmers are named.

The process is slightly different to most, with the use of mash kettles instead of mash tuns, fermenting in open-topped washbacks on-grain and using a warm, short fermentation to get fruity characters and a predominant grain profile. Distillation is in a four-plate hybrid still.

Releases are all from single virgin American oak (bar the single malt that's in refill), with 'no blending, chill-filtering or any spirit manipulation'. There's also considerable variety – up to 20 different styles. 'We approach distilling with eyes wide open and we're not afraid to learn,' says Jimmy. His approach means that he is also teaching.

LIMEBURNERS

★ **LIMEBURNERS PORT CASK STRENGTH** 61%

Light nose prickle, some butteriness, then Poire William and juicy plump fruits. Palate has light blackcurrant jam, chocolate and yellow fruits.

FRUITY/SOFT

NEW ZEALAND

Location can do much to dictate a mindset. If you are so distant from suppliers and markets, even with the benefit of modern communications, you are left with no choice other than to look locally for resources to draw upon. This can result either in a stolid conservatism, or it can expand the thinking and inspire creative practical solutions. The latter is key to the emerging New Zealand style – gently textured, already complex whiskies.

Imports of brandy and rum were the country's first favoured tipple, and then, as farming communities began to be established, whisky followed – perhaps due to the high number of Scottish immigrants. It was only in 1868 with the passing of the Distillation Act that legal distillation began, though it was to be short-lived. A series of steep hikes in taxation and a patchwork of 'dry' areas put paid to the legal producers and distilling went underground once more. Though one whisky distillery re-emerged (Willowbank – see page 341), it fell foul of global machinations and closed in 1997. What has happened recently is therefore a new start.

The enthusiasm that helps to drive the creation of a new whisky-making country also needs to be tempered to some extent by a legislative framework, not to restrict creativity but to reassure consumers and prevent less scrupulous individuals taking advantage of a lawless state of affairs. Whisky might like to think that it's the new Wild West, but it soon becomes obvious that every town needs a sheriff. Encouragingly, New Zealand's whisky producers were quick to create a set of guidelines for its production. In 2021, members of Distilled Spirits Aotearoa set out a set of guidelines and definitions, which is now legally binding.

If the legal side is now controlled, there remain hurdles for this new industry. Winning over an often-sceptical home market is one, but with a population the same as Scotland, export is vital. That puts extra pressure on the producers, as it means that stock needs to be laid down now in anticipation of that uptick in sales in five to seven years' time. The last thing any distiller wants is demand rising and only having half-empty warehouses and rationing supply.

That first taste of a 'new' country's whisky is often taken as being representative of every whisky from that place. Quality has to be high, character has to be different and not simply 'nice' because nice doesn't sell long-term. Any new whisky needs story, truth, transparency and a quality that makes that first sip one where a new drinker thinks, 'Hang on, this isn't just a whisky from a distillery in New Zealand, it's a New Zealand whisky.' What those defining characteristics will be is what's exciting. There is depth here. The next steps will be intriguing.

Right *New Zealand's barley is key to the expansion of the country's whisky industry.*

NORTH ISLAND

The aroma of manuka smoke might well come to be a signature New Zealand character (see box). This local option for smoking whisky was pioneered by Mathew **Thomson** at his eponymous distillery in Auckland. What was originally made with malt from a homebrew store and then smoked over manuka wood on the barbecue is now a small but commercial-scale distillery housed in Riverhead's Hallertau Brewery.

Both stills are homemade: a repurposed Hoga with a conventional neck and cooling tube in the lyne arm, and a 1,900-litre (502-US-gallon) straight-sided, flat-bottomed mash still with an induction jacket, rummager and a 19th-century 'Irish-style' head. Both are direct-fired.

The manuka smoked malt (from Gladfield – see box, page 341) adds herbal, resinous, clove-like elements and a hint of meatiness, bringing some depth to his signature gentle, elegantly sweet whiskies. All have a light oiliness. Ageing is in a mix of full-sized and small casks with the spirit entering at 56% ABV. By not filling the casks completely, oxidation is increased.

'Place influences flavour in many subtle and not so subtle ways,' says Thomson. 'Nature, climate, raw materials and landscape all add up to your unique flavour profile, but place also adds preciousness.

SMOKING NEW ZEALAND STYLE

Kiwi distillers have two local options for smoking their malt. *Leptospermum scoparium*, aka the manuka tree (actually a shrub), is native to New Zealand and Southeast Australia. Noted for the curative properties of its honey and essential oil, the wood has been used for carving and tool handles and for smoking foods. New Zealand peat is comprised of sphagnum moss, wire rush sedge (*Empodisma robustum*), flax, shrubs (Myrtaceae, or myrtle family, which includes both manuka and eucalyptus) and Ericaceae, or heath or heather family – ember-like, aromatic/resinous and earthy.

NEW ZEALAND

THOMSON

THOMSON LOCAL FOLK & SMOKE 46%

Lifted red fruits, strawberry chews/bubblegum, then rose and orange. Full-fat, oily palate. Ripe fruits with subtle clove-like manuka smoke in the back.

SMOKY/PEATY

★ THOMSON TRIPLE TIPPLE 46%

Green pears, peach juice, caramel flan, light barley and sweet spices. Thicker on the palate, poached pear now and light lime zest. Long.

FRUITY/SOFT

★ THOMSON FULL NOISE 54.2%

Hefty, resinous, oil of clove, tea tree oil, a tiny touch of meatiness at the back. There is a honeyed sweetness, caramelized pear, some pimentón. Defines manuka smoke.

SMOKY/PEATY

'All we wanted was for the New Zealand whisky world to be proud of something that didn't exist ten years ago. Not being Scotch, but finding our own style. There's no history of smoking food with peat or using it as fuel. That's why we use manuka. It's part of what we are.'

Auckland's tiny **1919 Distilling** was started by Soren Crabb in 2017 after he left his life as a sea captain. He is another who built his own still, whose sharp downward turn on the lyne arm helps to make an intentionally 'super-heavy spirit', but there's sufficient reflux to add fruits and make it gentle rather than sulphury.

The new make (which is bottled as White Barley) is estery, but with succulent almost sugary weight. There's light citrus and a hint of liquorice on the end.

WAIHEKE ISLAND

It's a 40-minute ferry ride from Auckland to Waiheke Island. Known for the high quality of its wines, whisky is now also being made there. **Waiheke Whisky** started with a blank sheet. 'There never was any thought about going to Forsyths and asking them to build a still,' says founder Mark Izzard. 'It was always, let's make it and do it ourselves.' The design was unusual, a geodesic dome made up of 67 square copper sheets welded together. It may have looked unlike any other still in the world, but the logic was sound, as the shape increased the amount of copper conversation.

The original, which ran at the Mudbrick winery, has been retired and a new distillery called The Heke has been built. There's a brewhouse with conversion vessel/lauter tun/whirlpool to get 'super-clear' wort, a half dozen temperature-controlled fermenters all running off the ethernet

Above *Waiheke Whisky has the only geodesic stills in the world.*

Above *Waiheke – where wine meets whisky.*

Above *Pōkeno – keeping farmers happy with draff.*

Above *Local barley is one of the key elements at Pōkeno.*

and, naturally, a scaled-up geodesic spirit still – this time built by Speyside Copper Works.

Mudbrick's winemaker Patrick Newton is now a distiller and there are plenty of winemaking techniques being used – the 'port' made from Petit Verdot fortified with Tempranillo spirit, Chardonnay casks with lees still in them. New Zealand peat is in most (see Smoking New Zealand Style box, page 336). There's a vinous feel to these gentle, balanced whiskies. None are overcooked. 'We're from here, we use local barley, we use local peat and wine casks,' says Patrick. 'You have to explore what is around you. We're not trying to be someone else.'

HEADING SOUTH

Searching for a site for their distillery, in 2019 Matt Johns and his wife Céline settled on the town of **Pōkeno**, 25 miles (40km) south of Auckland. With a two-decade career in Scotch behind him, his approach was always going to be different to his North Island colleagues.

This is a classic, modern distillery – all steel and copper, all of which, bar the stills (two wash and one spirit, from McMillan of Prestonpans, Scotland), was designed and built here. Very clear wort and 80-hour ferments, slow distillation and still rests to refresh copper all point to a specific style. 'I like sweet and clean fruit-forward whiskies,' says Matt. 'The process was designed around that.' At time of writing it can make 250,000 litres (66,043 US gallons) a year, but there is space to double capacity. With no specialized cooperage in New Zealand, he built his own where re-char, STR and repair can be done.

Its first releases arrived in 2023, thanks to Pōkeno's weird microclimate. 'We're actually below sea level,' Matt explains. 'There are heavy mists every morning and 93–97% humidity. The air pressure is so high that the bungs pop out of the casks. Maturation is quick, about 9–10% a year, but we take the losses. It's what makes our signature.'

The whiskies have a distinct citric element, with a generous, almost fat mid-palate weight – conceivably coming from New Zealand barley (see box, page 341). Experiments have also started with casks, coopered on site, made from native woods such as totara (*Podocarpus totara*).

Export is key and Pōkeno has been quick to sign up global distribution. 'The local market will be crowded in the next five years,' says Matt, 'so we

WAIHEKE

★ WAIHEKE MOSS 44%

Mix of flowers and bay leaf, Bounty bar, dried pineapple, cherry, raspberry and fresh bread. Light smoke. Light roasted barley, vanilla pod. Sweet-hearted.

FRUITY/SOFT

WAIHEKE SERIS 1 42%

Toffee Crisp, pineapple, soft fruits, lemon pepper, dried peach. A hint of smoke. Concentrated purple fruits on the palate, some florals. Butterscotch, myrrh, smoke.

FRUITY/SOFT

WAIHEKE BOG MONSTER 46%

Big. Earthy, heather root, gentian, pitch pine, resinous. Rich and generous with slightly medicinal cough syrup-like fruits. Well balanced.

SMOKY/PEATY

have to scale up and build an international brand market by market, measuring ourselves against Starward, Paul John, Amrut and Kavalan.'

SOUTH ISLAND

The west coast of the South Island has long been home to extraction industries: gold, coal mining and then forestry. All have boomed and subsequently crashed, leaving towns such as Reefton with few prospects for employment. 'I didn't want this place to die, so the solution was to create a business with jobs that weren't reliant on those old industries,' says Patsy Bass who founded **Reefton Distilling Co.** in 2018. Starting as a small gin distillery, it's now housed in a large site on the outskirts of the town. It's producing whisky from a 'clearish' wort, followed by a 96-hour fermentation and slow distillation in small stills to maximize copper conversation. Most of the barley (from Gladfield – see box, page 341) is unsmoked, but manuka-smoked malt is also being used (see box, page 336). Might whisky have another West Coast to rival Scotland, Ireland and the USA?

On the shores of Lake Dunstan, Daniel McLaughlin and Mark Neal bought 36 hectares (89 acres) of dirt on which they are building what will become the country's largest producer, **Scapegrace.** There will be two pots making a heavy new make for whisky as well as a rectifying column for vodka/grain neutral spirits (GNS). Early releases, made at Workshops Whisky in Christchurch, have been aged in 50-litre (13.2-US-gallon) and 100-litre (26.4-US-gallon) virgin French oak, but their punchiness will be calmed down as a shift to 200-litre (53-US-gallon) plus casks gets underway. Manuka smoke is used for Revenant releases.

In the 1860s, the Cardrona Valley was the centre of a gold rush. Seams long exhausted, in more recent times it's become known as a ski area...and since 2015, whisky. **Cardrona**'s founder Desiree Reid set out to become a perfumier, but a distilling course at Chicago's Koval distillery (see page 228) changed her mind.

A 1.5-tonne (1.7-US-ton) mash tun and six fermenters give wash (made from New Zealand

PŌKENO

PŌKENO ORIGIN 43%
Lemon blossom, and light florals. Sweet and approachable. Custard, white chocolate, sherbet lemon, passion fruit and light peach.
FRAGRANT/FLORAL

PŌKENO DISCOVERY 43%
Light green fig, some grassiness, then plum, black cherry, fruit tea and echinacea. Spicy palate, clove, haskap berry with citrus zest.
FRUITY/SOFT

★ **PŌKENO TOTARA CASK** 46%
Light resin elements along with roasted malt. Becomes softer and more coconut-accented, cranberry, cherry chocolate and a balancing bitter twang.
FRUITY/SOFT

Above *Pōkeno, looking out to the world.*

Above *A new sort of gold is coming from the Cardrona Valley.*

barley), which is shared between the column still (for gin and vodka) and the whisky side. Fermented for a minimum of 72 hours and up to 5 days, it's double distilled in Forsyth pots. Air-resting the stills between passes refreshes the copper, while slow distillation maximizes reflux.

It takes time for these parameters to be bedded in, but after they are, distillers can start to play. 'We're at that point now,' says master distiller Sarah Elsom – another former winemaker who has made the switch. 'I'm going to try different yeasts that are more tolerant to higher temperatures.'

Her wine background is reflected in the way she approaches texture, flavour, blending and location, and how to find the right balance for Cardrona's subtle power. While sherry and ex-Bourbon casks are used, it's ex-Pinot Noir that give Cardrona a clear signature going forward. 'Given where we are, it makes sense,' she says. It's a distillery that was built with one woman's determination, and whose team is taking it forward. In 2023, it was bought by International Beverage.

CARDRONA

★ CARDRONA THE FALCON 52%

Ripe and generous. Great balance stuck between ripe berry fruits, caramel toffee and light oak. Barbecued peach, baked banana and nutty black chocolate. Rich and long.

RICH/DEEP

CARDRONA FULL FLIGHT 62.7%

Rich with sticky date, and moist ginger cake. A touch of cordite on the palate (and with water), then Cardrona's stone fruits. Lingering.

RICH/DEEP

SOUTHLAND

In 2005, Southland farmer Rob **Auld** picked up a bottle of locally made Wilson's whisky. 'I realized that my barley might have been in there,' he says. 'That's when we began to think, why don't we just do it ourselves?' The dream came to fruition in 2015. Now, 66% of the 200-hectare (494-acre) farm he works with wife Toni is under crops. There's an 80kg (176lb) hand-stirred mash tun, and 60-litre (15.9-US-gallon) fermenters where the wort spends a week, urged on by a mix of 3 yeasts.

As he puts it, 'all of our R&D is in the ground.' There are four strains of barley (Laureate, black, red and purple), wheat, black and white oats, rye, triticale and maize. More will probably follow.

'Where is the terroir in whisky?' he asks. 'Southland barley is big and fat. The yield is low, but the flavour is rich. Now, with a few years under our belt, we can see the difference year to year. That's hugely exciting.'

All the grains have been distilled separately to the same regime, but now comes a period of fine-tuning where ferment length and cut points can be tweaked to find the sweet spots. Mixed mash bills could emerge. 'People know all about casks,' he says, 'but not about what went before.' Auld could well change that.

Above *The all-women team at Cardrona.*

Above *Dunedin's Speight's Brewery is home to NZWC's stills.*

Southland's Hokonui Hills were home to the McRae family, skilled moonshiners who produced significant amounts of illicit booze after the passing of the Alcoholic Liquors Sale Control Act of 1893, which turned the region dry, but Old Hokonui moonshine continued to be quietly made until the 1960s.

EAST COAST

After Repeal in 1969, the Baker family started Dunedin's Willowbank distillery (Wilson's blend and Lammerlaw single malt). Bought by Seagram in 1981, it was sold to Foster's in 1997, who demolished the site three years later. The remaining stock was bought in 2010 by Greg Ramsay, who began to eke them out as Milford single malt under the **New Zealand Whisky Collection** (NZWC) label.

Now Dunedin is producing whisky again. In 2021, NZWC entered into an agreement with Speight's Brewery, which makes the wash to NZWC's spec. Two recipes using different yeasts are used. Distillation is in the hands of NZWC's Mike Byars, another former winemaker.

'We run long ferments, then go gentle on the stills, otherwise the new make gets too chewy,' he says. It's sweet, lightly malty, red fruit-forward with a touch of oils.

It's aged and blended 70 miles (112km) north in Oamaru in a mix of casks ranging from 55 litres (14.5 US gallons) to 278-litre (73.4-US-gallon) barriques – mostly ex-American whiskey of different types, but also French oak, ex-wine and sherry.

GLADFIELD MALT

A fundamental element to the creation of a New Zealand style has been the use of locally grown and malted barley. Doug and Gabi Michael own a 202-hectare (500-acre) farm 25 miles (40km) south of Christchurch. Long-established barley growers, they noticed in 2004 the growing demand from craft breweries to access small batches of speciality malts. The result was Gladfield Malt. Distillers soon followed. As well as their own crops, they work with 150 contracted growers.

Even if Kiwi distillers are using the same variety as their colleagues in Scotland, their whiskies have a noticeable sweetness and thicker texture, which could be down to the higher levels of non-fermentable sugars that have been discovered in New Zealanc malt. Once again, conditions come into play.

Blending is key to the emerging style. 'For a winemaker, it's all about palate and nose. People think of age difference, but it's really about blending and balancing. We've been making Sauvignon Blanc here for 50 years and we still don't fully know what New Zealand wine is. Why should we already know what whisky is?' The answer will be fascinating.

It is a sentiment that has been articulated in a variety of ways throughout this world tour. Not just 'what is whisky?' but 'what can my whisky be?'.

The answer is 'whatever you want it to be'. Whisky in the 21st century is a sea of possibilties. It is about engagement with history, craft and place.

For all the concerns about over-capacity, the richness of the new thinking is thrilling.

GLOSSARY

Entries in SMALL CAPS are cross-references within the Glossary.

Age statement An age on the label refers to the youngest component. Remember that age is not necessarily a determinant of quality.
ABV (alcohol by volume) The alcoholic content of a whisky expressed as a percentage of the total volume of liquid. By law, Scotch whisky must be 40% ABV or more. *See also* PROOF.
Angel's share As a cask breathes during MATURATION, so some of the alcohol will evaporate. This is known as the 'angel's share'. In Scotland, it accounts for a 2% loss of each cask's volume every year.

Backset *See* SOURMASHING.
Barley Barley contains naturally occurring enzymes that, once MALTED, aid in the conversion of starch into fermentable sugars. A percentage of malted barley is therefore added to the MASH of cereals in the production of virtually all types of whisky, while single malt uses 100% malted barley.
Barrel Term used to define a 200-litre (53-US-gallon) American OAK CASK.
Beer (USA) Alcoholic liquid to be distilled, aka WASH.
Beer still (USA) The first still (normally a COLUMN STILL) in DISTILLATION.
Blended whisky A mix of GRAIN WHISKY with MALT (in Scotland) or BOURBON/RYE (USA). Some 85% of the Scotch whisky sold globally is blended.
Bourbon American WHISKEY style, which must conform to the following rules: be made from a MASH containing at least 51% CORN (maize); distilled to a maximum of 80% ABV (160° PROOF); and aged in new CHARRED OAK BARRELS at a strength of no higher than 62.5% ABV (125° PROOF) for at least 2 years.
Butt A 500-litre (132-US-gallon) ex-sherry CASK used for maturing SCOTCH WHISKY.

Caramel A permitted additive used in many whiskies (but banned in BOURBON production) to adjust the colour of the spirit to ensure consistency between batches. Heavy use dulls aroma and gives a bitter finish.
Cask All-encompassing term referring to the different types of OAK containers used for maturing whisky.
Charcoal mellowing This technique, which defines TENNESSEE WHISKEYS, involves passing the new spirit through vats of charcoal prior to ageing.
Charring All American BARRELS are charred prior to use, creating a layer of active charcoal that acts as a filter to help remove harshness and other unwanted immature aromas. CHARCOAL MELLOWING accelerates this process.
Clearic *See* NEW MAKE.
Condensing The final part of DISTILLATION where alcohol vapour is turned back into liquid.
Corn (maize) The main cereal used in BOURBON production, corn adds a fat sweetness to the final spirit. Also used in Canada and in the production of GRAIN WHISKY.
Corn whiskey American whiskey style. By law, corn whiskey must be made from a minimum of 80% corn. No minimum ageing requirements.

Dark grains The name for the mix of pot ale (high protein residue after first DISTILLATION) and DRAFF, which is sold as a nutritious animal feed.
Distillation The process that sets spirits apart from wine or beer. Because alcohol boils at a lower temperature than water, if an alcoholic liquid (BEER/WASH) is heated in a still, the alcohol vapour will be driven off in preference to the water, thereby increasing the alcoholic strength and concentrating the flavours contained within the WASH.
Doubler (USA) The simple POT STILL in which the alcohol from the first DISTILLATION is redistilled to produce the final spirit.
Draff The spent grains that are left after all the sweet liquid (WORT) has been extracted from the mash tun. It's sold as animal feed.
Dram Though widely thought of as a Scottish term meaning a drink of whisky, 'dram' is of Latin origin and refers to a small measure of any spirit.
Drum maltings The most common method of MALTING BARLEY. These huge plants contain large horizontal drums where the green malt GERMINATES.

Esters Chemical compounds created during FERMENTATION. Typically a floral and intensely fruity aroma.

Feints Final alcohol at the end of the second DISTILLATION (aka tails, aftershots).
Fermentation The process by which sugar-rich WORT is converted into alcohol by the addition of YEAST and vital in the creation of flavour.
First-fill A slightly confusing Scottish/Irish/Japanese term, referring to CASKS. When distillers refer to a barrel as being 'first-fill', they mean it's the first time it has been filled with Scotch (or Irish, Japanese etc.) whisky. Because these industries tend to use second-hand CASKS, it's not, however, the first time it has been filled. *See also* REFILL.
Floor maltings Traditional way of MALTING barley. The damp grain is spread on a floor and left to GERMINATE, periodically being turned by shovels or ploughs. Today, floor maltings have been mostly replaced by DRUM MALTINGS. *See also* SALADIN BOX.
Foreshots The first spirit to appear in the final DISTILLATION. Foreshots are high in alcohol, contain volatile compounds and are redistilled with FEINTS and low wines in the next DISTILLATION (aka HEADS).

Germination Process in which the BARLEY's growth is nurtured during MALTING.
Grain whisky Made from a mix of a small percentage of MALTED BARLEY and either CORN or wheat and distilled to under 94.8% ABV in a column still. The Scotch Whisky Act decrees that grain must possess the character of the cereal from which it is made.

Heads *See* FORESHOTS.
High wines (USA) The final spirit produced from the second distillation in the DOUBLER (aka doublings).
Hogshead Type of CASK, mostly made from American OAK, with a capacity of 250 litres (66 US gallons), aka hoggies.

Indian whisky A somewhat controversial term, since the Indian industry does not conform to the globally recognized definition of whisky being exclusively a cereal-based spirit and permits 'whisky' to be made from molasses.

Lincoln County Process The technique that separates TENNESSEE WHISKEY from BOURBON. This involves passing the new spirit through beds of CHARCOAL to remove harsh elements (aka leaching/mellowing).
Liquor (USA) The hot water used in MASHING.
Lomond still A POT STILL that contains adjustable plates in its neck, which increases REFLUX and gives the resulting spirit a characteristic oily/fruity quality.
Lyne arm/lie pipe Aka swan neck. The top part of a POT STILL, which leads from the body of the still to the CONDENSER. The angle of the lyne arm will have an impact on the spirit character. An upward angle encourages REFLUX and tends to make a lighter spirit; a downward angle tends to produce a heavier spirit.

Malting The process that makes the starch available to the distiller through steeping the dormant BARLEY in water, GERMINATING it to start growth and then arresting the barley's growth by drying it in a kiln. This can be done in FLOOR, DRUM or SALADIN MALTING plants.
Mashing Process by which cereal starch is converted into fermentable sugars.

Mash bill Term used to describe the mix and percentage of different cereals used in whisky making.
Maturation The final part of the whisky-making process takes place in CASKS and can provide up to 70% of a whisky's final flavour (and its colour).
Mothballed Term referring to a distillery that has been closed but has not been decommissioned.

NAS Shorthand for a whisky with No Age Statement on the label.
New make An alternative Scottish term for newly distilled spirit. Aka CLEARIC in Scotland and WHITE DOG in the USA.

Oak Legally all Scotch, American, Canadian and Irish whisk(e)y must be aged in oak BARRELS. During MATURATION, the whisky interacts with the aromatic extractives present in the wood. This interplay between spirit and oak adds to a whisky's complexity.

Peat(ing) Peat plays an important role in the aroma of many whiskies. Semi-carbonized vegetation laid down over thousands of years on wet acidic boggy ground, peat is cut, dried and then burned in the kilning process in order to impart a smoky aroma to the final spirit.
Phenols The chemical term for the aromatic compounds given off when PEAT is burned. They are measured in phenolic parts per million (ppm), and the higher the ppm, the smokier the whisky. The ppm measurement refers to the MALTED BARLEY and not the NEW MAKE spirit. Up to 50% of the phenols are lost in the distilling process.
Proof A measurement of alcoholic strength now only used (on labels) by American distillers. American proof is exactly double the ABV (ALCOHOL BY VOLUME) measurement, i.e. 40% ABV equals 80° USA proof.
Pot still The copper kettle-style stills used in batch DISTILLATION.

Quarter cask Contains 45 litres (12 US gallons); its use has been revived recently as a way to inject a large amount of fresh OAK into a young whisky.
Quercus The Latin term for OAK. The most commonly used varieties in whisky are: *Quercus alba* or American white OAK; *Quercus robur* or European OAK; *Quercus petraea* or *sessile*/French OAK and *Quercus crispula* or mizunara/Japanese OAK. Each has its own range of aromas, flavours and structure.

Rackhouse American term for a warehouse.
Rancio Tasting term used to describe the exotic, leathery/musky/fungal notes found in very old whiskies.
Refill Term given to CASKS that have been filled once already with SCOTCH WHISKY.
Reflux Technical term referring to the CONDENSING of alcohol vapour within the still (i.e. prior to it reaching the CONDENSING system), which turns back into liquid and is redistilled. Reflux is one way of lightening the spirit and removing unwanted heavy elements, and can be promoted through still shape as well as by the speed of DISTILLATION.
Ricks American term for the wooden supports that whiskey BARRELS lie on during MATURATION. Traditional, tall, metal-sided RACKHOUSES are also called ricked warehouses. Also used to describe the stacks of sugar maple that are burned to give the active CHARCOAL bed through which TENNESSEE WHISKEY is filtered.
Rye (USA) Cereal used in the production of RYE WHISKEY, BOURBON and Canadian whisky. Rye gives an acidic, mouthwatering effect, with aromas of sourdough, citrus and an intense spiciness.
Rye whiskey (USA) Legally a rye whiskey is one made from a MASH BILL containing a minimum of 51% rye, which conforms to the regulations governing (USA) STRAIGHT WHISKEY.

Scotch whisky Must be produced in Scotland at a distillery, from MALTED BARLEY (to which other whole cereals can be added) that is then MASHED, converted to a fermentable liquid through the BARLEY'S own enzymes, FERMENTED with YEAST, distilled to less than 94.8% ABV, matured in Scotland in OAK CASKS not exceeding 700 litres (185 US gallons) in size for a minimum of 3 years and bottled at no less than 40% ABV. Nothing other than water and spirit CARAMEL are permitted to be added.
Saladin box A method of MALTING that sits halfway between a traditional FLOOR MALTINGS and a modern DRUM MALTINGS, in which the GERMINATING BARLEY is placed in a large open-topped box and turned using a screw mechanism.
Single barrel (USA) A slightly confusing term; the whiskey in the bottle comes from a single BARREL, but each batch of a single BARREL whiskey may comprise more than one BARREL.
Single pot still An Irish speciality made from a mash of malted and unmalted barley to which, historically, oats, rye and wheat can be added.
Sourmash(ing) (USA) The non-alcoholic liquid residue left at the end of the first DISTILLATION, which is then added to the mash in the fermenter. This can make up to 25% or more of the total liquid in the fermenter. Adding this souring agent to the mash eases FERMENTATION. Every BOURBON/TENNESSEE WHISKEY is sourmashed (aka BACKSET, spent beer, stillage).
Straight whiskey (USA) Any whiskey made from a minimum of 51% of any one grain (corn, rye, wheat) distilled to 160° PROOF (80% ABV), aged at no more than 125° PROOF (62.5% ABV) for a minimum of 2 years in new CHARRED OAK BARRELS and bottled at a minimum of 80° PROOF (40% ABV). No CARAMEL addition or flavour enhancement is allowed.

Tennessee whiskey This whiskey is controlled by the same regulations as BOURBON, but distillers in Tennessee filter the new spirit through beds of maple CHARCOAL (aka the LINCOLN COUNTY PROCESS).
Thumper Another name for a DOUBLER. The thumper is filled with water through which the low wines pass – a process that helps remove some of the heavier alcohols. As it does, it makes a 'thumping' sound.
Toasting Involves heating the staves of the CASK over a fire to make them more pliable. The heat also caramelizes the complex wood sugars in the OAK. It's these sugars that interact with the spirit to produce a complex mature whisky. By varying the level of toasting, distillers can create a wide range of effects.

Uisce beatha/Usquebaugh Scottish/Irish Gaelic terms for whisky; translates as 'water of life' – the term that has long been given to distilled spirits. It's widely believed that 'uisce' was the root of 'whisky'.

Vatted malt Archaic term meaning a mix of single malts. *See* BLENDED WHISKY.
Vendome still Type of POT STILL with a rectifying column in the neck.
Viscimetry The coils and eddies that appear in whisky when water is added.

Wash The fermented liquid (aka BEER) that is distilled into whisky.
Wash still The first still in batch DISTILLATION, where the fermented WASH is DISTILLED.
Wheated bourbon A BOURBON whose MASH BILL contains wheat rather than rye. This gives a generally sweeter character.
Whiskey/whisky By law Scotch, Canadian, and Japanese whisky are spelled without an 'e', while Irish and American have an 'e', though not all American whiskeys conform to this.
White dog American term for NEW MAKE.
Worm (tub) The traditional manner of CONDENSING spirit. The 'worm' is a coil of copper that is immersed in a vat of cold water. Because of the lower level of copper interaction that takes place in this method, worm tub whiskies tend to be heavier in character.
Wort The sweet liquid that is drawn off from the mash tun.

Yeast The micro-organisms that convert sugar into alcohol (plus carbon dioxide and heat). Different YEAST strains (types) will have an impact on flavour production.

BIBLIOGRAPHY

Acar, Matthieu, *Une brève mais intense histoire du whisky Français*, Flammarion, 2023
Bryson, Lew, *Tasting Whiskey*, Storey Publishing, 2014
Van Eycken, Stefan, *Whisky Rising* (2nd edition), Cider Mill Press, 2023
de Kergommeaux, Davin, *Canadian Whisky*, Random House Canada, 2017
de Kergommeaux, Davin (with Blair Phillips), *The Definitive Guide to Canadian Distilleries*, Random House, 2020
Kuebler, Leon (editor), *The Distillation of Whisky, 1927–1931*, James Eadie Ltd, 2023
O'Connor, Fionnán, *A Glass Apart*, The Images Publishing Group, 2017
Ronde, Ingvar, *Malt Whisky Yearbook 2024*, MagDig Media Ltd, 2023
Tsuchiya, Mamoru, *Japanese Whisky Yearbook 2023*, Japan Whisky Research Centre, 2023

RESOURCES

New to whisky and want to know more? Why not find a local whisky club, or start one yourself. There are also whisky festivals pretty much every week. Go along, chat, share, enjoy.

The internet is swagged and festooned with a multiplicity of websites, blogs, podcasts and videos. Here are some of the best:

WEBSITES
whiskyfun.com *Essential tasting notes from Serge Valentin*
whiskysponge.com *The no.1 site for whisky satire*
canadianwhisky.org *Davin de Kergommeaux's vital site for Canadian whisky*
ozwhiskyreview.com.au *For anyone curious about the Australian whisky industry*
maltermagasin.se *Scandinavia's largest whisky magazine*
thedramble.com *Tastings and opinions*
whiskybase.com *Long-established site featuring reviews plus*
thewhiskywire.com *'Dramfotainment' for enthusiasts*
whiskyandwisdom.com *Whisky opinions, articles, feature pieces, info and reviews*
whiskyintelligence.com *The whisky industry and Scotch whisky news*
whiskynotes.be *Sincere whisky reviews since 2008*
potstilled.com *Irish whiskey news, reviews and everything in between*
whiskey4breakfast.com *Eating and drinking in Ireland made easy*
whiskeytribe.com *Unpretentious opinions*

BLOGS
blog.thewhiskyexchange.com *Billy Abbott's musings on all things spiritous*
masterofmalt.com/blog *Features from the online drinks retailer*
insidethecask.com *A blog about drinks, retail and travel by Andre de Almeida*
malt-review.com *Whisky and spirits blog*

PODCASTS
notanotherwhiskypodcast.com *Whisky banter and insight*
whiskytalk.fireside.fm *A podcast from The Scotch Malt Whisky Society*
whiskycast.com *Weekly podcast from Caskstrength Media*
whiskysisterspodcast.com *A lively weekly whisky podcast*
onenationunderwhisky.com *It's like, a podcast, people*
bourbonpursuit.com *The official podcast of Bourbon*
whiskey-lore.com *Finding the true stories behind the legends and lore*
bourbonlens.com *A podcast for every whiskey drinker*

YOUTUBE
Aqvavitae *The water of life – whisky evangelism*
It's Bourbon Night *Everything you ever wanted to know about Bourbon*
The Fred Minnick Show *Like whiskey? Like to laugh? Want to learn more?*
Whisky And Things *Whisky reviews and other whisky-related topics*

UK/US TERMS

UK	US
aniseed	anise
barbecue	grill
barbecued	grilled
biscuit	cookie
boiled sweets	hardy candy
candy floss	cotton candy
chilli flakes	red pepper flakes
chocolate honeycomb	chocolate cinder
coriander leaf	cilantro
corn	maize
corn cobs	ears of corn
crystallized fruits	candied fruits
green pepper	green bell pepper
icing	frosting
icing sugar	powdered sugar
lino	linoleum
porridge	oatmeal
rapeseed oil	canola oil
sweets	candy
vanilla pod	vanilla bean
wholemeal bread	whole wheat bread

INDEX

Page numbers in **bold** refer to main entries.

PICTURE CREDITS

The publishers would like to acknowledge and thank all the whisky distilleries, brands and their agents who have kindly provided images for this book.

Distillery photography:

Agitator Whiskymakare AB 177b; Backwoods Distilling Co.6ar, Rebecca Haycraft (@bechaycraft) 326a; Crafty Maltsters 108; distinguished vineyards.com 282; Liam McNulty, nomunication.jp 289; OurWhisky Foundation/Christina Kernohan 6c; Amalga Distillery/Nathan Kelly 243; Amrut Distilleries Ltd. 304; Archie Rose Distilling Co. 327; Armorik Distillerie Warenghem/Ariel Ramirez 155; ASW Distillery 220; Auld Farm Distillery 19, 334; Aurora Spirit Distillery for Bivrost 182; Bacardi/John Dewar & Sons Ltd. 76, 47; Bakery Hill Distillery/Ryan Wheatley 323; Basque Moonshiners/Adrian Ruiz de Hierro 185; Black Gate Distillery 329b; Boplaas 311a; Brown-Forman/John Paul 2, 42; Bushmills Irish Whisky/ Bradley Quinn 128, 129; Caldera Distilling Inc. 248; Cape Byron Distillery 330a & bl; Cardrona Distillery 340a; Chichibu Distillery 279; Cinco International 264b, 265; Clonakilty Distillery 141; Cooper King Distillery 147; Corowa Distilling Co./Jim Tomson 6bl, 328al, 328ar; Cotswolds Distillery/Lorentz Gullachsen 148; Crown Royal Distillery 252; Da Mhile Distillery/Heather Birnie @heather.birnie 151b; Destillerie Heinrich Habbel/Jeanette Mueller 167; DEVINE Distillery 2, 260a; Diageo 31, 32, 43, 102, 204al; Distillerie Castan 161; Distillerie Des Menhirs/Kelly Lamar Kalagayan 157; Distillerie Miclo 20r; Distillerie Tessendier 162; Domaine des Hautes Glaces/© Céline Clanet160; Don Michael 267b; Edrington/Ian Gavan 55; Eimverk Distillery 183; Fary Lochan Destilleri 179r, Lasse Hyldager 179 left; Filliers 168; Fleurieu Distillery 331b; Garrison Brothers Distillery 223a; Glenmorangie Plc. 24; Gordon & MacPhail/John Paul 66b; Heaven Hill Distillery Inc. 198, 200a; 255al & b; Hellyers Road 321; High Coast Whisky/Rania Ronntoft 177a; High West Distillery/Trevor Hooper 236a & b; High Wire Distilling Co. 219a; Hombo Shuzo Co 281; Hood River Distillers, Inc. 190b, 240a; Inchdairnie Distillery: Colin McPherson 109, Till Britze 81; International Beverage 31b, 45l; Irish Distillers Limited/© IDL 126, 126r, 135l; Jack Daniel's Distillery 212a, 213bl; Jeptha Creed 205bl & br, 229; John Distilleries Pvt Ltd./Arun Mathew 305l, 305r; Kanosuke Distillery 288; Kavalan Distillery, Taiwan 298l, 298r; Kentucky Peerless Distilling Co. 201al & 201ar; Kilchoman Distillery/Konrad Borkowski 104; Killowen Distillery 130; Kinglake Distillery 325; Kings County Distillery 194; Kyro Distillery 174; La Alazana Whisky 268a; Langatun Distillery 172; Last Straw Distillery 250a; Leopold Bros. 233a; Limeburners 333; Loch Lomond Group 80, 85; Lochlea Distillery 121l & r; Mackmyra Svensk Whisky 178; Madoc 269; Maison Lineti/Thomas Liaunet 163; Maiz Nation Whiskey 266b; Micil Distillery/ Sam Hunter 140r; Morrison Bowmore Distillers Ltd. 110b; Nantou 299; Nelson's Green Brier Distillery 215a; Nestville 186, 187; New Riff Distilling 210al & ar; Odd Society Spirits/Cause and Affect 261r; Overeem Whisky 320; Penderyn Distillery 150, Barbara Smok 151a; Pernod Ricard/Cathal Noonan 140l; Piccadily Agro Industries/Ankit Nandwani Photography 9, 307, Vishal Rana 302; PUNI Distillery 18, 173; Quaker City Mercantile 197a, 197bl; Rampur Distillery, Radico Khaitan Ltd./Dhrum Khurana 306; Rock Town Distillery 221a; Sakurao Distillery 287; Sazerac Company, Inc. 246, Buffalo Trace Distillery 206a; Shelter Point Distillery 259a; Shizuoka/Gaiaflow Distilling Co. 283; Shortcross Distillery 130; Sierra Norte Distillery 266a; Simon's Fine Distillery Alexander Heinrichs 166l; Sons of Vancouver 258; Spirit of Yorkshire/David Lindsay 146; St Kilian Distillers/ Male Gruener 165; St. George Spirits 238a; Starlight Distillery 227; Stauning Whisky/Jacobdue.dk 180; Stork Club/Katja Hiendlmayer 166r; Stranahan's 232b; Sullivans Cove Distillery 316, 319al, 319; Suntory Group/Koki Yamashita 280; Teeling Whiskey 136; Teerenpeli Distillery & Brewery 176; The Borders Distillery/ Alix McIntosh Photography 118; The Lakes Distillery 145; The New Zealand Whisky Collection 6al, 7, 20l, 341; The Pokeno Whisky Company 338al, 338ar, 339b; The Watts Whisky Company (Pty) Ltd. 310br, J Dreyer 310bl; Thomson Whisky/James Hutchinson 336a, Sait Akkirman 336b; Three Societies 301; Thy Whisky 181; Tin Shed Distilling Co. 332a; Tony Vengrove 196b; Traverse Image 234a, 234b; Uncle Nearest/Amy Haring 214b; Về Để Đi/mailan 21l, 22c & r, 300; Virginia Distillery Co. 218l; Waiheke Whisky Ltd. 337a & b; Waldviertel Distillerie Haider 170; Waterford Whiskey 138l & r; Westward Whiskey 239bl & br; Whisky Rozelieures 159; WhiskyENJOY享威297; WhistlePig Whiskey 192a.

Additional Photography:

Alamy Stock Photo: G. Dienst/Zoonar 110a, Iain Masterton 34, imageBROKER/ Markus Keller 30, Reiss McGuire 87, Ambling Images 270, 274, Andrew Milligan/ PA Images 45r, Andy Sutton 38, Ashley Cooper 29a, Avalon/Picture Nature 72, Axon 33, Bazancik 69, CNMages 202b, Dan Leeth 256r, Darren Robinson 46, David Chapman 27, David Chapman/imageBROKER.com GmbH & Co. KG 244, David Gowans 75, Denis Bringard/hemis.fr 158, Derek Croucher 102b, Erik Schelzig/Associated Press 213br, Euan Cherry/PA 119, Gael Fontaine 154, Greg Balfour Evans 127, Gregory Gerault/hemis.fr 88a, Hayley Pankhurst 211a, History and Art Collection 124r, Horst Friedrichs 35, 49a, 63l, 65, 73, 83, 84, 88b, 90, 91, 92, 99a, 99b, 103, 112r, 113, 116, hpbfotos 308, Iain Masterton 67, Ian Rutherford 28, 94, 95, 98, 100, imageBROKER/Konrad Borkowski 96, imageBROKER/ Robert Haasmann 23, imageBROKER/Roland Marske 50, JasperImage 40, 48, 49b, 60, 61, 64, 66a, Jeremy Graham 207b, Jeremy Sutton-Hibbert 105r, Jerry Larson/Associated Press 222a, Jimmy Morris 111, Joe Hendrickson 226a, John Bracegirdle 36, 54, 58, Julian Worker 105l, karenfoleyphotography 208b, Katharine Rose 149, Kay Roxby 79, Ketaro Aoyama/Associated Press 276, Kristian Buus 53b, LOOK-foto 139l, imageBROKER/Markus Keller 89, Malcro 39, Marcia Chambers 209a, McClatchy-Tribune/Tribune Content Agency LLC 195a, 228al, Melissa Kopka 230b, Michael Wald 135r, Mieneke Andeweg-van Rijn 139r, Mikel Bilbao Gorostiaga 184, Nancy Hoyt Belcher 237a, Peter Horree 13, Peter McCabe 122, Peter Sandground 71, Pictorial Press 57, Rajiv Dasan 211b, Reda & Co 133, Richard Wayman 131, Rik Hamilton 41, 78, Robert Perry/ PA Images 70, Sebastian Scheiner/Associated Press 312, Shane Pedersen 314, SiliconValleyStock 190a, Simon Price 8, SO-Photography 106, Tim Gainey 44, travelib 112l, Trevor Thompson 290, Tuul and Bruno Morandi 137, Universal Images Group North America LLC/DeAgostini 53a, 63r, World History Archive 52, ZUMA Press Inc. 235l; Cephas Picture Library Ltd./Nigel Blythe 101; **Dreamstime.com:** Aleksandar Todorovic 262, Brian Smith 62, Carrick4 115, David Massie 56, Dmitriy Melnikov 204b, In Sung Choi 277a, Mollynz 251b, Nalidsa Sukprasert 257a, Neven Vorkapic 228ar, Osa855 254, Paolotrovo 29b, Per Bjorkdahl 272a, Peter Austin 272b, Pkzphotos 241b; Flickr/Paul Sableman/ Middle West Spirits (CC by 2.0 DEED) 225; **Getty Images:** Qilai Shen/Bloomberg 293, Akio Kon/Bloomberg 286, Andrew Caballero-Reynolds/AFP 203, Andy Buchanan/AFP 117, Carl Court 275, Carla Gottgens/Bloomberg 318a, Emily Macinnes/Bloomberg 77, Hulton Deutsch/Corbis 124l, Qilai Shen/Bloomberg 296, Richard A. Brooks/AFP 285, SSPL 134, Tomohiro Ohsumi 277b, Westend61 120; iStock: agaliza 188, backpacker 278, DonNichols 224, jameslee999 292, loeskieboom 322a, Michael Conrad 142, Photosspeakathousandwords 152, searagen 132; **Shutterstock:** Angus Alexander Chisholm 74, Canadian Press 253a, EWY Media 249, Jane Hobson 93, keeble1337 37, PenToolCAD 313, PJ Parker 86, Pormezz 6br, Todamo 51.

ACKNOWLEDGEMENTS

To all of the distillers, whisky makers and PRs who answered my calls for help and obliged with generosity in terms of their time and indeed whisky. This book could not have been done without their help. I thank every one of them. Particular thanks to Dawn Davies MW for sparing, and then sending, samples.

The world of whisky is an increasingly busy and complex one and it's impossible for one person to keep tabs on everything that is happening, so I threw myself at the mercy of friends around the world who knew better than I what was going on in their backyards. Their help and assistance is hugely appreciated and the freedom with which their expertise was shared only goes to show what a remarkable community this is. This book was only possible as the result of their knowledge and writings. In book order...

Ireland: My thanks to Fionnán O'Connor, whose chats and writings helped immensely in the telling of Ireland's story.

Wales: Thanks to Jens Tholstrup for his insights.

France: To Christine Lambert and Philippe Jugé for contacting producers on my behalf; to François de Guillebon and the La Maison du Whisky team for organising samples, and to Matthieu Acar, whose book came out at just the right time.

Germany: To Bernhard Schäfer for initiating contacts.

The Netherlands: To Hans Offringa for initiating contacts.

USA: Sincere thanks to Chris Middleton for his research into James Crow and Native American corn. Also to Marty Duffy, Todd Leopold, Steve Hawley and Jared Himstedt for their insights and tips.

Canada: To Davin de Kergommeaux, who is forever generous in passing on information and thoughts, and helping to organize samples.

Mexico: To Alejandro Aispuro, who rode to my assistance as I was beginning to despair. *Un fuerte abrazo*.

Japan: The writing of this section would not have been possible without the information contained in the books of Stefan Van Eycken, and Mamoru Tsuchiya and Liam McNulty, the last of whom helped here with contacting producers and then translating their words.

China: To Chester Zhi for a fascinating evening in Beijing and for numerous whisky samples.

Australia: To Chris Middleton for his unsurpassed historical knowledge and to Luke McCarthy for organizing and then sending a massive box of samples. Cheers mate!

New Zealand: While in New Zealand, Michael Fraser Milne and Siona Collier drove me around backroads, hunting down distilleries. Long live Nightcaps!

South Africa: To Marc Pendlebury, for his assistance in South Africa.

And finally to my friend and colleague Ingvar Ronde, with whom I swapped news of new distilleries in different parts of the world. He gave me more than I gave him.

To sounding boards who listened patiently as I ranted, and who then came up with sensible solutions: the late and hugely missed Ailana Kamelmacher, The Luncheon Club, Ewan Morgan, Ollie Chilton, Charlie MacLean, Christina Kernohan, Lexi Burgess, Marcin Miller, Keith Law, Brendan McCarron, Billy Abbott, Arthur Motley, Leon Kuebler, Alan Winchester and Mark Diacono.

To my agent Tom for always being a voice of calm and reason.

To the Octopus team: Alison Starling, Leanne Bryan, Juliette Norsworthy, Jo Richardson, Giulia Hetherington, Jen Veall, Allison Gonsalves, Geoff Fennell, Matt Grindon, Ailie Springall, Megan Brown, Liz Hermann, Martin Darlison and Leanne Kelman. Thank you for your patience and understanding as this beast lumbered slowly towards completion and for then making it look so beautiful. The backroom teams never get the credit they deserve. Without them, there would be no book.

And, as always, to my ever-patient wife Jo. You are my rock and I a mere limpet upon it. And to Rosie, my light. I drive you both mad but you keep me sane. I love you.